Distillation
Control

Distillation Control

For Productivity and Energy Conservation

F. G. SHINSKEY

The Foxboro Company
Foxboro, Massachusetts

SECOND EDITION

McGRAW-HILL BOOK COMPANY

*New York   St. Louis   San Francisco   Auckland   Bogotá
Hamburg   Johannesburg   London   Madrid   Mexico
Montreal   New Delhi   Panama   Paris   São Paulo
Singapore   Sydney   Tokyo   Toronto*

Library of Congress Cataloging in Publication Data

Shinskey, F. Greg.
 Distillation control.

 Includes bibliographical references and index.
  1. Distillation.   I. Title.
TP156.D5S5   1984   660.2'8425   83-11301
ISBN 0-07-056894-4

1234567890   DOC/DOC   89876543

ISBN 0-07-56894-4

The editors for this book were Harold Crawford and Joan Cipriano,
the designer was Naomi Auerbach, and the production
supervisor was Thomas G. Kowalczyk. It was set in Primer
by Beacon Graphics.

Printed and bound by R. R. Donnelley & Sons Company.

Contents

Preface to the Second Edition

Since the first edition of *Distillation Control* was published in 1977, the technology of *controlling* separations has changed far more than the separations themselves or their economic criteria. As a consequence, the problems to be solved remain essentially the same, but the tools available to us have improved markedly. This has become especially evident to instructors and students who have used *Distillation Control* as a text in short courses. Although these courses have been extremely successful, instructors have had to supplement the text with a growing percentage of material.

Principally this supplementary material has been in the form of papers by the author and an associate, C. J. Ryskamp. The material reflects a deeper understanding of the distillation process and successes in applying new types of control systems. The understanding stems from achieving a complete analytical solution for all independent relative-gain arrays for a simple, yet accurate, multivariable column model. Much of the credit belongs to Professor T. J. McAvoy and his associates at the University of Massachusetts who developed the model. This made it possible at last to predict the performance of fairly complex control structures, verified by and supporting Ryskamp's results on real columns and McAvoy's results on simulated columns.

In essence, the design of distillation control systems has moved from what was much of an art to very nearly a science. It is now possible to enter the parameters of a two-product column into a programmable calculator and read all applicable relative gains. The designer can then select the two or three best numbers from the set, make some judgments about the dynamic responsiveness of their structures and likely constraints, and quickly arrive at a basic system structure. Yet the savings in time and effort are far less important than the confidence that the design will be as effective as possible within the given constraints. Previously, considerable controversy existed among proponents of various system structures. The new mathematical procedure for structure selection should eliminate most of this.

Not all columns are of the simple two-product type. In fact, many of the problems faced by industry are now in multiple-stream columns and multiple-column arrangements. Heat-recovery loops have introduced a few complications of their own. This edition has a separate chapter dedicated to multiple-product processes, applying both relative-gain analysis and solutions learned through field experience.

One of the important features of this second edition is its organization. Only the first and last chapters remain in their original location. Following Chap. 1, "Objectives," the reader is introduced to a discussion of thermodynamic concepts as they relate to distillation. This is important background material in that all inefficiencies in the separation process (which contribute to operating costs) are directly identifiable as irreversible thermodynamic operations.

Chapter 3 presents the fundamental relationships which determine product compositions. Both binary and multicomponent equations are given. They form the basis for all the analysis work leading to the selection of system structure. The next four chapters develop that structure, beginning with control over flow rates and inventories, both in material and energy. Reboiler and condenser control are covered here. Chapter 5 presents the relative-gain concept applied to the typical multivariable distillation column, culminating in the selection of an optimum system structure. Chapter 6 discusses implementation of the system with appropriate control algorithms for feedforward, feedback, and decoupling. Next, constraints are applied to permit operation under a host of unfavorable conditions.

The third part of the book is devoted to complex processes: multiple columns, sidestream columns, and unconventional separations. Here are covered absorption, stripping, extractive distillation, multiple liquid phases, and many of the unusual processes devised for separation of azeotropes. The book concludes with a chapter on optimization, followed by an appendix.

Those who are familiar with the first edition will find that the second follows the same manner of presentation, but it is more exact, comprehensive, and better organized. New readers should find it a clear, concise path to the implementation of effective and reliable distillation control systems.

If the text seems to overly emphasize the problem of structure selection, let me quote from *Critique of Chemical Process Control Theory* by Alan Foss of the Department of Chemical Engineering at the University of California in Berkeley:*

> Perhaps the central issue to be resolved by the new theories of chemical process control is the determination of control system structure.... There is more than a suspicion that the work of genius is needed here, for without it the control configuration problem will likely remain in a primitive, hazily stated, and wholly unmanageable form.

*AIChE Journal, vol. 19, no. 2, March 1973, pp. 212 and 214.

Distillation
Control

PART ONE

Distillation Fundamentals

CHAPTER ONE

Objectives

The principal objective of distillation is to separate a given feedstock into products which are more valuable. The value of the products depends entirely on their quality. This makes quality control a foremost consideration in operating a column. But meeting quality specifications is only one aspect of column control: the operation must also be profitable and meet whatever production goals are set for it. These three objectives, quality, profit, and production, are tightly interwoven. The purpose of this chapter is to identify these relationships and thereby determine the role of a control system in meeting them.

PRODUCT SPECIFICATIONS

There are many types of specifications which distillation products are required to meet. Most commonly, specifications are expressed in maximum percent concentration of designated impurities, and occasionally, minimum purity of the principal component. It is important to note the units placed on the specifications. Most *chemical* products are sold on a weight basis, in which case their specifications are expressed in *weight percent*. Most *petroleum* products are sold by volume, so that their specifications are typically in *liquid-volume percent* for liquids and *gas-volume percent* for gases. The most common chemical analyzers report results in units of mole percent, which is the same as volume percent for gases. Where conversion is required, the reader is directed to Appendix C for the appropriate formula.

Other specifications exist as well. For petroleum distillates, such properties as flashpoint, pour point, vapor pressure, and boiling range are commonly specified. On-line analyzers are available for some of these measurements, although they are all determinable by laboratory procedures. Where the on-line measurement differs in results from the accepted laboratory procedure, additional care must be taken to assure that the true specification is met.

Final Products   Products to be sold must meet certain minimum specifications as contracted between buyer and seller or determined by a third party such as a regulatory agency. There is usually no alternative allowed; a product failing to meet specifications is rejected, unless the contract is somehow negotiable. Good business practice therefore demands that the specification is never violated, with virtual disregard for the cost. This places an arbitrary but very real economic bound on a production facility. Figure 1.1 illustrates the economic problem associated with the manufacture of final products of guaranteed quality. As long as the critical impurity falls at or below the specification limit, the product will command a selling price established by the market for its grade. Observe, however, that reducing the impurity below the limit has absolutely no influence on the selling price. Yet the cost of operating the process is inextricably connected with the impurity level, along a smooth curve which is a mathematical function of the process itself, and the costs of labor, capital, and energy. Maximum profit will be realized at the point where selling price exceeds operating cost by the greatest margin. This is coincident with a product quality that exactly meets specifications and no better.

Failure to meet the specification limit changes the economic picture completely. The selling price drops stepwise to a lower value corresponding to the market for the lower-grade product, as shown in Fig. 1.1. At the point of transition, however, operating cost is essentially as high as required to make the higher-grade product. This then represents an extremely unfavorable operating point—one of minimum profit or loss, as illustrated in Fig. 1.1.

This assumes, however, that a market and a price exist for the lower-grade product. If this is not the case, the selling price would drop to zero at the specification limit, and operation to the right of it would not be tolerated. Such occurrences are not uncommon—they are simply not tolerated. At the point where the specification is exceeded, a decision must be made on the disposal of the product. These decisions are always painful and reflect the proficiency of the operating crew. Therefore it is understandable that great care is taken to avoid this situation.

Alternatives include diverting the product into another service, such as fuel. In essence, this drops the product's selling price to that of the current cost of fuel on

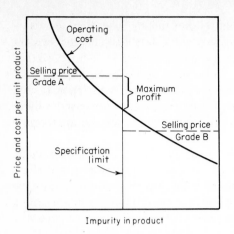

figure 1.1 *Cost of separation varies smoothly with purity, whereas the selling price of final products changes stepwise.*

an equivalent energy basis if burning it involves no other penalty such as increased maintenance. Another option is that of reprocessing the product. This action increases the operating cost stepwise as the specification limit is reached. The height of the step is proportional to the amount of product requiring reprocessing. Another alternative is that of blending below-grade product with above-grade product. This breaks the cost curve at the specification limit, raising its slope on the right-hand side of the limit. The effect on cost of alternative is investigated later in this chapter, under the heading "Operating Practices."

By-products   Many by-products have no rigid specifications and yet have values assigned to them. In fact the value assigned to a by-product is likely to be a function of composition, as contrasted to guaranteed products where price within a grade is fixed.

If a by-product's principal use is as a fuel, its value should be based on heat of combustion. This would apply whether the by-product is sold or is used in the plant that produces it. If sold, its heating value would have to be measured and multiplied by flow rate to arrive at a billing. Otherwise there would be no assurance of receiving value for money paid, unless the buyer insisted on a guaranteed energy content. Then the fuel would fall into the classification of a final product, and control would have to be applied to minimize energy "giveaway."

A value assigned to the energy content of a fuel does not necessarily remove any dependence upon composition. Consideration still must be given to the values of components as products relative to their values as fuel. Ethylene, for example, has a high value as a monomer, but it is not a good fuel since it tends to form soot. So if ethylene is lost to the fuel system because of limited recovery in distillation, it may not receive as much fuel credit as its energy content would indicate.

Wastes are products with negative value. In essence, the manufacturer pays to have them removed, as opposed to selling them for a profit. In times past, wastes were simply drained to the nearest waterway or burned at the pit or flare. That is, they were treated as having a value of zero.

Now, however, refineries and chemical plants are being forced to treat their effluents, and indiscriminate flaring is no longer tolerated. Treating effluents to the point where they meet environmental standards takes money. Therefore the treatment process, in upgrading an effluent to zero value, must have started with a negative-valued feed.

A methanol stripping column is a good example of this type of process. It is fed an aqueous stream containing perhaps 5 percent or less methanol. This stream cannot be discharged from the plant without treatment because its BOD (biochemical oxygen demand) is too high. So it must either be biologically decomposed or recovered.

As a waste product it has a negative value — essentially the cost of treatment required to bring its value to zero. But most of the methanol can be recovered by stripping with steam. The two products from the stripping column are typically a distillate of 90 percent or more methanol and a residue containing less than 0.1 percent. An economic analysis may be applied to the stripping operation. First, some specification must be set on the methanol product. Then the debit for methanol in the residue can be balanced against the cost of heating and cooling to arrive at an optimum residue concentration.

It is possible that a hard specification may be placed on the residue by a local regulatory agency. In that case, there is no opportunity for optimization, since the distillate also has to meet some purity limit in order to be useful in the plant.

Intermediate Products   Quite often, distillation is used to separate a product of a reaction from its unconverted feedstock. An example is the conversion of ethylbenzene to styrene by cracking. To maintain a reasonable yield of styrene, it is necessary to limit the conversion in the reactor to moderate levels. As a result, the reactor effluent may contain more unconverted ethylbenzene than styrene.

The two materials are separated in a vacuum column, with styrene leaving the bottom. This product must meet high purity specifications. An important factor in this separation is the tendency of styrene to polymerize in the column. This is the reason for vacuum operation: to minimize the temperature to which the product is exposed.

No standards exist for the ethylbenzene purity. If the styrene contained in it were to be recycled through the reactor without loss, the only penalty assignable to its presence would be a proportionate decrease in production rate. But a certain percentage of the recycled styrene is lost by overcracking. Its by-products include coke, whose value is negative in that it fouls the reactor. So substantial incentive exists for reducing the amount of styrene in the ethylbenzene.

But the cost of such reduction is also substantial. Because styrene quality must be controlled, improving ethylbenzene purity can be achieved only by increasing the energy input per pound of styrene produced. But as the energy input is increased, bottom temperature rises, promoting polymerization. The consequences are increased tars in the product, higher styrene losses, and more rapid fouling of the reboiler. These penalties must be combined with those arising from styrene in the distillate to arrive at the optimum distillate composition.

A similar relationship applies when separating butanes for alkylation to gasoline. Isobutane alkylates readily but n-butane does not. Consequently there is a penalty assigned to n-butane in the alkylation feed. However, the n-butane product from the separation is blended with gasoline to raise its vapor pressure for easy engine starting. If isobutane is used instead, less can be accommodated because of its higher vapor pressure. When the price of gasoline exceeds that of butanes (which is usually the case), n-butane has the greater value as a vapor-pressure additive because more of it can be used.

There is little reason to set any hard specifications on this butane separation. Instead, its operating cost should be minimized. That is, the products should be controlled at composition set points which minimize the sum of the penalties for n-butane in the alkylation feed, isobutane in the blending stock, and the costs of heating and cooling. Oddly enough, many towers ought to be controlled this way, but few are. Perhaps operators are more comfortable when they are given hard specifications to meet. Or perhaps the lack of optimization of these intermediate-product towers is due to the inability of engineers to define an optimum condition or the inability of management to demand it. Ample incentive exists — as little as 1 percent shift in composition between components differing in worth by only $1/bbl in a stream flowing at 10,000 bbl/per day raises profit by $100 per day. There are many streams meeting these requirements in a modern refinery.

Purity and Impurity   Specifications on *impurities* are unambiguous and usually lend themselves to direct control. The component in question is to be measured, inferred, or determined by laboratory analysis. Then control is applied to maintain its concentration below the specification limit. Even physical properties such as vapor pressure can often be associated with a particular component or group of components present in low concentration.

A specification on *purity* poses quite another problem. The principal difficulty is that purity is not readily measurable with any reasonable degree of accuracy. Analyzer accuracy is customarily given in percent of span. If it is necessary to measure a particular component over its full 100 percent span, the accuracy of the analysis will be much lower than if the impurity were measured instead over a much narrower span. As a consequence, even in those instances where purity may be specified, impurities will probably be measured and controlled.

Because few mixtures are truly binary, it may be necessary to measure the concentration of more than one impurity to determine the purity of the product. It is therefore not uncommon to control the sum of two or more components in a final product to maintain an acceptable purity. Recognize, however, that ordinarily only one of these components can be controlled at the column where the product is made.

Each column has the task of separating between the key components in the feed stream. The light-key component will have a higher concentration in the distillate than in the feed and a lower concentration in the bottom product. The reverse will be true of the heavy key. The impurities that can be controlled at a column are the concentration of heavy key in the distillate and light key in the bottom product.

For multicomponent feedstocks, there will be more than one impurity in each product. Components lighter (boiling at a lower temperature) than the light key all tend to appear in the distillate, and components heavier than the heavy key all tend to appear in the bottom product. Intermediate components, boiling between the heavy and light keys, cause special problems. They tend to concentrate in the middle of the column and should be withdrawn in a sidestream if they are not to interfere with separation of the keys. They also tend to appear in both end products, where their concentration can be controlled only by manipulating side-stream flow.

Off-key components may have to be controlled to satisfy particular specifications on impurity or related property such as vapor pressure. They may also require control to satisfy a purity specification. An extreme case would be a concentration of off-key component which in itself would violate the specification on product purity. Then there would be no way to meet it by adjustment of the key impurity. If the concentration of the off-key impurity can be reduced, then more of the key impurity can be allowed while still satisfying the purity specification. If both these impurities are adjustable to meet a single purity specification, then there exists a particular combination which will satisfy the specification at minimum operating cost. This topic is discussed under "Two-Column Optimization" in Chap. 11.

Controlling Off-Key Components   Consider the example of a depropanizer separating a multicomponent feedstock into a propane distillate and a bottom product consisting of butanes and heavier components, as in Fig. 1.2. (Distillation columns in a petroleum refinery are named after their light key.) There may be a specification on the ethane content (y_2) in the propane product. But because all the ethane in the depropanizer feed is forced into the propane product, its ethane content must be controlled at the bottom of the deethanizer, where ethane is a key component.

Observe, however, that the desired ethane concentration in the deethanizer bottom product is a function of the concentration of propane at that point. In other words, increasing propane content in the depropanizer feed will require increasing ethane content also if a constant propane product composition is to be maintained. This can be demonstrated mathematically with a material balance of both components. Let F, D, and B represent feed, distillate, and bottom flow rates in and out of the depropanizer, in the usual units of liquid volume:

$$F = D + B \tag{1.1}$$

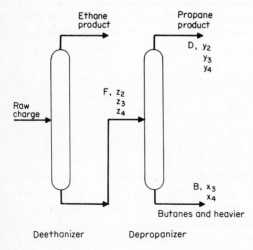

figure 1.2   *Control over ethane content of the propane product must be enforced at the deethanizer.*

Using z, y, and x as the volume fraction of components in those streams, with subscripts 2 and 3 representing ethane and propane, respectively, we have

$$Fz_2 = Dy_2 + Bx_2 \tag{1.2}$$

$$Fz_3 = Dy_3 + Bx_3 \tag{1.3}$$

Being interested only in the distillate product, let us substitute $F - D$ for B in Eqs. (1.2) and (1.3). Then

$$F(z_2 - x_2) = D(y_2 - x_2) \tag{1.4}$$

$$F(z_3 - x_3) = D(y_3 - x_3) \tag{1.5}$$

Dividing (1.4) by (1.5) and solving for z_2 yields

$$z_2 = x_2 + \frac{y_2 - x_2}{y_3 - x_3}(z_3 - x_3) \tag{1.6}$$

In the case of the depropanizer, x_2 (the ethane content in the bottom product) is small enough to be neglected, leaving

$$z_2 = y_2 \frac{z_3 - x_3}{y_3 - x_3} \tag{1.7}$$

A further simplification—which may not be justified—would eliminate x_3, which is small compared with z_3 and y_3, yielding

$$\frac{z_2}{z_3} \approx \frac{y_2}{y_3} \tag{1.8}$$

If Eq. (1.8) is valid, it can be exceptionally useful, for it states that control over the ethane-propane ratio in the feed will yield the same ratio in the product. Therefore, it is worthwhile examining how accurate this simplification is.

example 1.1

Let the following specifications apply: $y_3 = 95$ percent, $x_3 = 1.0$ percent. Solve Eq. (1.8) for values of $y_2 = 3$ percent and 4 percent, with $z_3 = 20$ percent and 30 percent. Then, using the estimated z_2, find corresponding values of y_2, using (1.7) as a test for the accuracy of the approximation.

Desired y_2, %	z_3, %	z_2, %	Resulting y_2, %
3.0	20	0.632	3.13
3.0	30	0.947	3.07
4.0	20	0.842	4.17
4.0	30	1.260	4.09

Observe that y_2 resulting from the approximation of z_2 is high in every case due to the omission of x_3 from the estimate. This deviation can be distributed more equitably by solving Eq. (1.7) to obtain a correction factor to apply to the estimate. First, factor y_3 and z_3 from Eq. (1.7) to fit the form of (1.8):

$$\frac{z_2}{z_3} = \frac{y_2}{y_3} \frac{1 - x_3/z_3}{1 - x_3/y_3} \tag{1.9}$$

Then the correction factor needed to match the approximation to the exact equation at one point is

$$C = \frac{1 - x_3/z_3}{1 - x_3/y_3} \tag{1.10}$$

The correction factor is then applied to improve the accuracy of the estimate:

$$\frac{z_2}{z_3} \approx C\frac{y_2}{y_3} \tag{1.11}$$

example 1.2

Repeat Example 1.1, using Eq. (1.11) instead of (1.8). Apply a correction factor based on a midrange value of z_3.

$$C = \frac{1 - 0.01/0.25}{1 - 0.01/0.95} = 0.97$$

Desired y_2, %	z_3, %	z_2, %	Resulting y_2, %
3.0	20	0.613	3.03
3.0	30	0.919	2.98
4.0	20	0.817	4.04
4.0	30	1.226	3.97

Observe that the correction factor has reduced the deviation in the final product quality to 1 percent of value. Since the deviation will increase with x_3 and variations in z_3, however, the approximation should be tested for accuracy in each application.

Obviously Eq. (1.7) can be used to arrive at an accurate set point for z_2 under all conditions. But being able to work with ratios of component concentrations simplifies the mathematical models of columns and the control systems applied to them. These ratios will appear again when material balances and separation factors are discussed.

Another point to be made under this discussion is that in most cases the off-key component does not have to be controlled rigidly. Typical specifications on the propane product of 95 percent propane would allow the balance to be any combination of ethane and isobutane. An *optimum* ethane content may exist, but the penalty for failing to control at that point would be smooth rather than abrupt.

Consequently variations in ethane content due to poor control of the deethanizer or inaccuracy in the estimate of its control point will not necessarily cause the propane to deviate from specification. If the depropanizer is controlled on the basis of ethane *plus* isobutane in the product, uncontrolled variations in the former can be compensated by controlled variations in the latter.

QUALITY VARIATIONS

The quality of a product leaving a distillation column tends to be quite variable since it is subject to influence from so many sources. There are environmental disturbances such as the effect of weather conditions on condensers. There are also changes in production rate imposed by the operator or by upstream units. Changes in feed composition may accompany production rate changes or may result from switching of feed sources or upsets in reactors generating the feed. All these are uncontrolled disturbances which may occur randomly.

The purpose of a controller is to overcome the effects of uncontrolled variations and restore quality to its desired value. In doing so, a controller tends to induce cyclical disturbances in its own controlled variable. Other controllers are also capable of upsetting product quality, owing to the extensive interaction existing among column variables. Also possible is complete loss of control, causing product impurity to rise upscale or approach zero. Each of these conditions is examined below.

Controlled Variations Cyclical variations in product quality are generally induced by the action of a feedback controller attempting to restore product quality to set point. A cycle may be expanding, undamped, or damped. Because the expanding cycle represents an unstable situation, it will not be tolerated by operators, who will respond by transferring the offending controller to manual. The undamped cycle is the limit of stability and could persist indefinitely. It is caused by a controller responding to a deviation in product composition by moving its output just enough to cause an exactly equal deviation in the opposite direction one-half cycle later.

If all elements in a feedback loop are linear (i.e., all gains are constant), the cycle induced by the controller will be sinusoidal. However, the true relationships between product compositions (or column temperatures) and manipulated variables, such as product flow, reflux, or reboil rate, tend to be nonlinear. Steady-state gains vary directly with the concentration of the controlled impurity. As a result, cycles caused by control action are often far from sinusoidal, as exemplified by Fig. 1.3. The steps in this record are caused by the sample-and-hold nature of the chromatographic analyzer presenting its results. The flat valleys and sharp peaks are caused by the gain variations noted above.

In the absence of external disturbances, a controlled cycle will have its area equally distributed about the set point. The reason lies in the integral mode of the controller. Its function is to change the controller output by an amount necessary to return the deviation to zero, following shifts in disturbing variables to new levels. The ideal *PID* controller algorithm is

$$m = \frac{100}{P} \left(e + \frac{1}{I} \int e \, dt - D \frac{dc}{dt} \right) \tag{1.12}$$

where m = controller output
P = proportional band, percent
e = deviation from set point
I = integral time constant

$$t = \text{time}$$
$$D = \text{derivative time constant}$$
$$c = \text{controlled variable}$$

While disturbing variables are constant, the steady-state value of m will not have to change; i.e., m will return to the same position at the end of each cycle. Consequently, the integrated deviation $\int e\, dt$ over each cycle is zero. This appears to be true in Fig. 1.3, as the areas above and below heavy-key set point y_H^* seem to balance.

Uncontrolled Variations   Uncontrolled excursions in product composition are the result of disturbances in the material or energy balance of a column. Compositions will be driven toward a new steady state unless some type of control action is applied, either manual or automatic. The dependence of gain on composition applies to all types of inputs. Therefore, the closer impurities are driven toward zero, the less they will be affected by disturbances. It is not surprising, then, that there is a tendency among operators to overpurify products, notwithstanding the economic penalties incurred thereby.

A column with controlled compositions that are also subject to disturbances will exhibit a combination of the above responses. The first reaction to a disturbance is the departure of product composition from set point toward what would be a new steady state. But an active feedback controller will begin a proportionate change in the manipulated variable to arrest the motion of the controlled composition. Proportional and derivative action cannot return it to set point, however, if the disturbing variable does not also return to its original position. Therefore, integral action is required, which will eventually return the deviation to zero.

Figure 1.4 describes the recovery of a controlled variable to set point by a *PID* controller, following a stepwise change in a disturbing variable. In this example the process is linear, so that a damped sinusoidal cycle results. Its period is proportional to the time delays inherent in the process, but it is also affected by the settings of integral and derivative times. The rate of recovery is a function of the integral time, and the maximum deviation varies with the size of the disturbance and with the proportional and derivative settings.

The area under the curve may be estimated from the controller settings. Equation (1.12) is evaluated first at t_0, when e and dc/dt are both zero, and again at t_1, when the same conditions have been restored. Subtracting the algorithm evaluated at the two steady states gives

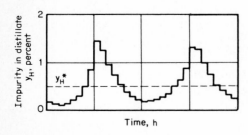

figure 1.3 *Product compositions tend to cycle nonsinusoidally because process gain varies directly with the level of the key impurity.*

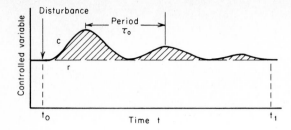

figure 1.4 *The integrated error following a step disturbance varies directly with its size and the product of the proportional and integral settings of the controller.*

$$m_1 - m_0 = \frac{100}{PI} \int_{t_0}^{t_1} e\, dt \tag{1.13}$$

Solving for the integrated error,

$$\int_{t_0}^{t_1} e\, dt = (m_1 - m_0)\frac{PI}{100} \tag{1.14}$$

The difference between m_1 and m_0 varies directly with the magnitude of the disturbance. One way to reduce the integrated error is to apply control action before a deviation develops, through the use of feedforward control (see Chap. 6).

Although the integrated error is related to the settings of proportional band and integral time, there are lower limits to these adjustments which cannot be passed without sacrificing stability. In effect, there exist optimum values of P and I which are related to the steady-state and dynamic characteristics of the process enclosed by the controller.

The deviation could lie on either side of set point, depending on the direction of change in the disturbing variable. Temporary disturbances will result in equal areas on both sides. If the intervals between disturbances is less than about three periods of oscillation, the controlled variable may never reach a steady state. This is not unusual for distillation columns, since the period of a composition loop may be long (1 to 2 h is common) and disturbances are readily propagated from upstream processes having similar or shorter periods. Likewise, interaction among loops in the same column is common, and random or diurnal changes in heating and cooling sources can be expected.

Considering all the above, controlled compositions often move in irregular patterns about the set point, rarely reaching a true steady state. If the process is reasonably linear, which happens when the controlled impurity is well away from zero or is sensed as temperature well away from the end of the column, the deviations above and below set point may be quite similar. If a concentration is controlled near zero, however, as in Fig. 1.3, there will be decidedly less variation on the pure side of set point than on the impure side.

Loss of Control Should control be lost altogether, impurity concentrations will follow the direction given by the disturbing variables. They may wander aimlessly either above or below set point, but if above set point, the variations will be far more pronounced.

There are three common reasons for loss of control:
1. Failure of the control system
2. Manual operation
3. Encountering of a constraint

The most likely point of failure is the product analyzer, as it is usually the most complex and demanding element in the control loop. In the absence of diagnostic logic, a controller will attempt to operate on the signal from a failed analyzer, driving its manipulated variable, without feedback. This open-loop situation will result in a valve ultimately reaching a fully open or fully closed position. In some analyzers, logic is available to signal failure, in which case the controller can be inhibited or placed in manual. Control is still lost but not at an extreme valve position.

Operators may transfer control to manual when suspicious of irregular performance either by the system or by the process. Whether manual control is superior or inferior to automatic control depends on both the effectiveness of the automatic system and the skill of the operator. Control systems are doomed to disuse even if they are ineffective only occasionally. Operators must have more confidence in the system than in themselves or they will not use it. The principal purpose of this book is to instruct in the design of controls that will be effective and reliable so that operators will have confidence in their performance.

Manual control does not have the characteristics of automatic control because it is discontinuous and somewhat random in execution rather than being continuous and calculated. Furthermore, different operators have different skills, habits, and misconceptions, so that performance under manual operation can vary considerably with time of day and day of the week. There is also a tendency to minimize adjustments, allowing the process to drift as long as specifications are not exceeded. Extreme conservatism results in relatively smooth operation, but is excessively costly in energy consumption per unit of product. This aspect is covered in more depth later in this chapter.

Loss of control also occurs whenever a constraint is encountered. The most common constraint is a fully open valve. This situation occurs whenever the capacity of an element is exceeded, owing to an overload or a falling capacity. The elements making up a typical distillation unit are the column, heat exchangers, and pumps. The failure of a pump to deliver sufficient flow will be signalled by a fully open valve. Similarly, if a reboiler cannot transfer the requisite flow of heat to achieve the needed separation, its heat-input valve will be fully open. Again, this could be due to an excessive load on the column, a fouled heat-transfer surface, or accumulation of noncondensible gas in the steam chest.

Condensers are also likely to present constraints, particularly owing to their dependence on atmospheric cooling, either directly or through cooling towers. Hot, humid weather can reduce their capacity significantly below that of other elements in the unit. Here again, open louvers and full-speed fans, or closed bypass valves indicate constrained operation.

Often, constraints must be imposed short of extreme valve positions owing to limitations inherent in some of the equipment. Excessive vapor loading through column trays can cause entrainment and flooding before reboiler or condenser capacity is reached. A low limit may also exist, below which weeping of liquid through vapor slots may reduce tray efficiency. Protecting a column from entering these adverse regimes is possible using constraint controls, and this procedure is developed in detail in Chap. 7.

Whenever a constraint, either natural or imposed by the system, is encountered, control over one variable must be sacrificed. Because it is essential to retain control over inventory variables such as pressure and liquid levels, a composition variable must be sacrificed. Loss of control will then cause the product composition to drift away from set point. The usual result is a violation of specifications because an upper limit on capacity has been reached. Operator intervention is then usually necessary, either to reduce the load, add capacity, or relieve the bottleneck. Chapter 7 presents some automatic mechanisms to help toward this end. Return of composition to set point is then possible, although overshoot is likely unless specific protection is provided, and this also is presented in Chap. 7.

PRODUCT RECOVERY

The principal economic benefit of improved control is usually enhanced recovery of the more valuable product. There are exceptions to this rule, since the optimum operating conditions depend as well on the cost of heating and cooling. However, most products are more valuable than the energy used to separate them. This is not to say that energy costs are to be ignored; in fact, they are rising and more significance is being attached to them. But at this writing, rising energy costs have forced product values upward at a similar pace, so that their relative worth may be scarcely changing at all.

Consequently, product recovery will be examined before the contribution of energy costs. Although this discussion applies principally to towers yielding a final product, the concepts will be useful when considering others as well.

Recovery Defined   For final products, recovery is defined as the amount of salable product generated per unit of that component in the feed. Mathematically, recovery of component i is defined as

$$R_i \equiv \frac{D}{F z_i} \tag{1.15}$$

where D is the product flow, F is feed rate, and z_i the fraction of the feed constituted by component i, in consistent units. Thus we speak of the "propylene recovery" of a propane-propylene column or "isobutane recovery" of a butane splitter.

Observe that recovery applies specifically to the *product* and not to the components of the product. If, for example, the *only* specification on a propane product were 95 percent purity, the remaining 5 percent could be *anything* else. Then ethane, butanes, carbon dioxide, etc., could all command the propane selling price as long as their total did not exceed 5 percent.

Note also that it is possible for recovery to exceed 100 percent, depending on the specification on the product. In fact, the upper limit on recovery is worth delineating at this point. Maximum recovery will be achieved when none of the component of interest is lost. For the depropanizer of Fig. 1.2, whose propane balance was given in Eq. (1.3), maximum propane recovery would occur at $x_3 = 0$. Combining Eqs. (1.3) and (1.15) under this condition gives

$$R_{3,\max} = \frac{D}{Fz_3} = \frac{1}{y_3} \qquad (1.16)$$

where y_3 would be the specification purity. For a 95 percent specification, recovery could be as high as 105.3 percent.

Recovery vs. Purity   Quantity and quality vary inversely with each other. Therefore, the cost of an increase in quality is frequently a reduction in quantity, or in the terms used here, recovery. Conversely, enhanced recovery is possible by operating closer to quality specifications.

Figure 1.5 describes the separation of a 50-50 mixture of two components having a relative volatility of 1.72 in a column with 30 theoretical trays. Recovery is plotted against impurity in the product for several values of V/F, which is the dimensionless ratio of boilup to feed. This ratio is significant in that it represents the index of energy consumed per unit of feed processed. (These curves were generated from the column model detailed in Chap. 3; the procedure is developed there.)

Consider an impurity specification of 5 percent, with V/F set at 1.5. Product recovery could be as high as 100 percent at that V/F, with the maximum attainable with perfect separation being $1/0.95$ or 105.3 percent. Recovery is very nearly linear with impurity above 5.0 percent, but it drops off sharply below that point. If it is necessary to operate at 3.0 percent to avoid violating specifications, recovery will decrease by 3.2 percent unless heat input is adjusted. This is the penalty for conservative positioning of the set point.

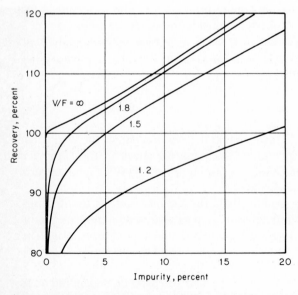

figure 1.5 *Product recovery is a function of both product purity and energy consumption.*

There is a second penalty associated with the amplitude of fluctuations, apart from their mean value (which is the set point). Because of the nonlinear relationships among column variables, cycling about a set point causes average recovery to be lower than it is at set point as a function of amplitude. For the column described in Fig. 1.5 with $V/F = 1.5$, cycling between 1.3 and 5.0 percent about a set point of 3.0 would reduce recovery by an additional 0.6 percent. (These numbers were derived from a simulation of the above column and can be expected to vary from one example to another and between real and simulated columns. Nevertheless, the general relationships will hold true for all, and columns with more nonlinear characteristics will exhibit greater variations in recovery with composition.)

Cost of Unrecovered Product   Unrecovered product leaves a column in the other stream; its value is thereby reduced to that of the other stream, or worse. In a propane-propylene separation, unrecovered propylene leaves with the propane and may be sold as propane. Occasionally, however, an additional penalty may be assigned, in that the unrecovered product may actually *lower* the value of the other stream. If the n-butane from a butane splitter is being used for gasoline blending, unrecovered isobutane leaving in that stream incurs an additional penalty, in that it is less valuable than n-butane for that service.

The cost of unrecovered product can be stated mathematically as

$$\$_D = \Delta v \, (D_s - D) \tag{1.17}$$

where D_s is the product flow achievable if specifications are just met, and Δv is the difference in value between product D and the presence of its principal component in the other stream. On a per-unit basis,

$$\frac{\$_D}{D} = \Delta v \left(\frac{D_s}{D} - 1 \right) = \Delta v \left(\frac{R_s}{R} - 1 \right) \tag{1.18}$$

where R_s is the recovery attainable by just meeting specifications. Equation (1.18) is useful in estimating the economic penalty associated with low recovery and therefore the incentive for improving control. The concept is illustrated in the following example.

example 1.3

Using the curve $V/F = 1.5$ in Fig. 1.5, estimate the cost of operating at an impurity level of 2.0 percent instead of the 5.0 percent specification, when the value difference between the products is $1.50/bbl.

$$\frac{\Delta \$_D}{D} = \$1.50 \left(\frac{100}{94.5} - 1 \right) = \$0.0873/bbl$$

At a nominal product rate of 10,000 bbl per day, estimate the daily loss.

$$\$_D = 10,000 \times \$0.0873 = \$873 \text{ per day}$$

Although perfect control is not to be expected, the estimate of savings achievable under perfect control is a worthwhile exercise in that it presents the maximum possible return on investment. In the author's experience, from 50 to 80 percent of this figure is typically obtainable by applying advanced control systems. Naturally, much depends on the present condition of the controls and the proficiency of the operating crew.

Maximizing Recovery   Figure 1.5 includes curves for higher ratios of energy input to feed rate (V/F). Where recovery is already high, there is little advantage in moving to a higher V/F ratio. However, low recoveries can be improved markedly by increasing the V/F ratio. Whether a given column will show a profit increase by such an adjustment is a subject which is explored later. The point to be made at the moment is that recovery can be maximized by operating a column at the highest attainable V/F ratio. This can be accomplished by maximizing V or minimizing F. But since feed rate F is determined by production requirements, maximizing V will maximize V/F for any given F. A point worth noting is that V/F being maximum implies that it is also variable. Hence when recovery is maximum it is also variable. This factor must be considered in the design of the composition controls because of the now variable relationship between flow rates and compositions.

Maximum recovery has historically been the operating mode for most columns. In general, V has been less expensive than feedstocks F, so there has been no reason not to maximize V/F. The principal exceptions have been sensitive products like styrene, where reducing energy input can reduce degradation and fouling.

Many engineers feel that in controlling the quality of their lower-valued product they will be maximizing recovery of their higher-valued product. Actually, however, controlling the quality of both products *fixes* the recovery. To maximize recovery means to minimize losses of the higher-valued product—in other words, to minimize the concentration of that component in the lower-valued product. It is achieved simply by controlling the quality of the more valuable product as closely as possible to specifications while holding energy input at its upper limit. No attention need be paid to the quality of the lower-value product: it will float with feed rate, etc., while always being as high as conditions permit.

COST OF HEATING AND COOLING

Rising energy costs and short supplies have spurred conservation programs from homeowner to government and industry. In Ref. 1 it is reported that in 1976 distillation accounted for about 3 percent of the total energy consumed in the United States. It lists several causes of excess consumption in any given distillation unit: leaks and hot spots, excessive reflux, low tray efficiency, improper feed point, excessive pressure drop, inefficient heat transfer, insufficient trays. Retrofitting of controls is listed first among the options recommended to reduce energy consumption.

The first step in an energy-conservation effort should be a survey of usage and costs. The following discussion will serve to place them on a quantitative basis.

Energy Sources   The primary energy sources in a plant include electricity and the direct combustion of fuels. Although using electricity for most heating purposes is grossly inefficient, as a motive force for pumps and compressors it does contribute energy to the separation unit.

Secondary sources include process steam and also oil that been heated by the combustion of fuel. A refrigeration unit which provides only cooling, or heating and cooling, would also qualify as a secondary source. In the past, these secondary sources have seemed—from the point of view of only one column—limitless in capacity. But this may have been simply because they serve several columns and could not be easily upset by an increase in demand by only one.

Tertiary supplies would have some energy already extracted from them. Examples are steam from turbine exhausts and waste-heat boilers, as well as hot effluent or coolant from reactors. Obviously, tertiary energy is much lower in cost than either primary or secondary supplies and therefore should be used when available. However, it does have two distinct disadvantages:

1. Its energy content tends to be low.
2. Both energy content and supply are variable.

The low energy content of tertiary supplies limits their usage to low-temperature operations. Water heated to 180°F by cooling a reactor effluent can supply heat only to towers separating close-boiling components (assuming a condenser temperature of about 110°F). Exhaust steam from a turbine could be used for preheating a naphtha feedstock, but it may not be hot enough for reboiling its heaviest components.   Variability of tertiary supplies creates other problems. As a reactor-effluent temperature varies with reaction rate, feed quality, poor control, etc., it will affect column boilup. Therefore, columns using these sources should have a limited amount of secondary energy available and a means of sensing and regulating the actual rate of heat being transferred.

Some heat-recovery schemes contain feedback loops within the process. Without tight control, they are capable of destabilizing a column or several columns. Here again, tight control can provide the needed stability while taking advantage of the energy savings. More is said on this subject in Chap. 8. A general discussion on energy efficiency also follows in the next chapter.

Energy Sinks   Rejecting energy to the atmosphere or to other streams within the plant is fraught with even more pitfalls. Energy rejection depends on the constancy of the sink. Primary sinks are environmental, i.e., atmospheric air and river water. Secondary sinks include cooling-tower water, refrigeration units, and waste-heat boilers, in that the cooling fluid is contained and dedicated to that function. The tertiary category includes such nondedicated sinks as other process streams.

Because all heat is ultimately rejected to the environment, its transfer rate depends on the environment. Atmospheric cooling is especially variable, since conditions change seasonally, diurnally, and abruptly with the weather. To a limited extent, all distillation columns reject heat to the environment through losses in column walls and piping. The reflux flow to a propane-propylene column of two 100-tray sections was seen to change 2.5 percent between day and night, with constant boilup and river-water cooling. This difference can only be attributed to increased heat losses to the surroundings during the night. Rain and high winds are certain to increase the losses far more, as are seasonal changes.

Although river and cooling-tower water may not change in temperature between day and night, their seasonal variations are still important. The most economical condensers — the air-fan units — are also the most variable. At this writing, much of the control effort in the distillation field has been applied to regulating the rate of cooling from these atmospheric condensers. Actually, this variability can be used to advantage, as described in the discussion on condensers in Chap. 2, "Thermodynamic Concepts."

Costs and Conversion Factors   Even at currently rising rates, the cost of the energy required to boil a unit of product is but a small fraction of the value of the product. Since this relationship is used in later chapters on control and optimization, it is worthwhile quantifying at this point.

Fuel costs at this writing are quite variable, particularly between producing and consuming nations. For example, distillate fuel oil in the United States retailed at about $1.28/gal in 1982, compared with about $0.15/gal in Venezuela. The latter price is regulated and therefore does not reflect market value. Fuel oil saved by a refinery in Venezuela could be sold for several times its regulated value.

It is customary to report fuel costs in dollars per unit of energy. Some care needs to be exercised, however, in comparing costs of different fuels. If the cost of a fuel is reported on the basis of higher heating value, it assumes that the latent heat of water produced in combustion is recoverable. In actual practice it is not, because condensation in a flue gas must be avoided or severe corrosion will result from all but the cleanest fuels. As a consequence, lower heating values should be used, which assume loss of the water as vapor.

The cost in fuel of vaporizing a unit of that same fuel is its heat of vaporization divided by its lower heat of combustion. This number is about 0.008 for light hydrocarbons. It can be useful in estimating the efficiency of refinery distillation units, particularly those separating naphthas and heated by direct combustion.

Another useful parameter is the ratio of steam condensed per unit of hydrocarbon vaporized. About 6.7 mass units of light hydrocarbons can be vaporized by condensing 1 unit of steam. Polar compounds require considerably more energy. For example, only 2.1 mass units of methanol are vaporized by 1 mass unit of steam.

Cooling-water costs tend to be independent of temperature rise since they are primarily a function of pumping rate. The cost can be converted to equivalent heat flow only by assuming or assigning a temperature rise. The same is generally true for atmospheric cooling by fans: the principal cost is drive power. Cooling costs are usually only a fraction of heating costs for the same unit and are therefore easily lumped with heating costs. The two costs are not always additive, however, in that coolant flow often remains relatively constant as the heat transferred to it is changed.

Minimizing Energy Consumption   Maximizing recovery requires a maximum rate of energy input to the process. Minimizing energy consumption requires that recovery be reduced to some minimum acceptable value. There are several policies by which some "minimum" consumption can be achieved, but they involve an arbitrary assignment of product quality specifications or recovery. If the qualities of both products are specified at some minimum acceptable limits, then the energy required to make that separation is thereby established. Fixing a limit on recovery of a final product accomplishes essentially the same result.

For any given set of compositions, energy consumption may be reduced by operating the tower more efficiently. This requires moving in a different dimension, such as adding more trays to the tower, improving tray efficiency, or increasing the relative volatility of the components. Still other ways would be to trade feed heat input for reboiler heat input where it is more effective or to relocate the feed to a more optimum tray. Let us examine these items one at a time:

1. Adding trays cannot be accomplished while the tower is operating, so this method is discounted for the purpose of this discussion.

2. Improving tray efficiency is possible in cases where the vapor rate happens to be outside the most efficient range. Varying boilup outside these limits may create other difficulties such as weeping and entrainment, which are discussed in Chap. 7.

3. Relative volatility can be increased by adding another component such as an extractant or by operating at minimum pressure. Although the former method is reserved for especially difficult separations, minimum-pressure operation is applicable to all. Energy consumption may be reduced as much as 5 percent at night and as much as 25 percent during winter operation by taking full advantage of atmospheric cooling to minimize tower pressure.

4. Heat introduced with the feed is only a fraction as effective as that applied to the reboiler because the vapor it generates passes through only a fraction of the trays. The principal justification for preheating the feed has to be usage of a lower-temperature (and hence less costly) source of heat or equalization of vapor loading between top and bottom sections of the tower.

5. If the composition of the feed differs markedly from that of the tray it enters, separation efficiency will be lost through blending. Most columns have several alternative feeding locations, but they are rarely used.

Of all the foregoing points, improving relative volatility through minimum-pressure operation is the most readily implemented by controls.

Although any of the means listed above could be used to maximize recovery as opposed to minimizing energy consumption, they seem more rightfully treated under this latter heading. Maximizing recovery is achieved through maximizing heat input in systems where heating and cooling costs are inconsequential. Furthermore, at maximum heat-input rates, savings achievable through minimum-pressure operation are reduced.

Energy Consumption vs. Purity   The same information relating recovery to impurity for various values of V/F in Fig. 1.5 is replotted as V/F versus impurity for selected values of recovery in Fig. 1.6. Because recovery cannot exceed the reciprocal of purity per Eq. (1.16), each of the curves is asymptotic to the purity which is the reciprocal of recovery, Attempting to achieve a high recovery and high purity at the same time can be very costly in terms of energy consumption.

Specific energy consumption expressed as boilup-to-feed ratio V/F has two effects on the economic performance of a distillation unit. Increasing V/F obviously

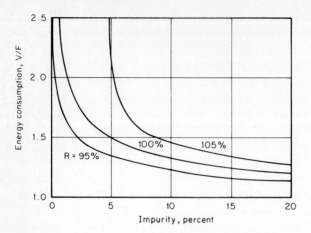

figure 1.6 *The desired recovery has a pronounced effect on the energy required to obtain a specified purity.*

increases the cost of energy per unit of feed. But it also limits production capacity and thereby increases the cost of labor and capital per unit of feed processed. This results from inherent limits within the distillation unit for handling vapor: The reboiler, condenser, and column itself all have limits on allowable vapor flow. The lowest of the three determines the vapor capacity of the unit at a given point in time. This natural limit on V imposes an upper limit on feed rate F which varies inversely with the V/F ratio needed to obtain the desired combination of recovery and purity.

In reducing production capacity by raising V/F, the rate of feed processed per column and per operator is reduced, both of which contribute measurably to production costs. During periods when market demand is low, these cost factors may not be important, but this cannot be considered an efficient operating mode for the long term.

Total operating cost is a combination of:
1. Unrecovered product
2. Energy usage
3. Labor cost
4. Capital cost

If the objective of the distillation unit is to deliver *one* final product meeting a given specification, the operating cost may be minimized by appropriately adjusting V/F and recovery. This is but one of the optimization problems discussed in Chap. 11. If the objective of the unit is to deliver *two* final products meeting given specifications, recovery will be fixed and energy usage minimized. In the first case, two variables (V/F and R) can be manipulated to control a single composition; this leaves a degree of freedom to permit minimizing cost. In the second, both compositions have to be controlled, eliminating the degree of freedom necessary for optimization.

Because of the nonlinearity of the curves in Fig. 1.6, cycling about a set point consumes more energy than operating at a steady state at the set point. For the same column simulated earlier at 100 percent recovery, operation at a set point of 3.0 percent impurity to meet a 5.0 percent specification brings an increase in energy use of 10.7 percent; but cycling between 1.5 and 5.0 percent about that set point adds another 3.2 percent to the energy consumption.

OPERATING PRACTICES

The principal function of the operator in the normal course of events is to ensure that all products satisfy the specifications imposed on them. There is a variety of ways to accomplish this end, depending on the effectiveness of the unit's automatic controls, the frequency and magnitude of disturbances, and the resources which may be available in the way of parallel units, storage tanks, etc. Not all options are equally cost-effective, however, as will be seen. Therefore, it is important to examine each of the common practices and weigh its advantages and disadvantages against the others.

Overpurification   If product composition could be perfectly controlled, its set point could then be positioned exactly at the specification limit. Because this is the most profitable operating point, as indicated by Fig. 1.1, there is substantial incentive for devising extremely effective controls. In actual practice, however, control is imperfect, and compositions are subject to fluctuations of varying intensity and duration. Because of the nonlinear response of composition to all input variables, excursions in impurities will be more pronounced above set point than below, as illustrated in Fig. 1.3. If specifications are not violated at all during the normal course of operation, the impurity set point must be positioned well below the specification limit, so that the most severe upsets likely to be encountered will not cause a violation.

Again, operators tend to be very conservative with set points. There are two basic reasons for this:

1. There are personal disincentives for violating specifications, but usually no personal incentives for reducing operating costs.

2. Columns are more stable as impurities are reduced toward zero, thereby requiring less care and attention.

Control systems that demonstrate effectiveness will encourage operators to move set points closer to the specification limit and thereby improve profit margins.

A device that could be useful in this role is an automatic set-point processor. It would take from the operator the responsibility of positioning the set point relative to the magnitude of fluctuations in product quality, and do it continuously in a calculated manner. Then control would be applied to keep the highest impurity level always below the specification limit. During intervals of little variation the set point would approach the specification limit, but the arrival of disturbances would then drive it farther away. The performance of such a system might appear as in Fig. 1.7.

Blending   When controlling composition with an algorithm containing integration, the integrated deviation from set point for product composition will be zero between any two steady states having the same controller output. This includes cycles in product quality caused by the composition controller and fluctuations

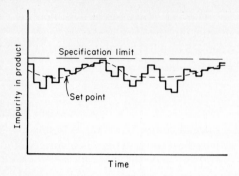

figure 1.7 *The set point can be automatically precessed below the specification limit in proportion to the amplitude of variations in product quality.*

created by transient disturbances. If the product is flowing at a constant rate during these excursions, accumulating it over time can result in a blend that is very close to the set point in composition.

This offers a reasonable alternative to overpurification: The set point can be positioned very close to the specification limit if the product is accumulated in a agitated storage tank. The effectiveness of the tank in smoothing fluctuations depends on its residence time, the degree of mixing, and the period of the fluctuations that are to be smoothed. The amplitude of composition variations leaving the tank (dy_o) compared with that entering (dy_i) is related to its time constant τ and the period τ_o of the cycle:

$$\frac{dy_o(t)}{dy_i(t)} = \frac{1}{\sqrt{1 + (2\pi\tau/\tau_o)^2}} \tag{1.19}$$

Where $\tau > \tau_o$, this relationship can be reduced to

$$\frac{dy_o(t)}{dy_i(t)} \approx \frac{\tau_o}{2\pi\tau} \tag{1.20}$$

A tenfold attenuation can be accomplished with a time constant only 1.6 times the period of the fundamental cycle.

The time constant of a mixed vessel approaches its resident time (i.e., volume divided by flow) if its contents are perfectly mixed. However, real levels of mixing typically produce a time constant that is only a few percent below the residence time, with dead time making up the balance. Then cycles of 1- to 2-h duration can be successfully smoothed in a mixed vessel of 3 h or more residence time.

Tankage can then allow a final product to be controlled at a set point very close to the specification limit and therefore reduce the operating cost associated with set-point positioning. However a penalty still exists proportional to the amplitude of fluctuations in composition leaving the column. Because of the nonlinear relationships among composition, recovery, and energy, as developed in Figs. 1.5 and 1.6, higher costs are incurred by overpurifying than are saved by underpurifying to the same extent.

This penalty can be accurately calculated where the products from parallel columns are blended or where an off-specification product is blended with overpure

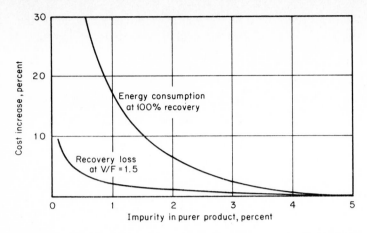

figure 1.8 *Cost of blending two products to meet a 5.0 percent specification.*

material. Figure 1.8 illustrates the economic penalties associated with blending for the column that has been used as an example throughout this chapter. Operated at constant 100 percent recovery, product containing 7.0 percent impurity could be blended with an equal amount of product containing 3.0 percent to meet the 5.0 percent specification. However the total energy consumed in making these two products is about 2.4 percent higher than if the specification were met without blending. As individual products deviate further from the specification prior to blending, the energy penalty increases exponentially.

Also shown in Fig. 1.8 is the cost increase incurred if the column is operated at constant V/F of 1.5, and recovery is varied to adjust composition. The penalty in lost recovery is not nearly as severe, principally because the variation of recovery with composition above 5.0 percent impurity is essentially linear. But of principal significance is the fact that the penalties incurred by blending are only a fraction of those resulting from overpurification to the same degree. Blending is then a reasonable practice if differences between the blended streams are kept small and centered about the specification limit.

Rerun Another alternative method of disposing of off-specification production is to return it to the feed tank for reprocessing. This is far more costly than blending it with overpure product because of the great difference in composition between feed and product. In essence, the cost of making the product very nearly acceptable is largely wasted when that product is rerun.

The penalty may be taken either as an increase in energy consumption or as a loss in recovery. If recovery is to be maintained, then the energy used by the column must increase in proportion to the increase in feed contributed by rerun product. If energy consumption is held constant, then increasing the feed rate by the amount of rerun product will reduce recovery by lowering the V/F ratio.

The worst option would be to mix the *two* products together for reprocessing. This increases the feed rate by *both* their rates and sacrifices all the energy and used in their first distillation. Also, any time the energy-to-product ratio rises, the capacity of the plant decreases and its labor and capital costs increase as well.

REFERENCES

1. Mix, T. W., and J. S. Dweck: *Conserving Energy in Distillation,* Industrial Energy-Conservation Manual No. 13, MIT Press, Cambridge, Mass., 1982.

CHAPTER TWO

Thermodynamic Concepts

When fuel scarcity first became apparent and costs began to rise, plant operations which were large energy consumers became a target for investigation. Distillation was singled out as the largest energy user in both the chemical and petroleum industries. But initially there was disagreement on how to evaluate the efficiency of a distillation unit. Some argued that the thermal efficiency of a column was in the range of 90 to 100 percent because nearly all the energy introduced at the reboiler passed through the mixture being separated, little being lost through column walls. Others pointed out that column thermal efficiency is zero because all the heat sent to the reboiler warms the environment, whether it is lost through column walls or transferred through the condenser.

Actually, neither of these arguments is correct. Both are oversimplified by failing to recognize the common denominator that relates energy flow to the composition change it produces. Thermodynamic principles need to be applied to relate the two and thereby permit a true evaluation of efficiency.

THERMODYNAMIC LAWS

The laws of thermodynamics govern all processes and set real limits on their performance. These laws are not understood particularly well, even by engineers who design and operate the processes which depend on them. In fact, thermodynamics has been considered little more than an intellectual curiosity by most students required to take a course in the subject to obtain an engineering degree. Perhaps the reason for this feeling was the failure of teachers and authors to apply thermodynamic principles beyond turbines and compressors to many of the common operations people encounter in their daily activities. In any event, the science has taken on new importance, and an understanding of the basic laws is of substantial help in decision making in many fields.

The First Law   The first law of thermodynamics is simply a statement of the conservation of energy. As shown in Fig. 2.1, the sum of all the energy leaving a process must equal the sum of all the energy entering, in the steady state. The laws of conservation of mass and energy are followed implicitly by engineers designing and operating processes of all kinds. Unfortunately, taken by itself, the first law has led to much confusion when attempting to evaluate process efficiency. People talk of energy conservation being an important effort, but in fact, no effort is required to conserve energy — it is naturally conserved.

The conclusions which can be drawn from the first law are limited because it does not distinguish among the various energy forms. Shaft work introduced by a reflux pump will leave a column as heat to the condenser just as readily as will heat introduced at the reboiler. Some engineers have fallen into the trap of lumping all forms of energy together in attempting to determine process efficiency. This is obviously not justified; the various energy forms have different costs.

Economic performance is a much more meaningful rating of a process than energy efficiency because it includes a degree of distinction among energy forms. Although it is really the standard by which processes are measured and investments are made, economic performance rating still has certain deficiencies. These deficiencies are related to artificial pricing policies, externalized costs, and an inability to predict future trends. Fortunately, it is possible to place true thermodynamic values on resources and thereby determine a true thermodynamic performance for a process. Although it will rarely be completely consistent with economic performance, it is useful in pointing out externalized costs and weakness in pricing, which could and probably will eventually be corrected by the market. To develop this insight, it is necessary to understand the second law.

The Second Law   There are many different statements of the second law as applied to cycles in which heat is converted into work. At this point, a more general

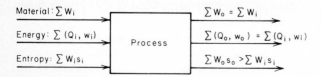

Material: $\sum W_i$ $\qquad \sum W_o = \sum W_i$

Energy: $\sum (Q_i, w_i)$ Process $\sum (Q_o, w_o) = \sum (Q_i, w_i)$

Entropy: $\sum W_i s_i$ $\qquad \sum W_o s_o > \sum W_i s_i$

figure 2.1   *Mass and energy are naturally conserved but entropy is created in every process.*

statement is desirable: The conversion of energy from one form to another always results in an overall loss in quality. Another is: All systems tend to approach equilibrium (disorder). These statements point out the difficulty in expressing the second law. It cannot really be done satisfactorily without defining another term describing quality or disorder.

That term is *entropy*. This property of state quantifies the level of disorder in a fluid, body, or system. Absolute zero entropy is defined as the state of a pure, crystalline solid at absolute zero temperature. Each molecule is surrounded by identical molecules in a perfectly ordered structure at rest. Motion, randomness, contamination, uncertainty, all add disorder and therefore contribute to entropy. Conversely, order is valuable, whether in the clarity of a gem stone, the purity of a chemical product, the cleanliness of a living space, or the freshness of air and water. Order commands a high price and can be created only by applying work. Most of our work is expended in creating or restoring order in the home, the workplace, and the environment. High entropy in the environment is what we recognize as pollution: chemical, physical, and thermal. It is one of the externalized costs of manufacturing.

The purpose of every productive process is to reduce entropy by separating mixtures into pure products, reducing uncertainty in our knowledge, or creating works of art from raw materials. In general, there is a progression of decreasing entropy from feedstocks to products. However, this is an uphill struggle inasmuch as the natural tendency is for entropy to increase as systems approach equilibrium.

The driving force for the *decrease* in entropy required of production is a concomitant *increase* in entropy by a greater amount in the rest of the universe. Generally speaking, this increase is sustained within the same plant and is therefore responsible for the decrease in product entropy. Whereas the entropy decrease resides in the conversion of feedstocks into products, the greater increase is indicated by the conversion of fuels, electricity, air, and water, into combustion products, wastewater, and waste heat.

The bottom line of Fig. 2.1 describes the second law, just as the middle line describes the first law. The total entropy of all streams leaving a process must always exceed that of all streams entering. If the entropy were to balance, as do the mass and energy, the process would be *reversible,* i.e., it could function as well backward as forward. Reversible processes are only theoretically possible, requiring dynamic equilibrium to exist continually — they are not productive. Furthermore, if the inequality were reversed, i.e., if there were a net entropy decrease, all the arrows would also be reversed and the process would be forced to run backward. In essence, it is the entropy rise that drives the process: It is the same driving force that makes water flow downhill, heat flow from hot to cold, vessels leak, glass break, metals corrode. In short, all things approach equilibrium with their surroundings.

Expressions of Entropy   Entropy has three dimensions: temperature, pressure, and composition. For an ideal gas, it can be expressed as

$$s = C_p \ln T - R \ln p - R\Sigma y_i \ln y_i \qquad (2.1)$$

where C_p = specific heat at constant pressure

T = absolute temperature
R = universal gas constant
p = absolute pressure
y_i = mole fraction of component i

Entropy rises with increasing temperature and falls with increasing pressure; it is a double-valued function of composition.

As concentration y_i approaches either zero or unity, $-y_i \ln y_i$ approaches zero; this function passes through a maximum at $y_i = e^{-1} = 0.368$. (Note that the negative sign is canceled by the negative value of logarithms of fractions.) However, the composition term requires the summation of this function for *all* components. For a binary mixture, the summation of the function for both components passes through a maximum at $y_1 = y_2 = 0.5$. In fact, equimolar mixtures always exhibit the highest entropy, which increases with the number of components. Consider an equimolar mixture of n components; each y_i would equal $1/n$. Then

$$-\sum_{1}^{n} \frac{1}{n} \ln \frac{1}{n} = -\ln \frac{1}{n} = \ln n \tag{2.2}$$

Entropy increases as heat is absorbed at a given absolute temperature:

$$ds = \frac{dQ}{T} \tag{2.3}$$

Therefore, a vapor will have a higher entropy than a liquid of the same composition by the latent heat of vaporization at the boiling point:

$$\Delta s_v = \frac{\Delta H_v}{T_b} \tag{2.4}$$

The same relationship holds true for change to or from a solid, using latent heat of fusion at the melting point. As can be seen from the preceding expressions, entropy carries units of specific heat; the gas constant R must have the same units.

Like temperature, entropy can be expressed either as an absolute value or as a value relative to any particular base. When using a table or graph giving entropy as a function of temperature and pressure, be careful to note the base. Zero absolute temperature and one atmosphere pressure are common reference points for the *ideal gas state*, in which case the entropy of the corresponding liquid will be negative. This is therefore not an absolute measure of entropy. Absolute data are available, however, for permanent gases.

In any case, absolute entropy is not of particular importance in distillation; changes between states determine thermodynamic efficiency and work to be done. Each of entropy's three dimensions is examined later in this chapter to see how it is increased by certain operations, particularly those required for control.

The process in Fig. 2.1 has two basic streams separated by a barrier. The product is undergoing a reduction in entropy, designated Δs_p. On the other side of the barrier, utilities are undergoing a rise in entropy Δs_e. The subscript e is used to denote that all utilities are resources that come from the environment, and the resulting wastes are all returned to the environment; Δs_e is then the environmental impact of the process. If all the costs of operating the process are true and internalized, Δs_e represents the thermodynamic cost of business, and as such is a legitimate extension of present escalation into a future where scarcities will force true costs to be borne.

Thermodynamic efficiency is the ratio of the reduction in product entropy achieved by the process to the cost of achieving it:

$$\eta = -100 \frac{\Delta s_p}{\Delta s_e} \tag{2.5}$$

Efficiency is here expressed in percent. The negative sign is required to cancel that of Δs_p. In the event that Δs_p is positive, the process would be destructive rather than productive and its efficiency negative.

Available Work   The reduction in entropy of a productive process is accomplished by applying work, using skill and intelligence. The latter factors are contributions of the designer, fabricator, operator, and the controls (insofar as they embody a degree of programmed intelligence). As more is learned about a process, design and operating skills can be improved, so that less work is required to achieve a given objective. Thermodynamic efficiencies of most processes are so low that there is much room for improvement in this area.

While material and energy flow in and out of a process must balance, there is no such requirement for work. Work is an energy form which makes up part of the energy balance but is easily converted to heat. Available work is related to both entropy and enthalpy as

$$w_a = T_0 \, \Delta s - \Delta H \tag{2.6}$$

Available work is the maximum amount of work which could be extracted from a fluid or body in taking it from its present state to equilibrium with the environment at absolute temperature T_0 in a reversible process. Terms Δs and ΔH represent differences in entropy and enthalpy between the initial and final conditions.

As an example, consider superheated steam at 2400 lb/in^2 abs and 1000°F (typical conditions in a power plant) at the inlet to a turbine discharging into a condenser at 60°F. Entropy would decrease from 1.5326 to 0.0556 Btu/lb · °F in being condensed, and enthalpy would decrease from 1460.6 to 28.1 Btu/lb. Available work in the steam under these conditions is then

$$w_a = 520(0.0556 - 1.5326) - (28.1 - 1460.6)$$
$$= 664.5 \text{ Btu/lb}$$

This is the maximum amount of energy contained in the steam that would be convertible to work in a perfect turbine; it is about 46 percent of the energy which the steam gives up. In practice, the power plant is far from perfect, resulting in conversion of only about 35 percent of this energy into work.

While the power plant produces shaft work or equivalent electricity, process plants consume work whether it appears in mechanical form or not. Heat introduced at the reboiler in a distillation column travels through it to exit at the condenser. That heat contains available work relative to condenser conditions. Some of the work is used in reducing entropy of the products, whereas most is lost in inefficiencies such as column pressure drop and temperature differences across the heat exchangers.

Steam at various temperatures and pressures ought to be valued based on its available work, this being a true thermodynamic measure of its worth in a process. Mechanical work is completely available as is electrical energy; in essence, their entropy is zero, so that $w_a = -\Delta H$. Compressed air, on the other hand, contains available work but no energy relative to atmospheric air at the same temperature. Its available work consists entirely in low entropy: $w_a = T_0 \, \Delta s$.

Observe also that available work is a function of sink conditions as well as source conditions. All processes operate against the environment represented by a heat sink of cooling water, air, etc. A reduction in environmental temperature at night or owing to a change in weather or season can increase w_a significantly.

Reversible processes are those that conserve entropy, as noted in the discussion on Fig. 2.1. Irreversible processes cause a net increase in entropy, and therefore a loss in available work per Eq. (2.6). While no real process is completely reversible, some are reversible in principle. These will be recognized as energy conversion processes which have reverse images, such as the motor and generator, compressor and expander, pump and waterwheel, fuel cell and electrolysis cell, heat pump and heat engine.

One statement of the second law is that heat will not flow from a cold to a warmer fluid or body without the application of work. Therefore, if heat is allowed to flow from a warm to a colder fluid or body *without the recovery of work*, that process is irreversible. It results in a loss in available work; none was recovered when it could have been. Heat transfer is, in principle, irreversible. The same is true of throttling and blending operations. They all create entropy and therefore lose available work. Each is examined below in some detail, pursuant to its use in achieving process control.

FLUID FLOW

Control over a process is exercised principally by throttling flowing streams with valves and dampers. Motive power is provided by pumps and compressors. Since throttling devices do not recover any of the work imparted by the prime mover, they introduce irreversibility and hence inefficiency to the process. This section examines the work loss associated with throttling, offering suggestions on how to reduce or eliminate it.

Pumping Liquids   The work transferred to an incompressible fluid by a pump is simply the product of volume and pressure rise:

$$w = v \, \Delta p \tag{2.7}$$

For flowing systems, volumetric rate times pressure rise gives hydraulic horsepower:

$$\text{hhp} = \frac{F \, \Delta p}{1714} \tag{2.8}$$

where F = flow, gal/min

Δp = pressure rise, lb/in^2

Consider a system consisting of only a pump, motor, and throttling valve connected in a recycle loop. Power from the motor produces a discharge pressure which forces flow through the valve and piping back to the pump suction. While power is intentionally introduced into the loop, there is no place where that energy is intentionally withdrawn. Yet an energy balance must be satisfied. Since none leaves the loop in the form of work, all must leave in the form of heat. The fluid does in fact warm until enough heat is transferred to the surroundings to balance the power introduced by the motor. In throttling a liquid stream, then, a valve does not remove energy from the stream. Instead, it reduces the available work in the fluid by an amount equal to the volume passed, multiplied by the unrecovered pressure drop across the valve.

The principal purpose of a pump is to force liquid through the fixed resistances of the process and against a static head where present. The control valve is incidental to this task and should absorb no more pressure loss than absolutely necessary. When the valve is closed, there is no flow and hence no pressure loss from fixed resistances in the process and in the pump. Then all the dynamic pressure drop exists across the valve. As flow increases, the fixed resistances begin to develop a pressure drop proportional to flow squared. The drop across the valve then decreases by the same amount, reaching a minimum level when it is fully open.

Figure 2.2 shows the power loss across the fixed load compared with that across a control valve as a function of flow for two different valve sizes. Power loss across the load varies with flow cubed. Power loss across the valve rises linearly with flow but falls with flow squared, thereby passing through a maximum in the range of 60 to 70 percent flow. Ironically, that is the very range where valves are selected to operate. Later in this chapter, schemes are devised to drive valves toward their open or closed positions where power losses are lower.

Throttling Characteristics   A significant saving in pump power is possible by resorting to larger valves as evident in Fig. 2.2; the valve having the pressure-drop ratio of 16 is twice as large as that having a ratio of 4. However, the valve must be sized together with the pump and motor for the savings to be realized. The disad-

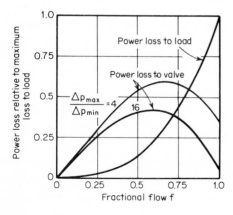

figure 2.2   *Power loss due to throt-*
tling is maximum just above mid-
range.

vantage of using larger valves is the effect of the variable pressure drop on their installed characteristics. A linear valve produces flow variations that are linear with stroke only when pressure drop is constant.

For liquids, the flow through a control valve is given by

$$F = aC_v \sqrt{\frac{\Delta p}{\rho}} \qquad (2.9)$$

where F = flow, gal/min
$\quad a$ = fractional valve opening
$\quad C_v$ = valve flow coefficient
$\quad \Delta p$ = pressure drop, lb/in^2
$\quad \rho$ = fluid density, g/ml

It can be seen that if Δp decreases as the valve opens, the change in flow per unit valve opening decreases. The result of this relationship is that fractional flow f (that is, F/F_{max}) varies nonlinearly with fractional valve opening, as shown in Fig. 2.3.

One of the most important factors affecting control-loop performance is the uniformity of the response of the controlled variable to valve position. If the controlled variable is a linear flow measurement or is linearly related to flow, then the nonlinear characteristics shown in Fig. 2.3 can cause trouble. Consider the example of fuel gas from a constant-pressure source throttled through a valve and a burner and then used to heat a constant stream of air. At low flow rates, when the pressure drop across the burner is nil, air temperature will respond sharply to valve position. But as full flow is approached, the sensitivity of temperature to valve position deteriorates since most of the supply pressure is dropped across the burner. If the temperature controller is adjusted for tight control at low temperatures, response will be sluggish at high temperatures (high flows). Or if adjusted for tight control at high temperatures, the increased sensitivity at low temperatures will result in oscillation there.

This undesirable relationship can be altered by using a valve with a suitably nonlinear relationship between valve opening and stem position. An "equalpercentage" characteristic exists in which the valve opening varies exponentially with the fractional stem position m:

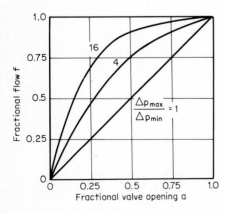

figure 2.3 *Variable pressure drop causes flow to respond nonlinearly to valve opening.*

$$a = r^{m-1} \tag{2.10}$$

Here r is the "rangeability" of the valve, i.e., its maximum flow divided by its minimum controllable flow under constant pressure drop. The name of the valve characteristic springs from the response of flow caused by a given incremental adjustment in stem position being an "equal percentage" of the *actual* flow for any position of the stem.

If an equal-percentage valve with a rangeability of 50 is operated under constant pressure drop, its fractional flow relates to stem position as the lowest curve in Fig. 2.4. But if it is used in a system that creates a variable pressure drop, a more linear relationship is achieved. If linearity is the only criterion of concern, then the equal-percentage valve performs admirably when the available pressure drop varies over 16 to 1.

However, variable pressure drop also reduces the effective rangeability of a valve. The maximum-flow capability of the valve will be reduced by the lower pressure drop available at that flow whereas the minimum-flow capability will not. Then the effective rangeability r_{eff} will always be smaller than the rated rangeability r by the pressure-drop ratio

$$r_{\text{eff}} = r \sqrt{\frac{\Delta p_{\text{min}}}{\Delta p_{\text{mzx}}}} \tag{2.11}$$

The rangeability of the equal-percentage valve described in Fig. 2.4 is reduced from 50 to 12.5 when the pressure drop varies by a factor of 16 between zero and full flow. Although this loss in rangeability can create a problem in some control loops, most of those on distillation columns will not be affected by it.

Some control loops contain nonlinearities other than the valve characteristic and can use a nonlinear valve to compensate for them. A head-type flowmeter such as an orifice or nozzle produces a differential-pressure signal h related to the square of flow:

$$h = kF^2 \tag{2.12}$$

The sensitivity of the differential-pressure signal is low at low flows, so the sensitivity of the valve should be high in that range. This characteristic corresponds to

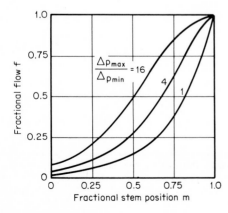

figure **2.4** *The equal-percentage characteristic compensates for variations in pressure drop.*

what was achieved with a linear valve having a variable pressure drop as in Fig. 2.3.

Most temperature-control loops require the opposite characteristic, even with constant pressure drop. This peculiarity of temperature loops, along with the need to compensate for variable pressure drop, explains why most control valves used in the process industries are of the equal-percentage variety.

Compression and Expansion   Compression is an important operation in many distillation units and an essential part of refrigeration systems. Yet it is not well understood thermodynamically. To illustrate, consider the air compressor shown in Fig. 2.5. The compressor raises the pressure and temperature of the air, with its enthalpy increasing by the amount of work introduced. If the air is then cooled back to ambient temperature, its enthalpy returns to its original atmospheric value because for an ideal gas (which air approaches at atmospheric pressure)

$$\left.\frac{\partial H}{\partial p}\right|_T = 0 \tag{2.13}$$

Its internal energy is also the same; the difference between enthalpy and internal energy is the $\Delta(pv)$ product, but for an ideal gas at constant temperature $p_1 v_1 = p_2 v_2$.

Consequently, compressed air at ambient temperature contains the same energy as atmospheric air, although it has available work. Its work content is a function of entropy alone. An ideal compressor operates isentropically. Differentiating Eq. (2.1),

$$ds = C_p \frac{dT}{T} - R \frac{dp}{p} = 0 \tag{2.14}$$

This relates temperature rise to pressure rise:

$$C_p \int_{T_0}^{T_1} \frac{dT}{T} = R \int_{p_0}^{p_1} \frac{dp}{p}$$

$$C_p \ln \frac{T_1}{T_0} = R \ln \frac{p_1}{p_0} \tag{2.15}$$

In practice, inefficiencies in the compressor will put additional work into the gas, causing T_1 to exceed that predicted by Eq. (2.15).

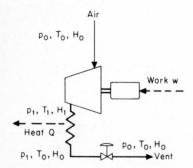

figure 2.5 *Cooling compressed air to its suction temperature removes all the energy introduced by the compressor.*

The cooling process reduces entropy following Eq (2.3):

$$\Delta s = \int_{T_1}^{T_0} \frac{dO}{T} = C_p \ln \frac{T_0}{T_1} \tag{2.16}$$

Because T_0 is lower than T_1, Δs is negative. This can also be determined by integrating the pressure term of Eq. (2.14):

$$\Delta s = -R \int_{p_0}^{p_1} \frac{dp}{p} = -R \ln \frac{p_1}{p_0} \tag{2.17}$$

The last element in Fig. 2.5 is a control valve which vents the compressed air to atmosphere. Because the energy content of the air is the same on both sides of the valve, there is no energy lost across it. This is characteristic of all throttling devices, whether the gas is ideal or real. In the case of an ideal gas, there is also no temperature change. However, real gases may exhibit a temperature change known as the Joule-Thompson effect:

$$\left. \frac{\partial T}{\partial p} \right|_H = \mu \tag{2.18}$$

where μ is the Joule-Thompson coefficient. It is a positive number for most gases so that temperature falls upon expansion, but the temperature of hydrogen and helium increase owing to negative coefficients.

Note especially that while the valve conserves energy, it loses work. The entropy removed by the sequence of compression and cooling is all restored by the valve. In sum, work has been converted entirely into heat without any net entropy reduction; because of venting, the process has become unproductive. If an expansion engine were to replace the valve, work could be recovered and the air would be exhausted below ambient temperature, thereby producing cooling. This is the basis of the air refrigeration cycle.

Figure 2.6 compares expansion through an engine against throttling for superheated steam. This is a phase diagram plotting temperature against entropy. Superheated vapor appears to the right of the saturated-vapor curve. Ideal expansion from point A to point C follows a line of constant entropy. By contrast, throttling from A to B follows a curve of constant enthalpy. (Because temperature falls somewhat, steam is seen to exhibit a positive Joule-Thompson coefficient.)

The ideal expansion process is reversible in that the work could be returned to the shaft and compress the steam back to point A. The throttling process is irreversible in that there is no mechanism by which the steam can be moved along a path of constant enthalpy from point B to A. The steam would have to be compressed and cooled, in which case it would follow a different path entirely.

Expansion in a *real* engine will follow a path that is a vector sum of the reversible and irreversible vectors. Some of the irreversibility is due to uncontrollable friction and some is due to throttling. Large turbines have multiple throttle valves so that only a portion of the steam is throttled at any time, with the balance expanding through more-efficient nozzles.

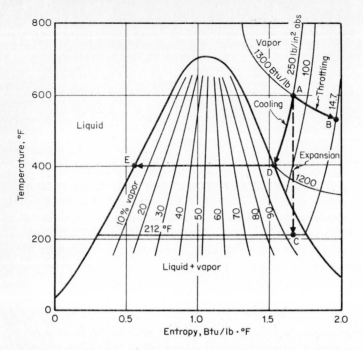

figure 2.6 *Expansion of steam through an engine produces work while conserving entropy; expansion through a throttling valve conserves energy while creating entropy and losing available work.*

Compressor Controls   Centrifugal compressors are quite similar to centrifugal pumps in that their discharge head falls as flow is increased. But owing to the compressibility of gases as opposed to liquids, compressors exhibit a characteristic instability in the low-flow range which limits their turndown. This instability is known as *surge*—a dynamic reversal of the pressure-flow characteristic that causes oscillations to develop within the machine. Unchecked, surge may destroy a compressor in very little time.

Because normal operation is impossible within the surge region, a compressor must be protected from low-flow conditions by its control system, even during start-up. This is usually accomplished by bypassing some of the discharged product back to the suction after cooling. The cooling is essential to remove the heat of compression—otherwise the gas will continue to rise in temperature until damage is done or protective devices trip. For maximum efficiency, only enough gas should be bypassed as is absolutely necessary for surge protection. To minimize bypassing, the surge region must be carefully defined and the controls programmed as to its exact location.

Compressor characteristic curves are usually plotted in coordinates of adiabatic head vs. volumetric flow at suction conditions. A typical plot appears in Fig. 2.7. The surge line describes what is essentially a parabola:

$$h_a = k_a F_s^2 \tag{2.19}$$

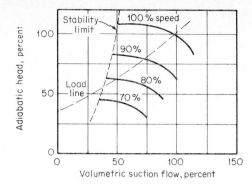

figure **2.7** *Characteristic curves for a typical variable-speed centrifugal compressor outlining the surge region.*

where h_a = adiabatic head, ft

   F_s = volumetric suction flow, ft³/min

   k_a = constant

Although this is a general model for centrifugal compressors, it is by no means all-inclusive. Some surge curves are nearly linear; others bend farther to the right at high heads. But there are further difficulties to be surmounted — neither adiabatic head nor suction flow is directly measurable. According to White [1], adiabatic head varies not only with compression ratio but also with molecular weight, temperature, supercompressibility, and specific-heat ratio. Compression ratio is nearly linear with adiabatic head for most machines, with the linearity deteriorating as compression ratio is increased. White then substitutes compression ratio for adiabatic head to develop his surge-curve model:

$$\frac{p_2}{p_1} - 1 = k_s F_s^2 \tag{2.20}$$

where p_2 = absolute discharge pressure

   p_1 = absolute suction pressure

   k_s = constant

Molecular weight, temperature, and specific-heat ratio primarily affect the coefficient k_s of the relationship.

   Suction flow is usually measured with an orifice or nozzle that produces a differential-pressure signal. The relationship between volumetric flow and measured differential pressure h_m for a gas is

$$F_s = k_m \sqrt{\frac{h_m T_1}{p_1}} \tag{2.21}$$

where T_1 is absolute suction temperature and k_m is the meter constant, which includes the specific gravity and supercompressibility of the gas. When Eqs. (2.20) and (2.21) are combined, we have

$$\frac{p_2 - p_1}{p_1} = \frac{k_s k_m^2 h_m T_1}{p_1}$$

Reduced to its simplest form, the surge line becomes

$$\Delta p = k_c h_m \tag{2.22}$$

where Δp is $p_2 - p_1$ and k_c is a control constant which includes all other constants and their variability. Some of the variables like suction temperature and super-compressibility affect the head-compression ratio relationship and the volumetric-flow relationship in a compensating manner. However, some residual remains such that

$$k_c = f(T_1, M, \gamma) \tag{2.23}$$

where M is the molecular weight and γ is the specific-heat ratio. Both M and γ are functions of gas composition and tend to change together. The value of k_c increases about 10 percent with a change in M from 16 to 20 for light hydrocarbon gases and about the same amount for a 120°F reduction in T_1.

The first step in designing an antisurge control system is to convert data from the manufacturer's surge curves into measurable Δp and h_m. Then plot Δp versus h_m as in Fig. 2.8 and construct the best straight line on the safe side of the data points. The slope k_2 and possibly the intercept k_1 should be adjustable in the field to allow a better fit to installed compressor characteristics and to correct for changing operating conditions. A control system which implements this function appears in Fig. 2.9.

The FFC in Fig. 2.9 is known as a ratio controller; it is intended to hold one input (Δp) in ratio (k_2) to the other $(h_m - k_1)$. If k_2 is set too high, a condition might arise which would place the compressor into surge; if set too low, gas may be bypassed needlessly, wasting energy. Therefore, accurate definition of the control line is important. Compensation may be applied for varying suction temperature if necessary. Constant-speed compressors controlled by adjustable inlet guide vanes have surge curves which vary with vane position, requiring compensation for their effect. With some compressors, locating the flow measurement on the suction line is not practicable so that a discharge measurement is needed. Then the control line must be compensated for variations in compression ratio, in effect referring volumetric discharge flow to suction pressure.

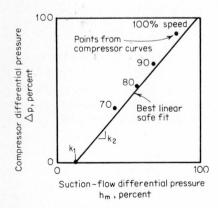

figure 2.8 *Data from the compressor's characteristic curves should be plotted as Δp versus h_m.*

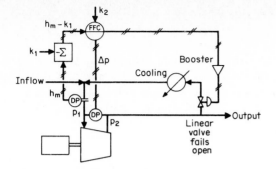

figure 2.9 *This system imple- ments the antisurge control func- tion described in Fig. 2.8.*

In addition to these steady-state requirements, the dynamic response of the surge-control system must be considered as well. The bypass valve should fail in the open position so that the compressor cannot be damaged by even a momentary air-supply outage. The valve should have a linear characteristic—an equal- percentage valve restricts flow too much at low stroke where flow is most needed in this application. Furthermore, the valve should be equipped with a volume booster to increase its stroking speed.

The controller itself must be especially equipped for this service. Its normal condition is saturated, since the bypass valve is normally closed. A conventional controller in this service would "wind up" with its integrating function keeping the valve closed until the control line is crossed and an unsafe condition exists [2]. An antiwindup device is available which can hold the controller output at the just- closed position of the valve by modulating the controller's integrating action. Then a change in the variables in the *direction* of surge will cause the valve to open by proportional action *before* the control line is crossed. Any unnecessary bypassing caused by this action is temporary, lasting at most for a few seconds while the controller integrates to the new load condition. Control action is then always on the safe side of the control line.

Start-up of a compressor is a precarious event. The bypass valve should be open, particularly if a parallel compressor is running, in which case a differential pressure is already established. However, an electrically driven compressor has been known to overload its motor when starting with the bypass *open*. These problems can be circumvented when enough thought is given beforehand to designing appropriate control logic.

Variable-Speed Drives Lost work caused by throttling can be avoided in many cases through the use of variable-speed drives for pumps and compressors. Figure 2.7 describes a centrifugal compressor whose head and flow can be reduced by reducing shaft speed. The potential for savings depends entirely on the nature of the load and the variability of demand.

If the delivered flow always needs to be high, then a constant-speed driver is adequate. But reductions in demand can be opportunities to save work if the load line is favorable. Figure 2.7 shows a load line having a moderate static head (the intercept at zero flow). Head varies with flow squared, while power varies with the product of head and flow. When the static head is zero, then no power is required at zero flow.

The absence of a static head is most favorable to energy savings. Half the rated flow is attained at half the rated speed and one-fourth the rated head, thereby requiring one-eighth full power. An ideal surge curve will also follow a load line through zero static head, eliminating the possibility of entering the surge region by flow reduction through speed variation. Flow reduction by throttling, however, swings the load line counterclockwise toward the surge line and even past it.

The elevated static head appearing in the load line of Fig. 2.7 shows that 80 percent speed is required to deliver 50 percent flow. The surge line is reached at about 40 percent flow, allowing no further savings in energy if demand falls below that point. In essence, the antisurge system must prevent flow from falling into the surge region by recycling whatever flow is not required by the process. The antisurge system thereby creates a false demand when required.

Variable speed can be achieved through several different mechanisms. Turbines, both steam and gas, are efficient drivers whose speed can be readily adjusted over their full range. There are also variable-speed couplings available to connect a constant-speed motor to a variable load. Some losses are associated with the coupling itself, as well as reduced motor efficiency when running lightly loaded at full speed.

Coupling loss can be avoided by using direct-current motors; however, they have commutators which can cause problems and they require a high-power controlled dc source. Recently, variable-frequency, variable-voltage ac supplies have been developed for excitation of standard one- and three-phase induction and synchronous motors. Because these are costly units, the potential for power savings must be carefully estimated to ensure an adequate return on investment.

Generally, any type of variable-speed drive provides more satisfactory control of flow than a typical throttling valve. In most cases, their range, linearity, and dynamic response are superior to valves [3].

When a single pump or compressor must serve several users, valves can be used in conjunction with a variable-speed drive, as shown in Fig. 2.10. It is imperative to control the pressure of the supply header, otherwise individual users will attempt

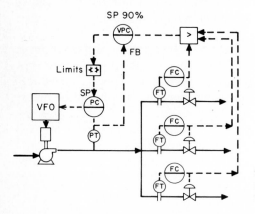

figure 2.10 *Optimum pressure is the lowest that can satisfy the most-demanding user.*

to satisfy their demands by taking from others. Yet the optimum pressure is not constant but the lowest value that will satisfy the most-demanding user.

Demand is indicated by the position of a control valve when under automatic control. The most-open valve then identifies the most-demanding user. The high-signal selector ($>$) compares the valve positions and transmits the highest signal to the valve-position controller (VPC). This device gradually reduces the pressure set point, causing pressure to fall and the flow controllers to open their valves. Eventually the most-open valve will reach the set point of the VPC, which is typically set at 90 percent. This is then the minimum pressure at which the most-demanding user can function, with a 10 percent margin to ensure stable control.

Raising the VPC set point will allow pressure to fall further but may cause loss of flow control for the most-demanding user. Reducing it allows more margin to accommodate upsets but at the cost of additional power. The VPC requires only the integral control mode, which minimizes the transmission of disturbances from the selected valve to the other users. Its time constant must allow for the complete response of both the pressure and the flow loops, which it encloses. An external feedback loop (FB) is shown to prevent the VPC from windup during intervals when the pressure controller cannot respond. These situations can develop when the operator must take over pressure control or when the pump is overloaded. Limits need to be placed on the pressure set point to keep it within a reasonable range when flow controllers are not in automatic.

BLENDING

Blending is an irreversible operation in that it is the opposite of separation which requires work. Blending of fluids having dissimilar compositions or temperatures conserves mass and energy but loses the work that was required to separate them. The thermodynamic inefficiency of blending is illustrated below for several common operations.

Blending Products It is possible to calculate the entropy rise associated with the blending of different mixtures having the same temperature by using only the composition dimension of Eq. (2.1). Consider mixing stream X having composition x with Y of composition y to form blend Z of composition z. To satisfy the material balance

$$X + Y = Z \tag{2.24}$$

$$Xx_i + Yy_i = Zz_i \tag{2.25}$$

The entropy rise per mole of blend caused by blending is

$$\Delta s = -R\left(\sum z_i \ln z_i - \frac{X}{Z}\sum x_i \ln x_i - \frac{Y}{Z}\sum y_i \ln y_i\right) \tag{2.26}$$

Dividing by the entropy of the blend will indicate a relative increase:

$$\frac{\Delta s}{s_z} = 1 - \frac{X\sum x_i \ln x_i + Y\sum y_i \ln y_i}{Z\sum z_i \ln z_i} \tag{2.27}$$

An enlightening exercise is to compare the relative entropy rise to the relative increase in energy required to meet composition specifications by blending. For the column described in Fig. 1.8, Eq. (2.27) was solved for equal values of X and Y to meet the specification of 5.0 percent impurity. The results are plotted in Fig. 2.11 along with the energy cost curve of Fig. 1.8.

The entropy model predicts the rise in energy use reasonably well, though conservatively, for moderate degrees of blending. It fails when one of the streams must approach zero impurity level due to a high impurity level in the other stream. The entropy model does not include any consideration of the number of stages required to reduce impurity to a particular level. If it must be reduced below the capability of the stages, no amount of energy can make the required separation.

Blending Feed   The same penalties associated with blending products apply to column feed as well, except that they could be more severe because there is more latitude. Ideally, the feed should be introduced at that point in the column where it encounters material of the same phase having the same composition. Significant differences shift the composition profile in the column at the feed point, producing a discontinuity. Composition differences per stage in such a pinch region are considerably less than elsewhere in the column, effectively reducing the number of theoretical stages and thereby increasing energy consumption.

While the entropy model can be used to estimate energy penalties associated with blending at the feed tray, it will probably give conservative estimates. Ryskamp's simulation [4] indicates an energy penalty of 45 percent when feeding a 50 percent feed at trays where the composition was 21 or 74 percent, and an 8.4 percent penalty when feeding at trays where the composition was 34 or 60 percent. The entropy model gives percentage increases well below these calculations. Ryskamp demonstrated that the separation load is uniformly distributed across all the trays when the composition profile falls in a straight line, when composition is plotted on a probability scale against tray number on a linear scale (see Fig. 6.5).

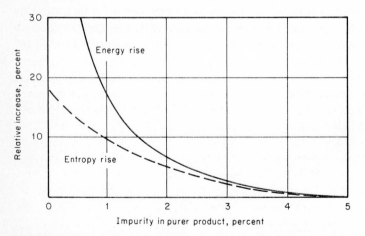

figure 2.11 *Comparison of entropy rise and energy requirements when two products are blended to meet a 5.0 percent specification.*

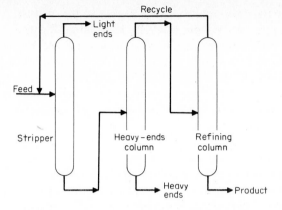

figure 2.12 *The recycle stream is used to recover whatever product may leave the top of the refining column.*

Recycling of unacceptable product back to the feed causes a large increase in entropy because of the great difference between the compositions of the two streams. Yet recycling is not uncommon. The usual reason for recycling is to satisfy a reactor of some kind in which complete conversion of a feedstock to a product is not feasible. Thus the unconverted feed may be separated from the product and returned to the reactor. If there is no reaction associated with the recycle stream, its purpose is usually to improve recovery of a final product. Figure 2.12 illustrates one possibility: the light ends from a refining column are returned to recover whatever product may be there. If the final product were taken overhead, the heavy cut would be recycled to the heavy-ends column to improve recovery.

When confronted with a system like that shown in Fig. 2.12, it is not easy to arrive at a sensible quality-control philosophy. The reason for the dilemma is that there are too many columns for the number of products. A three-product system ordinarily requires only two columns. Similarly, control of two specifications on the major product can be done with only two columns.

The third column and its recycle stream add a new dimension to the control problem. The question of control over the light-component content must be resolved. If it is controlled at the refining column, then what specification should be met in the bottom product from the stripping column? If that product already meets final specifications, the recycle flow can be zero. If it does not, a recycle flow will result, proportional to the excess of light component. Since the recycle returns to the stripping column, the two act like a double column having a single purpose. It is doubtful whether an optimum recycle rate exists — shifting the load from one column to another by increasing the recycle flow is not likely to have a significant effect on unit operating cost. In effect, this type of unit is usually more efficient if designed with the minimum number of columns. Even a double column is more efficient than two separate ones since it has a single reboiler and condenser. The refining column, as a rule, is added to allow trimming of the final-product quality and may not be required if control of upstream columns is adequate. A single refining column with major sidestream, as described in Chap. 8, can replace the three columns of Fig. 2.12.

Bojnowski et al. [5] describe a distillation unit of five columns in series used to separate a ternary feed. Overhead products from the last four columns were combined and recycled to the first. Owing to its obvious inefficiency, the unit was rebuilt by eliminating two of the columns and all but one recycle stream. After the revision, the same product purities were achieved at the same production rate, using only one-third as much steam as in the original configuration. Much of the excess steam consumption was due to blending streams that had already been separated.

Blending for Pressure Control   Most condensers reject heat into the atmosphere, either directly, or indirectly through a cooling tower, principally because these are the least expensive means of heat removal. When an overhead product has an atmospheric boiling point that is well above the temperature of the coolant, it will tend to draw a vacuum on condensing. This is generally true for all alcohols and for hydrocarbons containing more than 5 carbon atoms. For some, subatmospheric condensing may only take place during part of the year.

It is customary, however, to operate columns above atmospheric pressure unless there is a specific advantage to vacuum operation, such as minimizing degradation of a monomer like styrene. The principal consideration is probably safety. Leaks allowing product to escape are easier to detect than those allowing air to enter, and they are less likely to result in an explosion. As a consequence, the pressure in many of these columns is maintained above the vapor pressure of the overhead product at condenser temperature, by padding with gas or venting to atmosphere. Where the products are flammable, an inert gas such as nitrogen or carbon dioxide, or even natural gas, may be used. Other columns are simply vented. In either case, overhead product is contaminated by the blending, causing several minor problems:

1. Boiling point is depressed.
2. Relative volatility is decreased.
3. Vented gas carries away product.

The top two or three trays are required to strip the noncondensible gas from the reflux. While its concentration may be quite low, the effect of noncondensible gas on vapor pressure, and therefore boiling point, is considerable. Reflux is introduced at condenser temperature which may be well below the boiling point. This causes condensation of some of the vapors approaching the top tray, reducing overhead vapor flow and augmenting internal reflux, a point which is addressed in Chap. 4. Product quality cannot be inferred by measuring temperature either in the over-head vapor line or in the top two or three trays, owing to the depressed boiling point.

The relative volatility of most mixtures decreases as pressure is increased. Therefore, raising pressure artificially causes an unnecessary reduction in relative vola-tility and a concomitant rise in the energy required to make the separation. Consider, for example, a benzene-toluene column maintained at 14.7 lb/in^2abs with nitrogen at a reflux temperature of 80°F, where the vapor pressure of benzene is only 1.97 lb/in^2abs. Relative volatility for the mixture at the atmospheric boiling point of benzene is 2.6, whereas at 80°F, it rises to 3.2. Using the separation model described in the next chapter, it is estimated that the heat required can be reduced by 43 percent by operating at the lower temperature and pressure.

Every volume of noncondensible gas introduced into a condenser must even-tually be released. This is true whether the gas is added by a pressure regulator, or air is simply drawn into the condenser from a vent. If columns operated in an absolutely steady state, the ebb and flow of gas would be minimal. However, variations in the rate of heat transfer to the condenser affect the temperature difference between the coolant and the condensing vapor. The dew point can only be adjusted by changing the amount of noncondensible gas in the mixture. Therefore, every increase in heat load or coolant temperature will cause pressure to rise and gas to be expelled, while every decrease will cause pressure to fall and gas to be admitted.

While the consumption of gas over the course of time may become significant, losses of product with each venting are probably more costly. The concentration of product vapor in the vented gas is essentially its vapor pressure divided by the absolute pressure maintained in the column. For the benzene-toluene column described above, the vented gas could contain as high as 1.97/14.7 or 0.134 mole fraction benzene. Considering all the penalties associated with pressure control using noncondensible gas, the search for more-effective methods is very worthwhile. Some are described in detail in Chap. 4.

Blending for Temperature Control Streams that are blended must satisfy mass and energy balances, but entropy will rise. Consider the fired heater of Fig. 2.13 where unheated oil is blended with overheated oil to reach the controlled temperature. The mass balance is

$$F_1 + F_2 = F \tag{2.28}$$

If energy is all contained as sensible heat, an energy balance is simply

$$F_1 T_1 + F_2 T_2 = FT \tag{2.29}$$

The entropy of mixing per unit flow can be calculated using the temperature term of Eq. (2.1):

$$\Delta s = C_p \ln T - \frac{F_1}{F} \ln T_1 - \frac{F_2}{F} \ln T_2 \tag{2.30}$$

The entropy rise due to mixing can be evaluated relative to that caused by heating, to estimate performance:

$$\frac{\Delta s}{s - s_1} = \frac{\ln T - (F_1/F) \ln T_1 - (F_2/F) \ln T_2}{\ln T - \ln T_1} \tag{2.31}$$

The impact of the entropy rise is felt at the heater, where a higher-than-needed outlet temperature must be reached, while the flow is reduced thereby lowering its

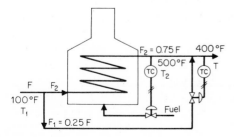

figure **2.13** *Blending the heated and unheated streams is an irreversible operation.*

heat-transfer coefficient. A model of the heater was developed in Ref. 6 to compare the heat lost to the flue gas caused by higher oil temperature and lower coefficient against the entropy rise calculated in Eq. (2.31). The results for the heater of Fig. 2.13 are plotted in Fig. 2.14.

There are additional considerations not accountable by heat-loss or entropy calculations. Both the hot oil and the heat-transfer surface are subject to degradation, the rate of which increases with temperature. The life of both can then be extended by operating the heater at minimum temperature, which precludes blending.

Networks of exchangers for heat recovery are very common in chemical plants and petroleum refineries. To maximize their performance, control is needed to balance flow rates, particularly between parallel paths. To minimize entropy rise, temperatures of streams should be equal before they are combined. Therefore, a control system to balance parallel networks effectively is simply a single controller comparing two exit temperatures and manipulating a three-way valve. This is shown in Fig. 2.15.

A similar problem exists in the balancing of parallel passes within a single heater. For maximum efficiency, all exit temperatures should be equal, regardless of the condition of the heat-transfer surfaces or the distribution of heat within the furnace. Figure 2.16 shows the mixed outlet temperature controlled by fuel flow, while individual pass temperatures are controlled by their own valves. There are only three independent temperatures, yet four valves are provided to control them.

The system shown drives one of the valves to an optimum position, eliminating it from participating in steady-state temperature control. It is the most-open valve which is selected and sent to the valve-position controller. The VPC then adjusts all pass-temperature set points to keep the most-open valve at the set position, typically 90 percent open. In the steady state, all temperatures will be balanced by positioning the other two pass valves relative to the most-open one. Whether the individual temperatures agree precisely with the mixed outlet temperature is not important.

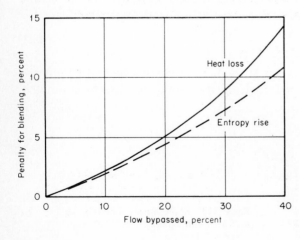

figure 2.14 *Heat loss and entropy rise from blending the two streams in Fig. 2.13.*

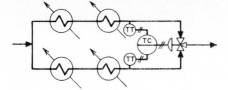

figure 2.15 *The temperature controller balances the streams so that there is no entropy rise when they are mixed.*

HEAT TRANSFER

As with other irreversible operations, the entropy rise associated with heat transfer is a function of the temperature difference between the two fluids. It can only be minimized by approaching zero temperature difference, which requires a heat-transfer surface that is far above the economic optimum. The purpose of this section is not to size a heat exchanger optimally but to control one optimally that is already sized.

Reboilers   Figure 2.17 is a pool-boiling curve for water, illustrating four distinct heat-transfer regimes. To initiate boiling, a sizeable temperature difference must be established across the heat-transfer surface. Once initiated, boiling will begin at a substantial rate if that same ΔT is maintained, because the development of turbulence establishes many nucleation sites. The rate of boiling can then be raised or lowered linearly with ΔT, although the line does not pass directly through zero. There is a residual ΔT_0 at which boiling stops and must be reinitiated; the residual is a function of the roughness of the heat-transfer surface. Specially prepared surfaces can reduce the residual significantly, as does forced circulation; this can be an important factor in heat-pumped columns.

Increasing ΔT will increase heat flux only to the point where all nucleation sites are occupied. Further increase causes vapor bubbles to converge into a film, actually forming an insulating barrier between the surface and the boiling liquid. As a result, heat flux falls with rising ΔT in proportion to the area thus blanketed. This transitional regime is unstable, exhibiting negative resistance: heat flux falls which tends to increase ΔT causing it to fall further. Eventually all nucleation sites are blanketed, and film boiling alone ensues. Heat flux again rises with ΔT;

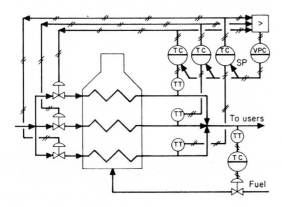

figure 2.16 *The pass-balancing controllers have the same set point, adjusted by the VPC to keep the most-open valve almost fully open.*

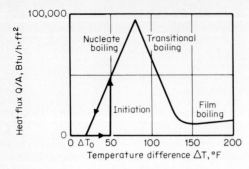

figure 2.17 *A typical pool-boiling curve indicates two unstable regions: the initiation of nucleate boiling and the transition to film boiling.*

however, ΔT is so high and flux so low that this represents an extremely inefficient operating mode.

If steam or another condensing vapor is the heating medium, both sides of the heat-transfer surface operate isothermally. In the nucleate-boiling regime, heat transfer Q is proportional to area A, and the temperature difference between condensing steam at T_1 and boiling liquid at T_2, less the residual ΔT_0:

$$Q = UA \, (T_1 - T_2 - \Delta T_0)$$ (2.32)

where U is the slope of the line representing nucleate boiling.

The steam undergoes entropy reduction in condensing:

$$\Delta s_1 = \frac{-Q}{T_1}$$

whereas the liquid entropy increases by boiling:

$$\Delta s_2 = \frac{Q}{T_2}$$

The net entropy change resulting from the transfer of Q units of heat is

$$\Delta s = Q \left(\frac{1}{T_2} - \frac{1}{T_1} \right) = Q \frac{T_1 - T_2}{T_1 T_2}$$ (2.33)

Considering that work loss is proportional to entropy rise, we have

$$w = T_0 \Delta s = Q(T_1 - T_2) \frac{T_0}{T_1 T_2}$$ (2.34)

This is essentially the Carnot equation relating work and heat in a heat engine or in a heat pump. Work w is the minimum that would be required to pump heat Q from lower temperature T_2 to higher temperature T_1, or what is lost in irreversible heat transfer from T_1 to T_2. While entropy and work are not considered important for conventionally reboiled columns, the above expression governs the efficiency of heat-pumped and refrigerated columns. Note the similarity to the expression for power consumption in pumping fluids, where flow was multiplied by pressure difference.

Nonisothermal heat transfer takes place when a liquid such as hot oil is supplied to the reboiler. Because its temperature falls as it passes through the reboiler, its average temperature must be higher than an isothermal fluid transferring the same heat. Therefore, nonisothermal heat transfer results in a greater entropy rise [7]. Additionally, heat-transfer coefficients tend to be lower, further increasing ΔT.

The efficiency of a hot-oil system can be improved using the controls shown in Fig. 2.18. Instead of supplying hot oil at a fixed temperature and a flow varied according to demand by the users, flow is maximized and temperature is set as low as is necessary to satisfy the most-demanding user. Again, the position of the most-open valve is selected and sent to the VPC, which slowly adjusts supply temperature. This will upset the users enough to reposition their valves, so that the most-open will reach the set point of the VPC in the steady state.

Maximizing flow in this way will maximize the heat-transfer coefficients in both users and heater. Minimizing oil temperature will reduce stack losses, extend the life of both the oil and the heat-transfer tubing, and reduce reboiler fouling. The additional cost of pumping is insignificant for a centrifugal pump operated at constant speed [3]. The differential-pressure controller (DPC) opens the bypass valve to maintain flow through the heater only during extremely low loads; it would rarely be required to act, as long as the VPC is functioning.

Condensers   Most condensers operate nonisothermally, because heat is transferred to air or water, thus raising coolant temperature. Many also have nonisothermal conditions inside, so that reflux is subcooled well below its dew-point temperature. This happens when columns are pressurized with gas or vented to atmosphere as described earlier. It also results when condensers are flooded intentionally with condensate to reduce heat transfer and thereby control column pressure.

Practices which maintain an unnecessarily high temperature difference between the coolant and the condensate are thermodynamically inefficient. Entropy rises in proportion to the temperature difference, resulting in lost work, or at least unrecovered work. There are two basic approaches toward this problem: One may attempt to recover the lost work for use elsewhere or reduce the work used by the process by increasing its efficiency. The second approach is generally more favor-

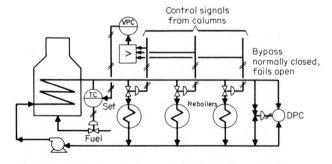

figure 2.18 *The valve-position controller maximizes flow and minimizes oil-supply temperature, both improving the efficiency of the system.*

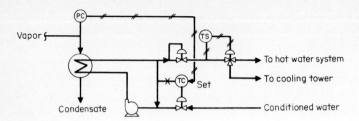

figure 2.19 *Circulation of cooling water allows its temperature to rise to a point where it may be useful for heating elsewhere.*

able, requiring less investment in equipment and not needing a partner willing to accept the recovered work.

A work-recovery system is shown in Fig. 2.19. Cooling water is circulated at a high rate through the condenser to minimize temperature differences. Pressure control is then achieved by adjusting condenser outlet temperature; addition of cold water to the loop causes an equal amount of hot water to be discharged. If it is sufficiently hot, it can be used elsewhere in the plant, where its available work is then recovered. The concept can also be extended to making the condenser a waste-heat boiler for generating steam to be used elsewhere. Higher column pressure will result in higher water temperature or generated steam pressure, making the recovered heat more useful, but the column separation more difficult.

Alternatively, column pressure can be reduced to its lowest controllable level, again by maximizing flow through the condenser and thereby minimizing temperature differences. A system to accomplish this is shown in Fig. 2.20. The VPC drives the condenser valve toward an extreme position which will maximize loading. If the valve admits cooling water, it is driven open; a condenser bypass valve would be driven closed. Again, the VPC should have a set point which allows about 10 percent margin from the end of the valve stroke, to maintain effective pressure control.

As in other applications, the VPC should have only the integral control mode to minimize transient upsets to the column. Its time constant should be long enough to allow the sensible heat stored within the column to be removed gradually as temperatures fall. Figure 2.21 describes how the pressure controller would move the valve quickly to maintain pressure at the onset of rain to its air-cooled condenser. The VPC then reacts by gradually reducing pressure set point until the valve returns to its minimum position.

The principal advantage of allowing the column pressure to float on the coolant is that it reduces energy requirements for the separation by raising relative vola-

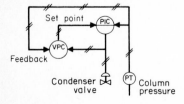

figure 2.20 *The valve-position controller slowly adjusts the pressure set point to keep the condenser fully loaded in the long term.*

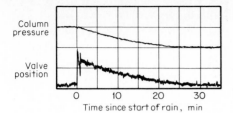

figure 2.21 *The column pressure moves exponentially to a new steady state as the condenser valve returns to its minimum position.*

tility. It can also increase column capacity. However, it does complicate column operation and controls by making pressure a function of coolant temperature and heat load. These details are covered in Chap. 4.

Refrigeration   Refrigeration is frequently used to pump heat from a column condenser to an atmospheric heat sink, in the manner shown in Fig. 2.22. Refrigerant boils in the column condenser at a low pressure, with its vapor then being compressed to a higher temperature. The atmospheric condenser removes superheat and latent heat from the compressed vapor, returning liquid refrigerant to the receiver under pressure.

Heat Q is removed from the column by work w applied to the compressor. To satisfy the energy balance, the heat rejected to the atmosphere is $Q + w$. The relationship between heat flow and work for the reversible process was given in Eq. (2.34), where T_1 and T_2 refer to the temperatures where the refrigerant boils and condenses. Boiling temperature will always be lower than the dew point in the column condenser by the ΔT required to transfer Q. By the same token, condensing temperature will also exceed that of the atmospheric heat sink by the ΔT required to transfer $Q + w$.

The actual temperature rise across the compressor is greater than $T_1 - T_2$ by the superheat developed during compression. This reduces the efficiency of the cycle somewhat, and attempts may be made to remove it by interstage cooling or injection of the cold vapor produced by flashing liquid across the expansion valve. A second source of inefficiency is the valve itself, whose entropy rise produces the flashed vapor, which is simply recirculated without contributing to heat removal.

From all the above considerations, it is clear that restrictions to heat transfer at both heat exchangers must be eliminated if the work required of the compressor is to be minimized. In many installations, the compressor suction pressure is controlled at a constant value, which fixes the boiling point of the refrigerant. Then the

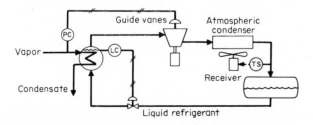

figure 2.22 *The column condensing tubes should be completely immersed in refrigerant to maximize heat transfer per unit power.*

rate of heat transfer is adjusted by manipulating the expansion valve. Because of the fixed boiling point, the mechanism by which the valve regulates heat transfer is by changing the level of boiling refrigerant and therefore the area of wetted heat-transfer surface. This is thermodynamically inefficient in that it restricts heat transfer; it is also dynamically slow in that refrigerant inventory must be moved to achieve control.

In the system of Fig. 2.22, the entire heat-transfer surface is immersed through the action of a level controller. Heat transfer is changed when the column pressure controller manipulates inlet guide vanes (compressor speed could be varied instead). The boiling of the refrigerant is then raised as high as possible to control column pressure, thereby minimizing compressor ΔT. Control response is rapid, too, because changing suction pressure induces or retards boiling immediately.

Most installations also provide control over discharge pressure or refrigerant condensing temperature by manipulating heat transfer from the refrigerant condenser. This is neither necessary nor desirable through the normal operating range, as it elevates the condensing temperature above the minimum attainable. That temperature should always be as low as the coolant will allow, down to the limit where operation of the compressor or expansion valve may be adversely affected. At this limit, simple on-off control over fans may be provided.

Where a single refrigeration system supplies several users, there is a common return header whose pressure is regulated at the compressor suction to provide a stable sink for all the users. Then each user will manipulate its own vapor return valve to adjust heat transfer. Compressor suction pressure should be set to minimize the pressure drop across those return valves. This can be accomplished by comparing all the user valve positions in a high selector and controlling that of the most-open valve by adjusting suction-pressure set point. The controls would appear as in Fig. 2.10, except that the control valves would be located at the suction of the compressor.

In those plants where several users are supplied with chilled water or brine, its supply temperature may be raised to a level which will just satisfy the most-demanding user. Here the control system shown in Fig. 2.18 applies.

THERMODYNAMIC ANALYSIS OF DISTILLATION

This section is devoted to the determination of thermodynamic efficiency of a column as a function of its heat source and sink. Whereas there are some opportunities for improving efficiency by selection of column internals, source and sink are seen to play dominant roles.

Column Analysis   Consider a distillation column separating a 50-50 mixture of propane and propylene into products each containing 1.0 mol % impurity, in 200 trays at 70 percent tray efficiency. At a reflux temperature of 100°F, column pressure is 225 lb/in² abs. Under these conditions, the relative volatility of the mixture is about 1.135. Using the column model developed in the next chapter, the required reflux ratio is found to be 15.0; the vapor-to-feed ratio is then 8.0, and the heat required is about 45,000 Btu/lb·mol of feed. Assuming a pressure drop of 2 in of water per tray, base pressure would be 240 lb/in² abs and base temperature 120°F.

If feed and products are all liquids at the same pressure and temperature, their entropy differences are only a matter of composition. Feed entropy is then

$$s_F = -R(2)(0.5 \ln 0.5) = 0.693R$$

Entropies of distillate and bottom products are equal:

$$s_D = s_B = -R(0.99 \ln 0.99 + \ln 0.01) = 0.056R$$

The entropy change per mole of feed is

$$\Delta s_p = 0.5\, s_D + 0.5\, s_B - s_F = -0.637R = -1.266 \text{ Btu/lb} \cdot \text{mol} \cdot {}^\circ\text{F}$$

(The value chosen for R in units consistent with this example is 1.987 Btu/lb·mol·°F.)

If there were no temperature drop across either reboiler or condenser, the entropy change related to the transfer of heat would simply be

$$\Delta s_e = 45{,}000 \left(\frac{1}{460 + 100} - \frac{1}{460 + 120} \right) = 2.80 \text{ Btu/lb} \cdot \text{mol} \cdot {}^\circ\text{F}$$

The thermodynamic efficiency of such a column would be

$$\eta = -100 \frac{-1.266}{2.80} = 45.3\%$$

Heat Source and Sink   The unusually high thermodynamic efficiency calculated above is primarily the result of a minimum temperature difference between the heat source and sink. It is simply the combination of 15°F boiling-point difference between propane and propylene, and 5°F related to the pressure drop of 15 lb/in² across the trays. If a temperature drop as little as 10°F is taken across both condenser and reboiler, the total difference between the heat source and sink is doubled, and the thermodynamic efficiency is cut in half.

In actual practice, the heat sink is the atmosphere. The heat source is more difficult to locate. It becomes necessary to trace the flow of heat from the reboiler back to its point of origin. If the heating medium is steam produced by firing a low-pressure boiler, then the heat source is the flame within the boiler. If the steam is exhaust from a turbine, however, then the turbine is the heat source and its exhaust temperature is source temperature. (In the same manner, the column becomes a heat sink to the turbine, with the condensing temperature of steam in the reboiler being sink temperature.)

The thermodynamic efficiency of a column is then principally a function of how well it is matched to source and sink. Using heat sources hotter than necessary makes poor use of resources, augmenting environmental impact signified by Δs_e. In an olefins plant where the propane-propylene splitter is typically found, reactor quench-water at 180°F is available for heating the reboiler. It is obtained by removing the sensible heat from the effluent of cracking furnaces, acting as the heat sink for those furnaces.

For purposes of comparison, the thermodynamic efficiency of the propane-propylene column has been calculated for a variety of heat sources and an atmospheric sink at 80°F. The results appear in Table 2.1.

TABLE 2.1   Thermodynamic Efficiency of Propane-Propylene Column Using Various Heat Sources

Source	T, °F	Efficiency %
Combustion at 80% efficiency	3000	1.4
Exhaust steam at 20 lb/in² abs	228	7.0
Quench-water at 180°F	180	9.6
Work-driven heat pump		13.6
Fossil-fueled heat pump		4.6
Exhaust steam at 20 lb/in² abs, double effect	228	11.1

Heat-Pumped Column   Rather than reject all the heat leaving a column to the atmosphere, some of it may be upgraded enough to be applied to the reboiler of the same column. Figure 2.23 gives an example for the propane-propylene column. Propylene vapor is compressed enough to condense against boiling propane. The work introduced by the compressor must be removed to an external sink to satisfy the column's energy balance. But by locating the heat sink at the compressor discharge, the column may be operated at a temperature below that of the sink. This offers an improvement in relative volatility which can be quite significant.

For the propane-propylene column at 70°F reflux temperature, relative volatility is increased to 1.152. This reduces reflux ratio to 12.2, V/F to 6.6, and the heat requirement to 42,000 Btu/lb·mol of feed. The minimum work required to pump that amount of heat can be calculated using Eq. (2.34). Propylene is boiling at 70°F, condensing at 110°F, and the environmental temperature is 80°F.

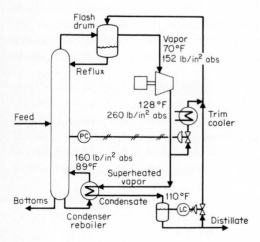

figure 2.23 *A heat pump is most useful when separating mixtures having a narrow boiling range.*

$$w_{min} = \frac{42,000 \, (110 - 70) \, (460 + 80)}{(460 + 110) \, (460 + 70)} = 3000 \text{ Btu/lb·mol}$$

Applying an efficiency of 60 percent to the compression cycle, actual work is estimated at 5000 Btu/mol feed.

Observe that this is only 12 percent of the heat flow (only 11 percent of the heat used by the column with conventional reboiler and condenser). The advantage of heat pumping is obvious. The work requirement can be reduced further by reducing the ΔT across the condenser-reboiler with an enhanced surface [8]. Removing 10°F from the ΔT could reduce actual work by one-fourth, since that represents one-fourth of the overall temperature difference.

Those separations featured by narrow boiling ranges and high reflux ratios offer the best opportunities for applying heat pumps. The propane-propylene separation is perhaps the most common example, but there are others as well. Null [9] made an economic study of a host of separations to outline the areas where heat pumps are cost-effective.

To complete the thermodynamic analysis of the heat-pumped column, it is necessary to investigate the work source. If electricity is used to drive the compressor, the entropy change in the conversion to work is zero. Then the entropic impact is simply that of converting work w to waste heat at ambient temperature:

$$\Delta s_e = \frac{5000}{(460 + 80)} = 9.27 \text{ Btu/lb·mol feed}$$

which gives an efficiency of 13.6 percent.

However, one additional factor needs to be considered — the source of the electric power. If electricity is generated in a fossil-fueled power plant, the efficiency of the plant is typically 33 percent. That is, only one-third of the energy in the fuel is converted into electricity, with the remainder being rejected to the environment as waste heat. Therefore, the total thermal impact on the environment is three times that estimated for the heat pump alone:

$$\Delta s_e = 3 \times 9.27 = 27.8 \text{ Btu/lb·mol feed}$$

which brings the efficiency of the heat-pumped column to 4.6 percent.

On an economic basis, the cost of power to drive the compressor may be five times as great per unit energy as low-pressure steam. The operating cost for the heat-pumped column should still be lower because its heat-to-work ratio is about 9. But if waste heat is available in the form of quench-water, for example, then *it* is the least-cost source, as well as the most efficient.

Double-Effect Distillation The reason for the wide variation in efficiencies for the same distillation column is the mismatch between the boiling range of the mixture and the source-sink temperature difference. Yet mismatching is quite common, owing to the lack of understanding of applied thermodynamics. As fuel costs continue to escalate, there will be a growing demand for increased efficiency. One way to better match the process to the source-sink ΔT is by cascading distillation columns.

Figure 2.24 shows cascaded propane-propylene columns. They are fed in parallel, but heat flows through them in series as in a double-effect evaporator. Because of the narrow boiling range of the mixture, the two columns easily fit between the low-pressure steam source and the cooling-water sink. The first column must operate at a higher pressure than the second to create enough ΔT across the common condenser-reboiler to carry the full flow of heat to the second column.

At the higher pressure, the relative volatility of the mixture is only 1.10 compared to 1.135. If the columns have the same number of theoretical trays, then the first column requires a higher reflux ratio than the second: 33 compared with 15. The latent heat of vaporization is also different: 106 Btu/lb in the first column overhead, against 132 Btu/lb in the second. Taking all these factors into account, the first column can only separate 37 percent of the feed while the second can accommodate 63 percent at the same heat flux. With single-effect distillation, the low-pressure column processed all the feed; but because now it needs to process only 63 percent of the feed, the remaining 37 percent is processed at no heat cost. In effect, the heat required per unit feed is reduced by 37 percent to 28,000 Btu/lb·mol.

To estimate the thermodynamic efficiency of the double-effect distillation, that of the single-effect column having the same heat source and sink is divided by the lower heat flow, 63 percent. This configuration is the most efficient of all those in Table 2.1 except for the heat-pumped column driven by a source of pure work (e.g., hydropower).

There are many possibilities for double-effect distillation. The benzene-toluene column described earlier could be operated at atmospheric pressure by condensing at 180°F in the reboiler of the propane-propylene column. The thermal properties of the two systems are well-matched, although they both might not be found in the same plant. Furthermore, their production rates would also have to match in both steady and unsteady states, if supplementary heat and cooling were to be avoided. Finally, in controlling product qualities, a degree of freedom is sacrificed to the common heat flux.

While there are obvious advantages to multiple-effect distillation, the obstacles are many and thereby limit the opportunities. The examples described of parallel

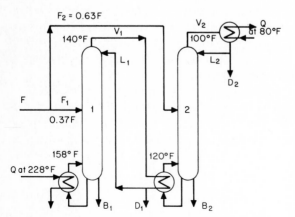

figure 2.24   *The cascading of distillation columns takes advantage of a source-sink temperature difference that is greater than a single column needs.*

columns in Fig. 2.24, and of independent separations such as the benzene-toluene and propane-propylene columns, are not the most likely candidates. The best opportunities are in serial separations of multicomponent mixtures, where there is a natural progression of boiling points, and feed rates are related. This arrangement becomes more like a true multiple-effect evaporator, in that both heat and feed flow serially. These applications are discussed in Chap. 8, "Multiple-Product Processes," where special consideration is given to column interaction.

REFERENCES

1. White, M. H.: "Surge Control for Centrifugal Compressors," *Chem. Eng. (N.Y.)*, December 25, 1972.
2. Shinskey, F. G.: "Effective Control for Automatic Startup and Plant Protection," *Can. Controls Instrum.*, April 1972.
3. Shinskey, F. G.: "Flow and Pressure Control Using Variable-Speed Motors," presented at the Control Engineering Conference, Chicago, May 18–20, 1982.
4. Ryskamp, C. J.: "Using the Probability Axis for Plotting Composition Profiles," *Chem. Eng. Prog.*, September 1981.
5. Bojnowski, J. H., J. W. Crandall, and R. M. Hoffman: "Modernized Separation System Saves More Than Energy," *Chem. Eng. Prog.*, October 1975.
6. Shinskey, F. G.: *Energy Conservation through Control,* Academic, New York, 1978, pp. 16, 17.
7. Reference 6, p. 19.
8. Wolf, W.: "High Flux Tubing Conserves Energy," *Chem. Eng. Prog.*, July 1976.
9. Null, H. R.: "Heat Pumps in Distillation," *Chem. Eng. Prog.*, July 1976.

CHAPTER 3

Factors Determining Compositions

A control system is a combination of mathematical functions using process output information to change process inputs. The closer its functions complement the mathematical relationships inherent within the process, the more effective that system will be in achieving its objectives. Therefore, it is imperative that the distillation process be modeled accurately to assist the designer in configuring a control system and to predict its performance.

The type of mathematical model presented in this chapter is fundamentally different from those used by column designers. The control engineer is not interested in estimating the optimum number of trays to perform a given separation, but rather how compositions can be changed by adjusting column inputs, given a fixed number of trays. Furthermore, a column is designed once, whereas its operating conditions may change almost daily. The type of model required for control purposes therefore cannot be a complex digital program requiring solution on a large computer. Instead, the model needs to be simple enough for analytical solution and

differentiation, yet accurate enough for purposes of optimization. After considerable search and evaluation, the characterization presented in this chapter has
been found to satisfy the preceding objectives.

MATERIAL BALANCES

The first and foremost relationships in any process are its material balances. They
have the dual advantage of being capable of precise statement and precise measurement. For some processes, material balances alone are sufficient to completely
relate all inputs and outputs; this is true where streams are blended. For many
others, a combination of material and energy balances is necessary; this is the case
for heat exchangers and evaporators. Unfortunately, distillation is more complex
than all of these, requiring a solution to liquid-vapor equilibria as well. Yet the
material-balance equations represent the first step in solving the problem.

   External Balances   External material balances are easier to write than to enforce; yet they constitute an essential factor in achieving control over product
compositions. At this point, it is only necessary to describe them mathematically.
Later, methods are presented to enforce them.

   Consider the two-product column appearing in its simplest form in Fig. 3.1. The
overall material balance is stated as

$$F = D + B \tag{3.1}$$

and the balance on each component i:

$$Fz_i = Dy_i + Bx_i \tag{3.2}$$

These two expressions may be combined by eliminating B to develop the D/F ratio:

$$\frac{D}{F} = \frac{z_i - x_i}{y_i - x_i} \tag{3.3}$$

or by eliminating D to develop the B/F ratio:

$$\frac{B}{F} = \frac{y_i - z_i}{y_i - x_i} \tag{3.4}$$

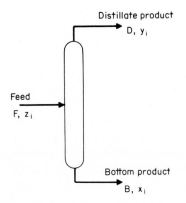

Distillate product

D, y_i

Feed

F, z_i

Bottom product

B, x_i

figure 3.1   *The external material
balance for a two-product
column.*

or by eliminating F to develop the D/B ratio:

$$\frac{D}{B} = \frac{z_i - x_i}{y_i - z_i} \tag{3.5}$$

Recognize that these three ratios are all interdependent: fixing one fixes the others as well through the overall material balance. In essence, there is only one independent material-balance relationship involving product compositions x_i and y_i. This single equation with two unknowns is insufficient to determine both compositions, but serves as half of the solution — the other half is provided by vapor-liquid equilibrium.

The equations above may be evaluated for any component i, and all evaluations must agree — the choice is inconsequential to the solution. For a multicomponent separation, however, there are some considerations which will prove helpful in expediting a solution. In general, it is assumed that all components lighter than the light key leave exclusively in the distillate and that all components heavier than the heavy key exit with the bottom product. These off-key components may then be lumped in two groups so that there are only four mathematically distinguishable components in the feed mixture.

Then Eqs. (3.3) to (3.5) may be written as shown for the heavy (H) and light (L) keys. But for the lighter components, $x_{LL} = 0$; then y_{LL} can be calculated from D/F and z_{LL} by reversing Eq. (3.3):

$$y_{LL} = \frac{z_{LL}}{D/F} \tag{3.6}$$

Furthermore, all mole fractions in the distillate must add to unity, so that

$$y_L = 1 - y_{LL} - y_H \tag{3.7}$$

In the same way, x_{HH} may be calculated from D/F and z_{HH}:

$$x_{HH} = \frac{z_{HH}}{1 - D/F} \tag{3.8}$$

and all mole fractions in the bottom add to unity:

$$x_H = 1 - x_{HH} - x_L \tag{3.9}$$

While D/F was used in Eqs. (3.6) and (3.8), it should be apparent that $1 - B/F$ could be substituted if desired.

The last four equations permit the complete solution for all components in the two products from a multicomponent separation, once the key concentrations have been found. The multicomponent two-product problem can then be solved as readily as a binary problem, regardless of the number of components in the feed. Examples of both are solved after the development of the separation model.

A graphical representation of the external material balance is helpful in appreciating the relationship. Figure 3.2 is a plot of y_L versus x_L for a particular ratio D/F (or B/F or D/B). Feed composition z_L is given; then adjustment to the D/F ratio

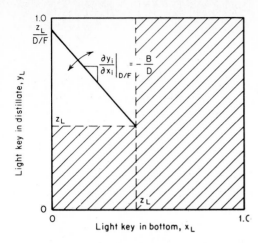

figure 3.2 *Product compositions are changed by pivoting the material-balance line about the feed composition.*

will cause the material-balance line to pivot about z_L. The slope of the line can be determined by solving Eq. (3.3) for y_i and differentiating at constant D/F:

$$y_i = x_i + \frac{z_i - x_i}{D/F} \tag{3.10}$$

$$\left. \frac{\partial y_i}{\partial x_i} \right|_{D/F} = 1 - \frac{1}{D/F} = -\frac{B}{D} \tag{3.11}$$

The y intercept can be determined by setting x_i to zero in (3.10). If D/F is set at z_L, then the line will pass through the point $(0, 1)$ representing pure products.

Increasing D/F will rotate the line counterclockwise, reducing purity y_L. Extreme values of D/F, i.e., zero and unity, correspond to vertical and horizontal positions, respectively. Operation is excluded from the shaded area of the plot.

Again, the material-balance relationship is insufficient in itself to establish the operating point; it could lie anywhere along the line. The graphical representation is developed more completely later in this chapter as other relationships between x and y are formulated.

Reflected Balances   A real column is, of course, much more complicated than the simple diagram of Fig. 3.1. Vapor generated at the reboiler is carried through the column, condensed, and split into distillate product and reflux, the reflux returned. Figure 3.3 shows the additions of vapor and reflux streams. The reflux is necessary to establish an equilibrium between liquid and vapor; without it, the trays above the feed point would be dry and therefore useless.

These internal flow rates tend to interfere with the establishment of a desired external material balance. If the flows of overhead vapor V and reflux L are constant, for example, then changing distillate flow D will have no effect on compositions but only cause the liquid level in the reflux accumulator to rise or fall. If the level is held constant, necessary to reach a steady state, then D is fixed by V and L.

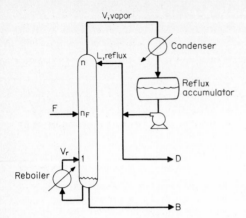

figure 3.3 *The relative values of vapor and reflux flows bear on the external balance.*

A steady-state material balance on the reflux accumulator gives

$$V = L + D \tag{3.12}$$

For the case of a total condenser as shown here, the compositions of the three streams are identical, so a component balance gives no new information. (Where a vapor product is taken, a component balance is necessary to relate the compositions of the different streams, and an equilibrium statement is also required.)

The principal factor affecting separation will be shown to be the reflux ratio L/D. At this point, it is helpful to see how it relates to ratios of D/F and V/F which have already been discussed. Dividing (3.12) by F gives

$$\frac{V}{F} = \frac{L}{F} + \frac{D}{F}$$

The reflux-to-feed ratio can then be factored into L/D and D/F, yielding

$$\frac{V}{F} = \frac{D}{F}\left(\frac{L}{D} + 1\right) \tag{3.15}$$

Another important consideration is the interrelationship of all ratios of these three overhead streams. Dividing Eq. (3.12) by D gives

$$\frac{V}{D} = \frac{L}{D} + 1 \tag{3.14}$$

Dividing it by L gives

$$\frac{V}{L} = 1 + \frac{D}{L} \tag{3.15}$$

It can then be seen that fixing any of the three ratios L/D, D/V, or L/V fixes them all. The result is similar to that observed in Eqs. (3.3) to (3.5), where D/F, B/F, and D/B were seen to be mutually dependent.

These two independent sets of three dependent ratios appear again and again as the principal factors through which product compositions are determined.

Internal Balances   Ultimately, it is the *internal* flow rates which determine product compositions. If the reflux returns to the column at the boiling point, then it will condense no vapor upon reaching the top tray, so that the internal vapor rate will be the same as the overhead vapor rate. Then internal reflux will also be the same as external flux.

However, subcooling of the reflux is quite common and should be considered. Material and energy balances at the top of the column in Fig. 3.4 will illustrate the difference between external and internal (subscript t) flow rates of vapor and reflux:

$$V_t - V = L_t - L \tag{3.16}$$

$$V_t(H_0 + \Delta H_D) + L H_L = V(H_0 + \Delta H_D) + L_t H_0 \tag{3.17}$$

where H_0 = enthalpy of product as liquid at boiling point
$\quad H_L$ = enthalpy of reflux
$\quad \Delta H_D$ = latent heat of vaporization of distillate
Rearranging (3.17) gives

$$(V_t - V)(H_0 + \Delta H_D) = L_t H_0 - L H_L$$

Then $L_t - L$ may be substituted for $V_t - V$ from (3.16) so that

$$L_t = L\left(1 + \frac{H_0 - H_L}{\Delta H_D}\right) \tag{3.18}$$

At this point, it is convenient to define *specific enthalpy* of a stream as

$$q \equiv \frac{H - H_0}{\Delta H_D} \tag{3.19}$$

Note that this definition has two very specific reference points, namely, enthalpy and latent heat of the distillate at its boiling point. Using the specific enthalpy of the reflux, (3.18) may be rewritten as

$$L_t = L(1 - q_L) \tag{3.20}$$

Because q_L will always be zero or negative, L_t will always equal or be greater than L.

If there is a substantial temperature gradient across a column, there will be an exchange between latent and sensible heats as vapor and liquid progress through the trays. Internal vapor rate tends to increase with its upward travel into cooler regions, and liquid rate tends to increase with downward flow into warmer trays.

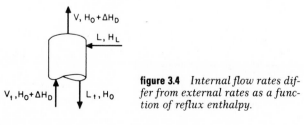

figure 3.4   *Internal flow rates differ from external rates as a function of reflux enthalpy.*

Only where extremely large gradients are encountered, as in crude-oil distillation, is this likely to cause a problem. Then, side coolers are provided to reduce vapor rates and increase liquid rates at points high in the column.

A shift in internal balances occurs at the feed tray, where vapor upflow is augmented by the vapor in the feed and liquid downflow by its liquid fraction. A subcooled feed will condense vapor as did a subcooled reflux, and a superheated feed will boil some of the liquid.

Rather than analyzing conditions at the feed tray in detail, it would seem more profitable to examine an overall energy balance. Consider the column with its condenser and reboiler in Fig. 3.5. An overall energy balance is

$$FH_F + Q_r = Q_c + DH_D + B_{HB} \tag{3.21}$$

Heat input to the reboiler generates a proportional flow of vapor:

$$Q_r = V_r \, \Delta H_B \tag{3.22}$$

Heat removed from the condenser is related to overhead flow rates:

$$Q_c = V(H_0 + \Delta H_D - H_D) \tag{3.23}$$

Each of these expressions may be substituted into (3.21); in addition, FH_0 may be subtracted from both sides. Next, all terms may be divided by ΔH_D, allowing specific enthalpies to be substituted for absolute enthalpies. The result of all this manipulation is

$$Fq_F + V_r \frac{\Delta H_B}{\Delta H_D} = V(1 - q_D) + Dq_D + Bq_B \tag{3.24}$$

If q_D and q_L are identical, Lq_L may be substituted for $(V - D)q_D$:

$$Fq_F + V_r \frac{\Delta H_B}{\Delta H_D} = V - Lq_L + Bq_B \tag{3.25}$$

Equation (3.25) provides an estimate of reboiler vapor rate, and therefore heat input, as a function of the enthalpies and rates of feed, reflux, and bottom product.

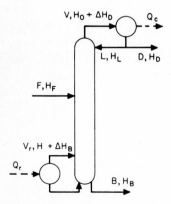

figure 3.5 *An overall energy balance is required to compare reboiler and condenser duties.*

Note particularly that the ratio $\Delta H_B/\Delta H_D$ may be significantly different from unity. For a methanol-water column, for example, it will be about 2.1. Latent heats vary inversely with temperature, so that ΔH_D tends to exceed ΔH_B when the two products are similar. For the propane-propylene separation at 100°F reflux temperature and 120°F base temperature, $\Delta H_B/\Delta H_D$ is 0.947.

Interactions between Energy and Material   Notwithstanding the rigorous relationships existing between external flow rates and product compositions in Eqs. (3.3) to (3.6), compositions are actually established by internal rates. If, for example, the relationship in Eq. (3.3) is desired, then control over the D/F ratio must be directly imposed. If it is not, then D will simply become the difference between L and V.

This creates two problems: Relatively small variations in V and L for columns with high L/D ratios will produce percentagewise large variations in D/F and hence in product compositions. Second, V is not capable of being controlled directly and therefore is subject to variations in V_r, L, q_L, F, and q_F:

$$V = V_r \frac{\Delta H_B}{\Delta H_D} + Lq_L + Fq_F - Bq_B \tag{3.26}$$

If D is made dependent on V and L,

$$D = V_r \frac{\Delta H_B}{\Delta H_D} - L\,(1 - q_L) + Fq_F - Bq_B$$

Substituting for B produces

$$D = \frac{V_r(\Delta H_B/\Delta H_D) - L(1 - q_L) + F(q_F - q_B)}{1 - q_B}$$

Variations in D caused by changes in other variables can be examined by differentiation:

$$dD = \frac{dV_r(\Delta H_B/\Delta H_D) - dL\,(1 - q_L) + L\,dq_L + dF\,(q_F - q_B) + F\,dq_F}{1 - q_B}$$

(Enthalpy q_B was not differentiated owing to its low and nearly constant value.)

Next, the differential may be divided by D to express inputs and outputs in relative terms; then V_r/D and the ratio of latent heats is replaced by $1 + L/D$:

$$\frac{dD}{D} = \frac{\dfrac{dV_r}{V_r}\left(1 + \dfrac{L}{D}\right) - \dfrac{dL}{L}\dfrac{L}{D}(1 - q_L) + \dfrac{L}{D}dq_L + \dfrac{dF}{F}\dfrac{F}{D}(q_F - q_B) + \dfrac{F}{D}dq_F}{1 - q_B}$$

$$\tag{3.27}$$

In the preceding expression, there is one variation that could be favorable: If $(q_F - q_B)/(1 - q_B)$ is close to the desired value of D/F, then feed variations dF will produce *desired* changes dD. All other variations may be considered unfavorable. They may be combined as if they were random using the root-mean-square technique:

$$\frac{dD}{D} = \sqrt{\sum\left[\left(\frac{dm}{m}\right)\left(\frac{\partial D}{\partial m}\right)\left(\frac{m}{D}\right)\right]^2 + \sum\left[\left(\frac{m}{D}\right)dq\right]^2} \tag{3.28}$$

where m is each flow rate and q each enthalpy. The method is best illustrated by example.

example 3.1

Consider a column having $L/D = 10$, $D/F = 0.5$, and $q_L = -0.1$. Let V_r and L be controllable to within 1.0 percent of value, and let q_L and q_F vary by 0.02. Estimate the effects of these variations on D. Ignore variations in F.

$$V_r: \quad [(0.01)11]^2 \quad = 0.0121$$

$$L: \quad [(0.01)10(1.1)]^2 = 0.0121$$

$$q_L: \quad [10(0.02)]^2 \quad = 0.0400$$

$$q_F: \quad [2(0.02)]^2 \quad \quad = \underline{0.0016}$$
$$0.0658$$

$$\frac{dD}{D} = \sqrt{0.0658} \quad \quad = 0.2565$$

These small variations cause 25 percent variations in D and therefore in the D/F ratio, principally through amplification by the L/D ratio. If D is controlled directly to 1.0 percent, its variations can be reduced 25-fold.

The largest disturbance in the example above was in reflux enthalpy:

$$q_L = \frac{H_L - H_0}{\Delta H_D} = \frac{C_p}{\Delta H_D}(T_L - T_0) \tag{3.29}$$

where C_p is the heat capacity of the reflux and T_L and T_0 are its temperature and boiling point, respectively. Their difference is the degree of subcooling of the reflux. For propylene at 100°F, the ratio $C_p/\Delta H_D$ is $0.00541°F^{-1}$. Then a change in q_L of 0.02 as in the above example would be caused by a change in reflux temperature of only 3.7°F.

The above discussion explains why columns having very high reflux ratios have not been controlled satisfactorily by manipulating boilup and reflux independently [1, 2]. They are too sensitive to heat losses and coolant disturbances, until either V or L is placed under level control allowing D to be set directly.

VAPOR-LIQUID EQUILIBRIA

Separation between components of a mixture takes place by virtue of the equilibrium that exists between liquid and vapor phases. When a liquid boils, the vapor formed tends to be richer in the more-volatile component(s) of the mixture. Fortunately, the relationship between the compositions of the two phases can be described mathematically, at least for ideal mixtures. These basic equations still hold true (with correction), for many nonideal mixtures, even with multiple components.

Ideal Mixtures   A review of the fundamental relationships existing in ideal mixtures will be helpful in laying the groundwork for later derivations. To begin, Dalton's law gives the total pressure in a gaseous mixture as the sum of the partial pressures of the components:

$$p = \sum_{1}^{n} p_i \tag{3.30}$$

In an ideal mixture, the partial pressure exerted by a component is proportional to its concentration in the vapor:

$$p_i = y_i p \tag{3.31}$$

Here y_i is the mole fraction of component i in the vapor.

Raoult's law states that in an ideal mixture the partial pressure exerted by a component at equilibrium is the product of its molar concentration x_i in the liquid times the vapor pressure p_i° of the pure component at that temperature:

$$p_i = x_i p_i^\circ \tag{3.32}$$

Combining the last two relationships allows us to express the concentrations of vapor and liquid relative to one another:

$$\frac{y_i}{x_i} = \frac{p_i^\circ}{p} = K_i \tag{3.33}$$

The ratio y_i/x_i or K_i is known as the *equilibrium vaporization ratio* or *K factor* for component i. For nonideal systems, experimentally determined values of K should be used instead of vapor-pressure data to evaluate Eq. (3.33).

The K factors change with temperature and pressure. They can be used to establish the composition of a vapor in equilibrium with a liquid mixture at a specific temperature and also to determine the vapor pressure of the mixture. At equilibrium, the set of y_i generated from a set of x_i and K_i must sum to unity. If they do not, then the pressure is not the vapor pressure of the mixture. The following example shows how to estimate vapor pressure using K factors in a trial-and-error procedure.

example 3.2

Given a liquid mixture containing 5 mol % propane, 90 percent isobutane, and 5 percent n-butane, estimate its vapor pressure and equilibrium vapor composition at 100°F. As a starting point, assume that $p = 72$ lb/in² abs, the vapor pressure of isobutane.

	x	$p = 72$ lb/in² abs		$p = 76$ lb/in² abs	
		K	y	K	y
Propane	0.05	2.25	0.113	2.13	0.107
Isobutane	0.90	1.00	0.900	0.95	0.855
n-Butane	0.05	0.74	0.037	0.70	0.035
			1.050		0.997

Relative Volatility   Equation (3.33) may be solved for each component in a mixture to allow comparison of their contributions to the liquid and vapor at equilibrium. Dividing (3.33) in terms of component i by the same equation in terms of component j yields the *relative volatility* of component i with respect to j:

$$\alpha_{ij} \equiv \frac{y_i/x_i}{y_j/x_j} = \frac{p_i^\circ}{p_j^\circ} \tag{3.34}$$

Again, for nonideal systems, actual values of K will give more exact results:

$$\alpha_{ij} = \frac{K_i}{K_j} \tag{3.35}$$

References 3 and 4 are recommended sources for K factors for mixtures of light hydrocarbons and common volatile nonhydrocarbons. Since these factors vary with temperature, pressure, and composition, their tabulation requires extensive nomography. As a result, they are not tabulated here, although values taken from Refs. 3 and 4 are used to enumerate examples. Since this book is not intended as a column design guide, exact equilibrium data are not required. However, the role of α in relating compositions, and the degree of its variation with pressure and temperature, are crucial.

If component i in Eqs. (3.34) and (3.35) is more volatile than j, then K_i will exceed K_j and α_{ij} will exceed unity. When $\alpha = 1.0$, separation of the two components by simple distillation is impossible.

In a binary system, graphic column design methods are based on an equilibrium curve of y versus x. For ideal binary mixtures, the points on the equilibrium curve may be calculated by substituting $1 - y_i$ for y_j and $1 - x_i$ for x_j. The subscripts may then be dropped:

$$\alpha = \frac{y/x}{(1 - y)/(1 - x)} = \frac{y(1 - x)}{x(1 - y)} \tag{3.36}$$

Then y may be found in terms of x:

$$y = \frac{\alpha x}{1 + x(\alpha - 1)} \tag{3.37}$$

or x may be evaluated in terms of y:

$$x = \frac{y}{\alpha - y(\alpha - 1)} \tag{3.38}$$

For purposes of illustration, Fig. 3.6 gives equilibrium curves for $\alpha = 1.2, 1.5, 2,$ and 4.

Unfortunately, α usually varies somewhat with composition, indicating nonideal behavior. When Eqs. (3.37) or (3.38) are used to make tray-to-tray calculations or to construct an equilibrium curve, α should be adjusted for compositions at each calculation.

To evaluate deviation from ideality, α was calculated for the butane-isobutane system from vapor pressures interpolated from Ref. 5 and K values taken from

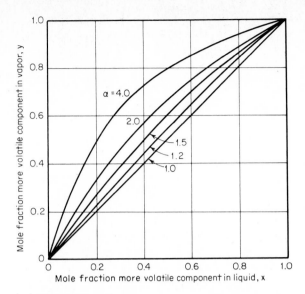

figure 3.6 *The equilibrium curve is symmetrical if* α *does not vary with composition.*

Ref. 4. The results appear in Table 3.1. Relative volatility calculated as K_i/K_n is only given to three significant figures since K values cannot be read more accurately from the charts given in Ref. 4.

Although α does not seem to vary significantly with composition, it is lower than the ratio of the vapor pressures. Use of the ratio of vapor pressures should be restricted to close-boiling homologs below 30 lb/in² abs where vapors do not appreciably deviate from ideal-gas behavior.

Variation of α with Temperature   Vapor pressure is related to temperature by

$$\ln p = a - \frac{b}{T + c} \tag{3.39}$$

where a, b, and c are constants specific to each substance. Vapor pressure is typically plotted on a logarithmic scale against temperature on a special scale devised to linearize the $-1/T$ function. Vapor pressures for a series of light hydrocarbons on these coordinates are shown in Fig. 3.7.

Observe that the lines tend to converge with increasing temperature. If relative volatility can be approximated by the ratio of vapor pressures, then its logarithm is approximately the difference in their logarithms:

TABLE 3.1   Relative Volatility Calculated for the Butane-Isobutane System at 100°F

p_i°, lb/in² abs	p_n°, lb/in² abs	p_i°/p_n°	x_i	K_i	K_n	K_i/K_n
71.9	51.7	1.391	0	1.35	1.00	1.35
			1.0	1.00	0.74	1.35

Subscripts i and n refer here to isobutane and normal butane.

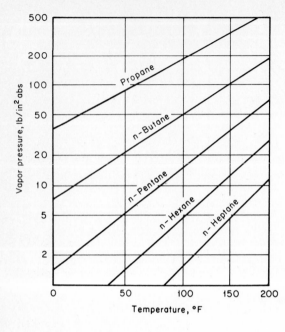

figure 3.7 *Vapor-pressure lines converge with increasing temperature.*

$$\ln \alpha_{ij} \approx \ln p_i^o - \ln p_j^o \tag{3.40}$$

Then convergence of the logarithms of vapor pressure indicates that relative volatility decreases with increasing temperature and pressure. This is a general rule which is followed by components of similar nature. Components having diverse natures, such as water and alcohols, may exhibit crossing vapor pressure curves — a sign of azeotrope formation.

Doig [6] reports variations in α for the butane-isobutane system as proportional to $p^{-0.055}$, as measured by the ratio of vapor pressures of the components. Table 3.2 gives relative volatilities for the butane-isobutane system as a function of temperature. The calculations were made using K factors from Ref. 4 evaluated at the vapor pressure of the more volatile component. Smuck [7] reports α for the propane-propylene system to vary from 1.15 at 175 lb/in² abs to 1.135 at 225 lb/in² abs and 1.12 at 300 lb/in² abs, as obtained from actual plant operating data. Although these variations seem small, examples in Chap. 2 show their profound effect on operating costs.

Single-Stage Equilibrium   Having developed an equilibrium model for a single stage, it is now possible to evaluate a complete solution with a material balance.

TABLE 3.2   Relative Volatilities of Butane-Isobutane as a Function of Temperature

T, °F	p, lb/in² abs	α
60	38.1	1.41
80	53.1	1.38
100	71.9	1.35
120	95.2	1.32

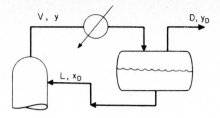

figure 3.8 *The partial condenser is a single equilibrium stage.*

Consider the partial condenser in Fig. 3.8 as a single-stage separator. Vapor V leaving the top of the column at composition y is condensed to the point where only reflux L forms. Distillate D is withdrawn as a vapor whose composition y_D is richer in the more-volatile component than y.

A material balance yields

$$\frac{D}{V} = \frac{y - x_D}{y_D - x_D} \tag{3.41}$$

Observe that this is in the form of Eq. (3.3). Additionally, an equilibrium model requires that

$$\alpha = \frac{y_D(1 - x_D)}{x_D(1 - y_D)} \tag{3.42}$$

where α is the relative volatility established at the temperature and pressure within the condenser.

For the binary system such as described here, the heat duty of the condenser would have to be regulated in order to condense only the desired fraction L/V, leaving D/V as a vapor. The resulting compositions y_D and x_D can be found by simultaneous solution of Eqs. (3.41) and (3.42), given values of y, D/V, and α.

One possible method would be to solve them graphically. An equilibrium curve of y_D versus x_D (Fig. 3.6) could be drawn for the appropriate value of α. Then the material balance line (as in Fig. 3.2) could be superimposed on the same coordinates, originating at point (y, y) and having a slope L/D. The point where the line and curve intersect would represent terminal compositions x_D and y_D.

A more exact method is to substitute Eq. (3.42) into (3.41) to produce a quadratic:

$$ay_D{}^2 + by_D + c = 0 \tag{3.43}$$

where $a = \dfrac{D}{V}(\alpha - 1)$

$$b = -\left[\left(\frac{D}{V} + y\right)(\alpha - 1) + 1\right]$$

$$c = \alpha y$$

It may be solved by applying the quadratic formula

$$y = \frac{-b - \sqrt{b^2 - 4ac}}{2a} \tag{3.44}$$

example 3.3

What are the terminal compositions for partial condensation of vapor containing 90 percent of the more-volatile product, when $\alpha = 4.0$ and $L/D = 2.0$?

$$a = \frac{1}{3}(4 - 1) = 1.0$$

$$b = -[(\tfrac{1}{3} + 0.90)(4 - 1) + 1] = -4.70$$

$$c = 4\,(0.90) = 3.6$$

$$y_D = 0.9635$$

$$x_D = \frac{0.9635}{4 - 0.9635\,(4 - 1)} = 0.8683$$

Partial condensers are used to allow lower column pressure when separating light hydrocarbons such as ethane and propane overhead. The distillate vapor may be condensed against a sink colder than the reflux condenser or left as a vapor. However, the split between reflux and distillate condensers can pose control problems, which are faced in Chap. 4.

Partial condensers often require the removal of distillates in both phases. Since this practice results in more than two products from the column, its discussion is deferred to Chap. 8.

The single-stage separator is a real representation of the interaction between the material and energy balances across an entire column. The larger problems are solved in the same way. But first, the single-stage separator needs to be expanded to include all the trays in the column.

Total Reflux Operation   Total reflux describes that operating condition in which all vapor is condensed and returned to the column as reflux; feed and product rates are zero. At total reflux, all trays may reach equilibrium since there is no net flow of product either upward or downward. It is possible to find the relationship between the compositions of the vapor leaving the top tray and the liquid leaving the bottom tray by solving the single-stage equilibrium Eq. (3.34) successively for as many theoretical trays as there are in the column:

$$\frac{y_i/x_i}{y_j/x_j} = \overline{\alpha}^{n_T} \tag{3.45}$$

Here $\overline{\alpha}$ is the average relative volatility across the column and n_T is the number of theoretical equilibrium stages. Equation (3.45) is a more general equilibrium statement, whereas (3.34) applies to the specific case where $n_T = 1$.

For a binary system, the multistage equilibrium equation allows y to be evaluated in terms of x, $\overline{\alpha}$, and n_T:

$$y = \frac{x\,\overline{\alpha}^{n_T}}{1 + x(\overline{\alpha}^{n_T} - 1)} \tag{3.46}$$

A multistage equilibrium curve can then be plotted, similar to the single-stage curves of Fig. 3.6 but with n_T as a parameter. Several multistage equilibrium curves appear in Fig. 3.9.

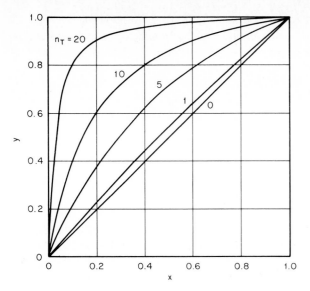

figure 3.9 *Multistage equilibrium curves for α = 1.2.*

The number of theoretical stages differs from the number of actual trays n by the average tray efficiency E:

$$n_T = nE \tag{3.47}$$

The definition of E suffers from the same problem as the definition of $\bar{\alpha}$. To be precise, $\bar{\alpha}$ and E should be evaluated at every tray for those systems where they tend to change significantly from tray to tray. While this may be necessary for column design, it becomes prohibitive for system analysis, where mathematical simplicity is more important than absolute accuracy.

If $\bar{\alpha}$ varies linearly with composition, then an arithmetic average of its value at the ends of the column should be adequate. If the variation is nonlinear, a geometric average — the square root of the product of the terminal values — is probably more representative. For example, the relative volatility between methanol and water varies from about 9 at low methanol concentrations to about 1.1 at high concentrations. The geometric average of $\sqrt{9(1.1)} = 3.15$ is probably a better estimate than the arithmetic average of 5.05.

For an operating column, E can be estimated from product compositions if $\bar{\alpha}$ and n are known. In most cases, the control engineer is interested in variations in operating conditions about a known point, so that exact values of $\bar{\alpha}$ and E are less important than the form of the model.

Another problem, however, is the variation of E with liquid and vapor rates. Too little vapor flow cannot adequately support a liquid head, and liquid will tend to drip downward through vapor slots, reducing contact. Too much vapor flow can carry drops or foam upward, causing mixing of dissimilar liquids and loss in efficiency. Valve trays were specifically devised to vary the vapor openings and thereby extend the flow range over which efficiency would be maximized.

In general, one can assume a constant tray efficiency — 70 percent is typical. If efficiency begins to fall due to entrainment or foaming, purities cannot be improved by increasing boilup or reflux — their upper limit has been reached. Similarly, if efficiency begins to fall due to weeping at low boilup, there is no possibility of saving energy by reducing it further — the low limit has been reached.

Equation (3.45) is the Fenske-Underwood equation [8]. It is useful in estimating the number of theoretical trays required to effect a given separation at total reflux. This is the minimum number of stages within which that separation may be achieved:

$$(nE)_{min} = \ln \frac{y_i/x_i}{y_j/x_j} \Big/ \ln \bar{\alpha}_{ij} \tag{3.48}$$

Total reflux operation is incapable of making any product. However, when feed is introduced and products are withdrawn, purities will decrease proportionately. For any real production condition then, n must exceed n_{min} if desired product compositions are to be maintained.

THE SEPARATION MODEL

The key to obtaining explicit solutions to distillation problems is the development of an analytical model relating compositions to reflux and boilup. In particular, for purposes of control system design and optimization, the model must be differentiable; numerical solutions simply cannot provide the necessary guidance. Over the course of the years, the author has used several models, each more accurate than the last. In this edition, a still more accurate model is followed; this model was developed by J. M. Douglas, A. Jafarey, and T. J. McAvoy at the University of Massachusetts [9].

Separation Factor   The separation factor is defined as the ratio of the light to heavy key in the distillate divided by the same ratio in the bottom product:

$$S \equiv \frac{y_L/x_L}{y_H/x_H} \tag{3.49}$$

Its maximum value has already been determined at total reflux:

$$S_{max} = \bar{\alpha}^{nE} \tag{3.50}$$

Its minimum value must be unity, where product and feed compositions are equal.

For a binary system, the multistage equilibrium equation (3.46) can be reduced to a more general form:

$$y = \frac{xS}{1 + x(S - 1)} \tag{3.51}$$

Because S is always much greater than unity, Eq. (3.51) may be approximated

$$y \approx \frac{1}{1 + 1/Sx} \tag{3.52}$$

Furthermore, we are principally interested in controlling impurities, i.e., x and $1 - y$:

$$1 - y \approx \frac{1}{1 + Sx} \qquad (3.53)$$

Then given a set of x and y, it is possible to calculate S, and from it determine how x and y vary with each other under conditions of constant separation.

Explicit Solutions An explicit solution for the composition of the products from a column can be found, given the separation factor and the D/F ratio. The procedure is the same followed before for the partial condenser. For the binary system, Eqs. (3.51) and (3.3) are combined to develop a quadratic,

$$y = \frac{-b - \sqrt{b^2 - 4ac}}{2a} \qquad (3.54)$$

where $a = \dfrac{D}{F}(S - 1)$

$$b = -\left[\left(\frac{D}{F} + z\right)(S - 1) + 1\right]$$

$$c = Sz$$

Having found y for given values of D/F and S, the corresponding value of x may be calculated from the separation equation,

$$x = \frac{y}{S - y(S - 1)} \qquad (3.55)$$

The multicomponent problem may be solved explicitly in the same way, although more equations are required. Again, the solution takes the form of a quadratic, where

$$y_H = \frac{-b + \sqrt{b^2 - 4ac}}{2a} \qquad (3.56)$$

and

$$a = \frac{D}{F}(S - 1)$$

$$b = \frac{D}{F} - z_{LL} + z_H + S\left(z_{LL} + z_L - \frac{D}{F}\right)$$

$$c = -z_H\left(1 - \frac{z_{LL}}{D/F}\right)$$

Composition y_{LL} is found by Eq. (3.6), and y_L by (3.7). Similarly, x_{HH} is found by (3.8). Then the separation equation may be used to determine x_L:

$$x_L = \frac{y_L(1 - x_{HH})}{y_H S + y_L}$$

(3.57)

Finally, x_H is found by Eq. (3.9).

example 3.4

Consider a debutanizer having the following sets of compositions:

	z, %	y, %	x, %
Propane (LL)	1.06	4.2	
Isobutane (LL)	7.4	29.1	0.11
n-Butane (L)	17.6	64.0	2.0
Isopentane (H)	16.2	2.0	21.0
n-Pentane (HH)	29.3	0.73	38.9
Isohexane (HH)	14.8		19.8
Heavier (HH)	13.6		18.2

Calculate D/F and S, using n-butane and isopentane as the keys.

$$\frac{D}{F} = \frac{0.176 - 0.020}{0.640 - 0.020} = 0.252$$

$$S = \frac{(0.640)(0.210)}{(0.020)(0.020)} = 336$$

Increase D/F by 0.01 at constant S, and estimate the new compositions:

$$a = 0.262\,(335) = 87.77$$

$$b = 0.262 - (0.074 + 0.0106) + 0.162 = \qquad 0.3394$$
$$+ \, 336\,[(0.074 + 0.0106) + 0.176 - 0.262] = -0.4704$$
$$-0.1310$$

$$c = -0.162\,[1 - (0.074 + 0.0106)/0.262] = -0.1097$$

Isopentane: $y_H = \dfrac{0.1310 + \sqrt{(0.1310)^2 + 4(87.77)(0.1097)}}{2(87.77)} = 0.0361$

Propane: $y_{LL} = \dfrac{0.0106}{0.262} = 0.0405$

Isobutane: $y_{LL} = \dfrac{0.074}{0.262} = 0.282$

n-Butane: $y_L = 1 - 0.0405 - 0.282 - 0.0361 = 0.641$

n-Pentane: $x_{HH} = \dfrac{0.293}{1 - 0.262} = 0.397$

Isohexane: $x_{HH} = \dfrac{0.148}{1 - 0.262} = 0.201$

Heavier: $x_{HH} = \dfrac{0.137}{1 - 0.262} = 0.184$

n-Butane: $x_L = \dfrac{0.641[1 - (0.397 + 0.201 + 0.184)]}{0.0361(336) + 0.641} = 0.011$

Isopentane: $x_H = 1 - 0.011 - (0.397 + 0.201 + 0.184) = 0.207$

The above example neglects the small amounts of off-key components which are lost to the opposite products, i.e., isobutane in the bottom and n-pentane in the distillate, as they are usually not significant to the solution. However, their exact values can be determined by first calculating separation factors for these components against the keys and then solving the resulting quadratic equations using those values of S.

Separation vs. Reflux Ratio   Having identified the separation factor, it is now necessary to evaluate the mechanism by which it is adjusted to control the compositions of the products. For feed entering the optimum tray at $q_F = 0$, the separation model of Douglas et al. [9] relates separation to reflux ratio and feed composition:

$$S = \left(\frac{\overline{\alpha}}{\sqrt{1 + D/Lz}}\right)^{nE} \tag{3.58}$$

Conversely, if compositions are known, reflux ratio may be determined from separation:

$$\frac{D}{L} = z\left[\left(\frac{\overline{\alpha}}{S^{1/nE}}\right)^2 - 1\right] \tag{3.59}$$

If $q_F \neq 0$, then a more complicated formula is required since the internal reflux is affected:

$$S = \left[\alpha\sqrt{1 - \frac{1 - q_F + L/D}{(1 + L/D)(1 - q_F + zL/D)}}\right]^{nE} \tag{3.60}$$

As L/D increases, the effect of q_F becomes less significant.

With these equations, it becomes possible to evaluate the effects of changes in any of the operating variables on product compositions. As an example, the column used to generate the economic data for Figs. 1.5 and 1.6 is examined below.

example 3.5

The column described in Chap. 1 separated a 50-50 mixture of two components having a relative volatility of 1.72 in 30 theoretical trays. Calculate the V/F ratio needed to produce products containing 5.0 percent impurities and the recovery achieved.

$$S = \frac{0.95/0.05}{0.05/0.95} = 361$$

Using Eq. (3.59),

$$\frac{D}{L} = 0.5\left[\left(\frac{1.72}{361^{1/30}}\right)^2 - 1\right] = 0.50$$

From Eq. (3.3),

$$\frac{D}{F} = \frac{0.50 - 0.05}{0.95 - 0.05} = 0.50$$

From Eq. (3.13),

$$\frac{V}{F} = 0.50 \left(\frac{1}{0.50} + 1\right) = 1.50$$

Recovery is calculated from Eq. (1.15):

$$R = \frac{0.50}{0.50} = 1.00 \text{ or } 100\%$$

Next, decrease the recovery to 96 percent and calculate the resulting distillate composition while holding energy flow constant.

$$\frac{D}{F} = Rz = (0.96)(0.50) = 0.48$$

$$\frac{L}{D} = \frac{V/F}{D/F} - 1 = \frac{1.50}{0.48} - 1 = 2.125$$

From Eq. (3.58),

$$S = \left(\frac{1.72}{\sqrt{1 + 1/(2.125)(1.05)}}\right)^{30} = 556$$

Using quadratic formula (3.54),

$$a = 0.48\,(555) = 266.4$$

$$b = -[(0.48 + 0.50)(555) + 1] = -544.9$$

$$c = 556(0.50) = 278$$

$$y = \frac{544.9 - \sqrt{544.9^2 - 4(226.4)(278)}}{2(266.4)} = 0.9738$$

With the recovery at 100 percent, increase the V/F ratio to 1.8 and calculate the resulting distillate composition.

$$\frac{L}{D} = \frac{1.8}{0.5} - 1 = 2.60$$

$$S = \left[\frac{1.72}{\sqrt{1 + 1/(2.60)(0.5)}}\right]^{30} = 2234$$

Following the same procedure as above,

$$y = 0.9794$$

Appendix A contains a BASIC program and one for an HP-11C calculator to solve for x and y given z, D/F, V/F, α, and nE. Other helpful programs are also found there.

Graphical Representations While numerical solutions tend to be more exact, graphical methods aid comprehension and facilitate extension of operating conditions into unexplored regions. They are therefore extremely useful devices in understanding the interrelationships among distillation variables.

To focus attention on product impurities, it is helpful to plot $1 - y$ versus x. This inverts the traditional equilibrium diagram but allows the scales to be expanded in the area of interest. Figure 3.10 plots the impurities in a binary column against

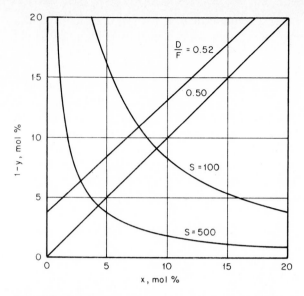

figure 3.10 *Product compositions are determined by the intersection of a material-balance line and a separation curve.*

each other for two values of S and two of D/F, with $z = 0.50$. It can readily be seen that changing D/F shifts the impurity from one end of the column to the other. In other words, it makes one product more pure at the expense of the other. Changing separation, however, causes both impurities to increase or decrease together.

Separation is not normally one of the variables manipulated to control composition: reflux and boilup are the favorite choices. Therefore, a useful exercise is to prepare a plot of $1 - y$ against x, adding L/F and V/F to the parameters of D/F and S. This has been done in Fig. 3.11 for the column described in Example 3.5. Conditions corresponding to the curves are: $D/F = 0.5$, $V/F = 1.5$, $L/F = 1.0$, and $S = 361$.

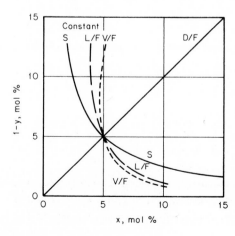

figure 3.11 *Each of the manipulated variables affects compositions differently.*

The curve of constant separation is symmetrical about the diagonal, but the others are not. The curve for constant reflux was developed by holding L/F at 1.0 and varying D/F. As D/F increases, L/D will decrease, causing separation to fall. Therefore, values of D/F above the diagonal produce compositions above the separation curve; similarly, lower values of D/F result in higher L/D and improved separation.

When V/F is held constant, increasing D/F causes reflux to fall, reducing L/D and hence separation more than when reflux is constant. Consequently, the curve for constant V/F departs still further from the curve of constant S, even turning back at high values of $1 - y$. If $1 - y$ is maintained above 5 percent, x is determined almost exclusively by boilup: If V/F is held constant, x will tend to stay at 5 percent, regardless of changes in other variables. This is an ideal situation, promising self-regulation of x within that operating range. But it is also uncommon, and one must search for it. The self-regulating property does not reveal itself except through the type of modeling done here.

Because these curves determine the steady-state response of product compositions to the available manipulated variables, they form the basis for control-loop selection. They are therefore reexamined in greater detail in Chap. 5 where the procedure for structure development is presented.

There is a wide variety of columns in existence, so that a considerable variation in operating curves may be expected. Columns having high reflux ratios will have their curves of constant S, L/F, and V/F more tightly compressed than those in Fig. 3.11, simply because variations in D have less effect on L and L/D. Nonetheless, the three curves always appear in the same order. Furthermore, the material-balance line of constant D always has a slope of opposite sign to that of the curve of constant S. As is seen in Fig. 3.11, it is possible for the slope of the constant-V/F curve to change signs, and less likely, that of the constant-L/F curve.

DYNAMIC EFFECTS

A control loop must contend with both steady-state relationships and dynamic elements in the path of information flow. A favorable steady-state relationship, such as that between boilup and bottom composition in Fig. 3.11, can guarantee success of a control system almost without regard to dynamic responsiveness. However, such good fortune is not to be expected, and less favorable relationships are the rule. Then, compositions are more easily upset by disturbances from a host of sources, including each other's controllers. Then the loop with faster dynamic response will hold composition closer to set point.

Hydraulic Lags   Small variations in liquid and vapor rates have similar effects on product quality in the steady state. However, the speed with which compositions respond to their manipulation varies widely.

Each tray may be covered with perhaps an inch or two of liquid (excluding froth) with up to 2 ft of vapor space between trays. Since the ratio of liquid to vapor densities may be upward of 30:1, each tray contains much more liquid than vapor. To increase the vapor flow leaving a tray, it is only necessary to increase the vapor flow approaching it. Because the pressure drop may be less than 2 in of water per tray, the compression of the vapor brought about by increasing its flow (and hence pressure drop) is insignificant except in vacuum towers.

For a column separating close-boiling components, an increase in reboiler heat input will therefore cause an increase in vapor flow entering the condenser within a few seconds. The dynamic response of the reboiler (5-s dead time) is probably of the same order of magnitude as that of a 50-tray column. When there is a large difference between component boiling points, however, considerable sensible heat may be stored in the temperature gradient across the trays. This has the effect of both delaying and reducing the effect of boilup on overhead vapor flow.

Considerable delay is encountered in transferring liquid from tray to tray, however. To increase the liquid flow leaving a tray, the head on the tray must first be raised by increasing liquid rate entering. Thus higher flow rates cause more liquid to be stored on the trays. Then the change in storage required for each new liquid rate delays the transfer of the new rate to trays farther down the column. Each tray is fitted with a weir which establishes its minimum liquid head. If the weir is rectangular in shape, the flow of liquid over it is related to the head above it by the $\frac{3}{2}$ power:

$$f = 3.33wh^{3/2} \tag{3.61}$$

where w and h are the width and head in feet, and f is in cubic feet per second. The volume of liquid retained above the weir is the tray area multiplied by the head:

$$v = Ah = A\left(\frac{f}{3.33w}\right)^{2/3} \tag{3.62}$$

When the flow of liquid entering the tray is raised, the overflow will momentarily remain the same. This difference between inflow and outflow increases the volume stored on the tray at a rate dv/dt. But this increases the head proportionally, which begins to raise the overflow at a rate df/dt. If this rate of rise of overflow were to be maintained, overflow would equal inflow in time

$$\tau_h = \frac{dv/dt}{df/dt} = \frac{dv}{df} \tag{3.63}$$

This is identified as the hydraulic time constant of the tray. Increasing overflow decreases the difference from inflow, reducing dv/dt and df/dt, so that a new steady state is approached exponentially. The overflow would follow the time response shown as tray 1 in Fig. 3.12 for a step increase in reflux.

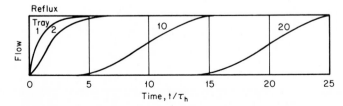

figure 3.12 *The response of liquid flow to a step change in reflux is delayed by the number of intervening trays.*

Because the overflow from tray 1 is already exponential, that from tray 2 must follow a second-order exponential, also shown in Fig. 3.12. The tenth tray from the top will show no perceptible change in overflow at all until almost 5 time constants have elapsed, and its change will be complete after about 15 time constants. At tray 20, the response is delayed by 15 time constants, but still requires about 10 more to approach completion. Then, for a large number of trays, the response at the nth tray can be approximated by an equivalent dead time of $(n - 4)\tau_h$ and an exponential lag of $5\tau_h$.

The time constant for an overflowing weir is not really constant. Differentiation of Eq. (3.62) gives

$$\tau_h = \frac{dv}{df} = \frac{2A}{3(3.33w)f^{1/3}} = \frac{0.2A}{wf^{1/3}} \tag{3.64}$$

Doubling the flow will cause τ_h to decrease by 21 percent, and reducing flow by half will raise τ_h by about 26 percent. In other words, the hydraulic time constant does appear to change somewhat with flow, but less than proportionately—less than 50 percent for a 4:1 flow change.

Beaverstock and Harriott [10] measured the hydraulic lag of liquid flow down a 12-in diameter, 15-sieve-tray column at an average of 6.5 s per tray. This was over three times as great as results predicted using the weir equation. The time required for a selected tray to respond to a liquid flow change introduced n trays above it was approximately $6.5n$ s. This verifies that the trays act essentially as described in Fig. 3.12.

From the foregoing discussion, it follows that the liquid-vapor ratio at any point in a column except the top tray may be changed much more rapidly by manipulating boilup than reflux.

Inverse Response   Inverse response is that dynamic characteristic by which a variable reverses direction partway through its reaction to a step disturbance as illustrated in Fig. 3.13. It is caused by a combination of two parallel reactions whose signs differ.

Consider, for example, a forced-circulation reboiler located at some distance from its column. Only part of the liquid is vaporized, so that the line returning to the column contains both liquid and vapor. If the heat input to the reboiler is raised, the rate of vapor generation will increase, so the return line must carry a higher percentage of vapor than before. To reach this new steady state, the return line will reduce its liquid content by expelling excess liquid into the base of the column,

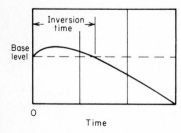

figure 3.13   *On an increase in boil-up, base level may rise before falling, exhibiting inverse response.*

raising the liquid level there. However, the higher vaporization rate will cause the total liquid inventory to begin moving from the bottom of the column to the reflux drum. The shift in inventory between the return line and the column base is only a transient event, after which base level will fall as in Fig. 3.13.

If a controller is to regulate base level with heat input, it must wait for the inverse response to elapse before a true reaction appears. In effect, inverse response delays the control loop in much the same way as dead time. A dynamic analysis by the author in Ref. 11 shows that it can be approximated quite accurately by a dead time of the same value as the observed inversion time. Buckley et al. [12] also described inverse response arising from displacement of liquid by vapor within the trays of the column itself. An increase in vapor flow can lower the froth density, lifting more liquid over the weirs and into the column base. In essence, the column liquid holdup varies inversely with vapor flow. Again, if the change in holdup is comparable to the capacity of the column base, level control by heat-input manipulation may be unsuccessful.

The inverse response centers not so much on the actual froth density as on how much it changes with boilup. Van Winkle [13] indicates that it changes more sharply with boilup at low rates than high, so inverse response observed at low boilup rates may disappear at higher rates. Also, perforated trays are more likely to give inverse response than bubble-cap trays.

The column described in Ref. 12 had 100 valve trays, which gave an inversion time of 11 min in response of base level to heat input. A forced-circulation reboiler was used. Level control by heat-input manipulation was impossible, and bottom flow was too small to be used. Although 3 min dead time elapsed in the effect of reflux on base level, reflux was selected as the best variable for its control. The authors determined that valve trays give inverse response at all vapor rates, whereas perforated trays give inverse response at low rates, direct response at high rates, and effectively dead time at intermediate rates.

Composition Time Constants   Changing the flow of liquid or vapor within the column will begin changing compositions immediately. However, their rate of change is a function of the mass of material in the column and the flow rates passing through it. Then as compositions begin to change, they induce further changes in composition, extending the time response of the column.

As an illustration, consider the response of the single tray and reflux drum in Fig. 3.14 to a change in reflux flow. Assume vapor flow V and composition z entering the tray are constant. Equilibrium exists between liquid and vapor compositions leaving the tray:

$$\frac{y\,(1-x)}{x\,(1-y)} = \alpha$$

Furthermore, the material balance must be satisfied:

$$Vz + Lx_D = Vy + Lx$$

Here a distinction is made between the composition y of the vapor and x_D of the reflux (and distillate) to allow for a time delay in the reflux drum. In the steady

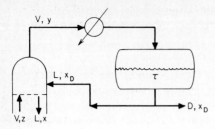

figure 3.14 *A change in reflux flow causes a change in vapor composition that returns later as a change in reflux composition.*

state, $x_D = y$, so that the above equations may be solved using the quadratic formula. If x_D is held constant, however, then the coefficients of the quadratic formula change.

These equations were solved for an initial steady state of $z = 0.9$, $\alpha = 2.0$, and $L/V = 1.0$; then $1 - y = 5.263$ percent. When L/V is reduced to 0.9 but x_D held at the above steady-state value of y, a new vapor composition is generated: $1 - y = 5.444$ percent. If the reflux drum were not mixed at all, x_D would remain at its original value until all the liquid was replaced by condensed vapor of the new composition. Then x_D would change and cause a second change in vapor composition to 5.506 percent. This procedure would continue to occur at intervals of the residence time τ in the reflux drum, with each step smaller than the last, as shown in Table 3.3 and Fig. 3.15. This exponential series of steps in the composition of the distillate product is the incremental equivalent of the first-order lag function $1 - e^{-t/\tau}$, as Table 3.3 shows, whose time constant is the residence time τ of the reflux drum, which is its volume divided by vapor flow.

If the reflux drum were perfectly mixed, the stepwise response of product composition would be replaced by the smooth exponential curve beginning immediately after the change in reflux. While it is not customary to agitate reflux drums, some mixing can be expected due to ordinary turbulence and diffusion, so that the steps in Fig. 3.14 are rounded or undiscernable.

Recognize that the preceding example is representative only for a single tray and reflux drum, neglecting the mass of liquid stored on the tray. In any real column, the change in reflux flow will begin to change compositions on all trays, after being delayed by the hydraulic lag described earlier. This will promote a series of sequential changes in vapor composition entering the top tray, which will then alter

TABLE 3.3   Dynamic Response of Top Tray and Reflux Accumulator

L/V_n	t/τ	$1 - y$, %	Response, %	$1 - e^{+t/\tau}$
1.0	0	5.263	0	0
0.9	1	5.444	65.6	0.632
	2	5.506	88.0	0.865
	3	5.527	95.7	0.950
	4	5.535	98.6	0.982
	5	5.537	99.3	0.993
	6	5.538	99.8	0.998
	∞	5.5385	100.0	1.000

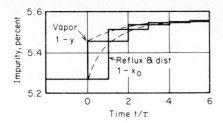

figure 3.15 *Distillate composition traces an exponential response following a change in L/V; the steps would appear in absence of mixing, whereas the smooth curves would be the result of perfect mixing.*

y and eventually x_D. Changes in reflux composition x_D also are propagated down the column, producing their own changes in vapor composition.

Composition changes will travel through the column more slowly than flow changes, however. Each tray acts like a first-order lag, but its composition time constant can be considerably longer than its hydraulic time constant in that *all* the liquid participates rather than only the amount above the weir. As a result, two waves will proceed through the column, the first caused by a change in reflux flow and the second, slower wave caused by a change in reflux composition. However, concurrent changes in vapor composition return and also travel across trays, resulting in a very complex set of interactions. These interactions are so pervasive that as long as compositions are changing anywhere in a column, compositions are changing or will be changing everywhere. A steady state is approached at about the same time everywhere.

This is confirmed by the observation that the dominant composition time constant for a column seems to be the same, regardless of where a disturbance is introduced or where composition is measured. McNeill and Sacks [2] reported essentially the same 20-h time constant for both feed-rate and distillate-rate changes when overhead vapor composition was measured; however, dead time in response to feed-rate changes was nearly 8 h, while that due to distillate changes was only about 15 min.

The only general relationship that seems to apply to the dominant composition time constant is that of the total liquid volume divided by the column feed rate:

$$\tau_x = \frac{\Sigma\,v}{F} \tag{3.65}$$

For a given tray design, liquid volume will be proportional to the tray area which in turn is proportioned to the vapor-carrying capacity:

$$v = kV_{max} \tag{3.66}$$

Combining these two equations for n trays then reveals the composition time constant to be proportional to the product of n and V_{max}/F:

$$\tau_x = \frac{knV_{max}}{F} \tag{3.67}$$

The reboiler and reflux accumulator will also add to the liquid holdup.

Control-Loop Response   An important observation to be made from Fig. 3.15 is that overhead vapor composition leads liquid product composition. The exponential response traced by the vapor composition is that of a lead-lag function [14]. Its lag time is that of the reflux drum, but it also contains a lead which is 65.6 percent of the lag time, as evidenced by its immediate response to 65.6 percent of the total change. By locating a product composition analyzer in the vapor line rather than on the distillate itself, the lag associated with the reflux drum can be almost eliminated from the control loop.

Figure 3.16 describes a typical response of product composition to a step change in reflux or boilup. The composition time constant τ_x dominates the response, and is estimated as the time required to reach 63.2 percent response $(1 - e^{-1})$, once a significant departure has been observed. Dead time τ_d is that time during which no significant response to stimulus appears. It is usually contributed by a combination of elements, such as the series of hydraulic lags in Fig. 3.12, inverse response, and other lags that are small in comparison with the dominant time constant. An example of a small lag would be that of the reflux drum when compared to the composition time constant for the entire column. Therefore, sampling the vapor instead of the liquid as described above can reduce the effective dead time seen by the controller.

The importance of minimizing dead time cannot be overstated. The period of oscillation of a control loop is directly proportional to its dead time. Because the optimum settings of integral and derivative time constants are related to the period, they are also functions of the dead time. The dynamic gain of a process varies with the ratio of dead time to its dominant time constant, which then determines the allowable proportional band of the controller. Equation (1.14) describes the integrated deviation from set point for a controller to be linearly related to the product of proportional and integral settings. Then the integrated deviation varies directly with dead time *squared*.

The composition time constant seems to be the same everywhere in the column, but the dead time is not. A change in reflux flow may cause overhead vapor composition to start changing almost immediately, but bottom composition response will be delayed by the combined hydraulic lags of all the trays. Consequently, reflux

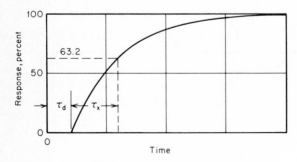

figure 3.16   *The response of composition to a step change in reflux or boilup typically reveals a dead time followed by a first-order lag.*

is not recommended to control compositions at the bottom of the column. Vapor, however, travels quickly through a column, particularly if the temperature gradient is small, so that it could be used to control composition at the top or bottom.

The dynamic response of a column is altered by the arrangement and gain of its level and pressure loops. If these loops manipulate only exit streams, i.e., distillate, bottoms, and heat removal, manipulating reflux and boilup will change compositions along dynamic paths like that described in Fig. 3.16. Any other configuration will significantly modify the response, however. Suppose, for example, that reflux is manipulated to control the level in the accumulator. Then increasing boilup will raise the level, causing the level controller to increase reflux by the same amount. The result would be the propagation of a reflux wave down the column largely canceling the changes induced by the boilup increase. Compositions or temperatures that originally started to rise would begin to fall again, coming to rest much closer to their original values than when accumulator level was controlled by distillate flow.

At this point in the development of the subject, the reader has insufficient information to quantify these dynamic interactions. It is first necessary to be able to quantify the steady-state interactions among the various control loops, which is shown in Chap. 5. Then the dynamic interactions will become evident.

REFERENCES

1. Van Kampen, J. A.: "Automatic Control by Chromatographs of the Product Quality of a Distillation Column," presented at the Convention on Advances in Automatic Control, Nottingham, England, April 1965.
2. McNeill, G. A., and J. D. Sacks: "High Performance Column Control," *Chem. Eng. Prog.*, March 1969.
3. Van Winkle, M.: *Distillation,* McGraw-Hill, New York, 1967.
4. *Technical Data Book — Petroleum Refining,* 2d ed., American Petroleum Institute, Washington, D.C., 1970.
5. *Matheson Gas Data Book,* 4th ed., The Matheson Company, East Rutherford, N.J., 1966.
6. Doig, I. D.: "Variation of the Operating Pressure to Manipulate Distillation Processes," *Aust. Chem. Eng.,* July 1971.
7. Smuck, W. W.: "Operating Characteristics of a Propylene Fractionating Unit," *Chem. Eng. Prog.,* June 1963.
8. Fenske, M. R.: "Fractionation of Straight-Run Pennsylvania Gasoline," *Ind. Eng. Chem.,* May 1932.
9. Douglas, J. M., A. Jafarey, and T. J. McAvoy: "Short-Cut Techniques for Distillation Column Design and Control, Part I. Column Design," *Ind. Eng. Chem. Process Des. Dev.,* vol. 18, no. 2, 1979, pp. 197–202.
10. Beaverstock, M. C., and P. Harriott: "Experimental Closed-Loop Control of a Distillation Column," *Ind. Eng. Chem. Process Des. Dev.,* vol. 12, no. 4, 1973.
11. Shinskey, F. G.: *Process Control Systems,* 2d ed., McGraw-Hill, New York, 1979, pp. 236, 237.
12. Buckley, P. S., R. K. Cox, and D. L. Rollins: "Inverse Response in a Distillation Column," *Chem. Eng. Prog.,* June 1975.
13. Reference 3, pp. 515 and 516.
14. Reference 11, pp. 181–183.

PART TWO

Control System Structure

CHAPTER FOUR

Controlling Flow Rates and Inventories

Chapter 3 develops the relationships between product compositions and various flow ratios in a column. Figure 3.11 provides an excellent summary of these relationships for a typical column, by comparing the variation of y with x when certain flow ratios were maintained constant. There are actually six different curves representing six independent flow ratios, only two of which are necessary to determine and therefore control product purities. The choice of which two are most effective will be determined in Chap. 5, following a presentation on control-loop interaction.

However, even after selection of the appropriate set of curves, there remains a wide choice of variables which may be controlled to establish the position of each curve. There are five principal streams, F, D, B, L, and V, which can be combined into ten ratios. If one wishes to regulate separation as an operating variable, this can be accomplished by fixing any ratio L/D, D/V, or L/V. To regulate the material balance requires fixing D/F, B/F, or D/B. The remaining selections are V/F, V/B,

L/F, and L/B. Choices or streams should include those whose flow rates can be measured and controlled with maximum precision. Therefore, this chapter is devoted to the task of examining how these flow rates can be measured and controlled, both instantaneously, and in their accumulation.

CONTROLLING FLOW RATES

Any vessel has a material balance and an energy balance, both of which must close. There may be any number of streams entering and leaving, some of which may be unmeasurable. The largest of them will tend to have the greatest influence over product quality and should therefore be controlled to the highest possible accuracy. Fortunately, it is not necessary to measure them all because in the steady state their algebraic sum must be zero. The stream that is largest or otherwise measurable with the least absolute accuracy should be assigned to close the material or energy balance so that its flow is determined by difference. But the first order of business is to measure those that can be measured.

External Liquid Rates   Fortunately, the flow rates of most of the liquid streams in a distillation unit are not difficult to measure. They are generally clean fluids that can be metered readily with an orifice and differential-pressure (DP) transmitter. (An exception would be the feed and bottom product of a beer still, which contain about 10 percent grain from the fermentation process wherein alcohol is produced from sugars.)

Feed, reflux, distillate, and bottom-product streams are usually liquids at or slightly below their boiling point. If at the boiling point, some flashing will develop with pressure drop across a metering orifice, producing a noisy and erratic signal. However, most of these streams are pumped, and the orifice is located between the pump and the control valve so that the boiling point is elevated by pump pressure well above the temperature of the liquid. Only in gravity-flow systems is the head loss across an orifice likely to present problems, and subcooling can be introduced where it becomes necessary.

The principal limitations of orifice meters are their low rangeability and sensitivity to density variations. The latter is a problem more common with gases than liquids, but rangeability is the same for both.

The rangeability of an orifice meter is limited principally by its nonlinear nature. The differential pressure h produced across an orifice is proportional to the square of flow f:

$$h = f^2 \tag{4.1}$$

where both variables are expressed here in terms of fraction of full scale. Differentiating (4.1) allows the flow error to be expressed in terms of error in the measured differential pressure:

$$df = \frac{dh}{2f} \tag{4.2}$$

In the field, differential-pressure transmitters are probably not more accurate than 0.5 percent. This error would be equivalent to a 1 percent flow error at

$$f = \frac{dh}{2df} = \frac{0.5}{2(1.0)} = 0.25$$

Therefore, the range of flow rates over which measurement could be made within 1 percent error is about 25 to 100 percent or a 4:1 range.

For many applications, this rangeability is adequate. Consider for example, that the efficiency of perforated trays in a distillation column falls sharply when vapor flow changes beyond a range of about 3:1 [1]. Orifice meters are adequate for measuring both boilup and reflux in these columns. They also appear satisfactory for most feed and product flows as well. Where greater accuracy and range are needed, turbine flowmeters have been used. Where solids are present, as in feed to a beer still, a magnetic flowmeter can be used.

The ability to measure and control *external* flow rates may be insufficient to hold compositions steady, however, as they are functions of *internal* flow rates. The difference between external and internal reflux is shown in Chap. 3 to be a function of reflux enthalpy or subcooling below the boiling point. By combining Eqs. (3.20) and (3.29), an expression can be derived for determining internal reflux from external reflux and temperature:

$$L_t = L\left[1 + \frac{C_p}{\Delta H_D}\left(T_0 - T_L\right)\right] \tag{4.3}$$

Figure 4.1 shows how this calculation is implemented. A differential-temperature measurement is made between the overhead vapor at T_0 and reflux at T_L. This difference is then multiplied by the coefficient $C_p/\Delta H_D$ for the product and then the resulting correction factor multiplied by the external flow rate [2].

Although this system compensates adequately for subcooling, it reacts incorrectly to variations in overhead composition. Should reflux temperature fall, $T_0 - T_L$ will increase and cause external reflux to be reduced, which it should. However, should vapor temperature rise owing to an increase in higher-boiling impurities, $T_0 - T_L$ will rise also, again reducing external reflux. This action is counter to the desired control action and will therefore contribute to the disturbance through positive feedback. Negative feedback from a composition or temperature controller is always required to ensure stability with this system.

Occasionally it may be desirable to control liquid flow in the bottom section of the column. Then the effect of feed rate and enthalpy will have to be included. The principal requirement for this information is in stabilizing a column having a dominant liquid sidestream. Special consideration is given to this problem in Chap. 8.

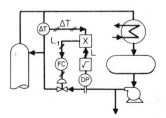

figure 4.1 *The flow of internal reflux in the top section of the column may be calculated by correcting the external reflux for subcooling.*

Measuring Vapor Rates Orifice meters are rarely inserted into column vapor lines, owing to their large size and the penalties associated with additional pressure drop. As a result, a direct measurement of vapor flow is not usually available.

One common alternative is to use the entire column or a section of it as an orifice plate. Each tray produces a pressure drop which is a function of vapor velocity, although it is also affected by liquid loading and the characteristics of the tray.

In an ideally functioning sieve tray, all the liquid will overflow the weir, and all the vapor will bubble through the holes. The vapor will therefore encounter a minimum pressure drop which is the product of weir height h_w and liquid density ρ_L, shown as the intercept in Fig. 4.2. In actual practice, however, liquid tends to drip through the holes at low vapor rates so that the lower portion of the curve is outside the viable operating range. At higher velocities, the pressure drop increases with the square of vapor flow owing to the resistance of the perforations. In the normal range of flows, differential pressure is a responsive and reliable indication of vapor flow, although lacking in absolute accuracy owing to its sensitivity to liquid flow as well. At still higher rates, downcomers may fill with liquid, restricting down-flow below the rate imposed on the upper trays so that the column begins to fill with liquid. This flooding condition is also detectable by the differential-pressure mea-surement. Its use in the control system is described in detail in Chap. 7.

To extend the range of efficient operation, columns are being fitted with valve trays. They are similar to perforated or slotted trays, but each aperture is covered by a valve which is lifted by upflowing vapor and closed by gravity. In the closed position, the valves reduce the leakage of liquid through the slots and offer a high resistance to vapor flow. In their throttling range, they act to regulate differential pressure, just like a weight-operated regulator. At high vapor rates, the valves open fully and the tray then behaves like a simple perforated tray.

Because of the self-regulating nature of a valve tray, the sensitivity of the differential-pressure measurement to vapor flow in the throttling range is poor. However, the measurement remains useful in the high-flow situation to indicate or prevent flooding.

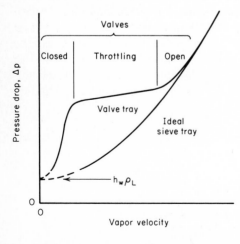

figure 4.2 *Pressure drop is an indication of vapor flow through sieve trays but is regulated by valve trays.*

The pressure-drop curves of Fig. 4.2 also indicate the likelihood of inverse response of base level to boilup. As boilup is increased, the density of the froth on the trays falls. If this changes the froth height more than it changes the pressure drop, the excess froth will spill over and temporarily increase internal reflux. This happens to sieve trays operating at low vapor rate but not at high vapor rate where a greater change in pressure drop is produced. Valve trays operating in their regulating range feature little change in pressure drop with boilup and therefore produce inverse response there.

The most serious obstacle in the way of measuring differential pressure reliably is in the installation of the transmitter. The two pressure taps are often located over 100 ft apart, and the lines typically contain vapor which will condense at ambient temperatures. If the transmitter is located below a pressure tap, liquid will accumulate in the line and give a false reading. Normal boilup may generate only 100 in of water differential pressure across a 100-ft-high column. Thus a few feet of liquid head, or even its variation with temperature, would cause serious error.

The line leading from the bottom of the column has problems, too. Even if it is straight and vertical, heat losses can develop a reflux within, which also causes an error. Therefore, vapor lines leading to the transmitter should be insulated. Purging them with noncondensable gas is not recommended since the gas will accumulate to the point where condensing is interfered with and pressure control becomes affected.

For most columns, the overhead condenser is not located above the column but instead is at grade or first level. The vapor line typically runs alongside the column most of the way down. If the top pressure tap is made where the vapor line enters the condenser, much of the elevation difference between the taps will be eliminated, as shown in Fig. 4.3. The transmitter will be as accessible as the condenser, and only the bottom line need be insulated. For double-column installations, as shown in Fig. 4.4, the transmitter for the bottom section may be mounted at grade level between the sections. Plug-style valves should be used, and liquid traps avoided.

Column differential pressure responds rapidly to heat input, controlling with a period of 20 s or thereabouts—essentially limited only by the dynamics of the reboiler. It is therefore quite capable of correcting for temperature or flow variations in the heating fluid as well as ambient disturbances.

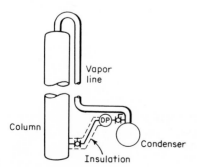

figure 4.3 *Connecting the low-pressure tap to the condenser inlet minimizes the difference in elevation between the taps.*

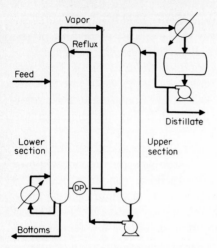

figure 4.4 *Differential pressure across the lower section can be measured to the bottom of the upper section of a double column.*

Differential-pressure control is recommended for the top section of a two-section column in order to overcome variations in heat losses in the long vapor and liquid lines connecting the sections. Differential-pressure control will be found very helpful in regulating vapor flow above a vapor sidestream, an issue which is discussed in Chap. 8.

A method of controlling overhead vapor flow with greater accuracy although with slower dynamic response is shown in Fig. 4.5. Because all the vapor leaves as reflux and distillate, it is possible to control vapor flow as their sum. Here vapor-flow set point V^* is introduced to a summing amplifier where measured distillate flow is subtracted. The result is the reflux set point L^* needed to hold the constant vapor rate.

But because this only sets *outflow* from the accumulator, an additional loop is needed to transfer control to *inflow*. The liquid level in the accumulator is controlled by manipulating heat input to the reboiler. Then changes in V^* will affect liquid level, causing the controller to change heat flow to restore the balance. The response of V to V^* in most installations is reasonably fast, with a time constant less than 1 min.

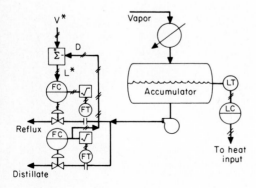

figure 4.5 *Overhead vapor flow may be controlled as the sum of reflux and distillate.*

The complement to this system would have distillate flow set from V^* and L; however, it is important that vapor flow should be adjustable over the widest possible range so that the larger of the two liquid flows should be selected for manipulation.

Because this system uses external flow rates, it also controls an external flow rate. If it is desired to control internal vapor rate V_t, then internal reflux L_t should be used in place of external reflux L. From Eq. (3.16),

$$V_t = V + L_t - L = L_t + D \qquad (4.4)$$

Closing Material Balances   The liquid-level control loop in Fig. 4.5 closes the material balance around the accumulator, forcing the boilup rate to equal the total liquid withdrawal rate in the steady state. Material balances always have to be closed at the base of the column and at the accumulator (when it contains two or more phases). There are three possibilities here:

1. If the accumulator is flooded, the liquid level will rise into the condenser, affecting column pressure; liquid level then becomes part of the energy-balance closure.

2. When both vapor and liquid are present, liquid level must be controlled by manipulating one or two of the streams V, L, D.

3. If the vapor condenses into two phases, as in steam stripping, the level of each phase must be controlled to close their individual balances.

The message given in Fig. 4.5 is an important one for it shows how closure of the material balance can be used to control the flow of a stream (V) that is not measured. Furthermore, it can also be used to overcome fluctuations in that variable if the level control loop is sufficiently fast. Suppose, for example, that the system of Fig. 4.5 is applied to the propane-propylene column described in Chap. 2, where the heat source is quench-water from the reactor effluent. Controlling quench-water flow is insufficient to maintain a constant boilup rate because its temperature will tend to vary. When this happens, the resultant changes in boilup will affect liquid level and be countered by the level controller's manipulation of quench-water flow.

The stream that is largest as well as most-variable should be the one selected to close the material balance for two reasons. First, liquid level will be affected most by the largest stream and therefore can best be controlled by it. Second, the largest flow is usually measurable with the least absolute accuracy of all the streams entering or leaving a vessel. By setting the flow rates of the smaller streams directly, the largest stream is forced to the same level of accuracy by the controller which closes the material balance. If material-balance closure were assigned to a smaller stream, it would then become subject to the larger errors associated with the larger streams. This point is made in Example 3.1 where boilup and reflux flows were set and distillate used to control accumulator level. Owing to the 10:1 reflux ratio, variations in distillate flow were estimated at about 25 percent, whereas it could have been controlled directly within 1 percent.

The system in Fig. 4.5 was devised specifically for the case where precise control of overhead vapor flow (as well as distillate flow) was desired. Again, modification to control internal vapor flow by setting internal reflux is possible. But consider the possibility that boilup can be controlled with acceptable accuracy by controlling the flow of steam to the reboiler. This is the practice that is most commonly followed; it is described later under "Manipulating Energy Inflow." In this case, V is set independently and accumulator level must be controlled by L and/or D. Several recommendations are given at the end of this chapter under "Inventory Controls."

The liquid level in the base of the column represents the liquid inventory in that vessel. In most columns, it is controlled by manipulating the discharge of bottom product, thereby closing the overall material balance. There are two basic reasons why B might *not* be used to control base level:

1. Bottom product is feed to a reactor and needs to be held at a fixed rate.
2. Its flow is so small that it cannot control level well.

In either of these instances, the external material balance must be closed by manipulating either feed or distillate flows. Both alternatives present problems, however, in that they are dynamically removed from the column base. Furthermore, feed rate is usually not available for manipulation. The usual practice is then to control base level with boilup, which transfers the external balance closure to the accumulator where level must be controlled by distillate flow. (A special case is examined in Chap. 8 for columns with a major liquid sidestream.) The principal stumbling block to this strategy is the possibility of inverse response of base level to changes in heat input. When faced with this dilemma, Buckley et al.[3] controlled base level with reflux and accumulator level with distillate. The various possibilities for level-control assignments are summarized in Table 4.1.

Closing the Energy Balance   Because of the equilibrium between liquid and vapor in a distillation column, changes in the energy balance manifest themselves in shifting inventory between the two phases. If heat is being added to the column faster than it is being removed, there will be a net conversion of liquid to vapor. Liquid inventory is sensed as liquid level in column base and accumulator. Vapor inventory is indicated by the pressure developed in the fixed vapor volume.

If one of the products from a column is a vapor, then vapor inventory and hence column pressure can be controlled by manipulating that stream. This is indeed the case wherever partial condensers are employed. By manipulating a product flow, the pressure controller is thereby closing the external material balance.

In many units with partial condensers, the flow of vapor product may be quite small — a noncondensible gas that has accumulated. Although it may represent a small part of the vapor inventory, it may have a very great influence over heat transfer in the condenser. If not released, it will reduce heat transfer, causing

TABLE 4.1   Assignments for Material-Balance Closures

Column	Accumulator	Comments
B	D	D depends on $V - L$
B	L or V	D is smallest flow
V or L	D	B is smallest flow
F	$D, L,$ or V	B is independent

pressure to rise. It is therefore capable of controlling column pressure if it is being fed to the column continuously; if not, then the pressure-control valve must occasionally close, during which time control will be lost. A system designed to accommodate both total and partial condensing situations is described in Fig. 7.16.

Closing the energy balance through pressure control should be accomplished through the manipulated variable that is large and most uncertain. The two largest energy streams are always the heat flow into the reboiler and the heat flow from the condenser. Usually, heat flow into the reboiler is easier to measure as the rate of steam being condensed. Furthermore, the steam supply is usually regulated so that the source is reliable.

Heat removal from the condenser is much more difficult to determine, particularly with air-cooled condensers. Furthermore, it tends to be more variable, changing with weather conditions, time of day, and season of the year. But by using heat removal to control pressure, the energy-balance closure forces it to match the heat inflow, thereby regulating it to the same absolute accuracy. Refrigerated condensers are easier to regulate, in that heat removal is proportional to refrigerant flow which is measurable.

For the heat-pumped column in Fig. 2.23, heat flow from the condenser is the same as that to the reboiler. Manipulation of that rate cannot close the energy balance, but only sets its throughput. A proportional flow of energy enters the unit as work to the compressor and must also be removed to close the energy balance. This is accomplished by the cooler shown at the compressor discharge, which therefore must be manipulated for pressure control.

In some distillation units, control over pressure is achieved by partially flooding the condenser with condensate. With the condenser above the accumulator, this can be accomplished either by closing a drain valve between the condenser and the accumulator, or by restricting flow out of the accumulator so that it fills completely. In the latter case, the liquid level in the accumulator does not require control, and there is no valve specifically provided to control pressure; instead, pressure is controlled by manipulating either V, L, or D (which would have otherwise been selected for level control). In effect, level is controlled within the condenser, closing both the energy and overhead material balances at once.

MANIPULATING ENERGY INFLOW

Chapters 1 and 2 include discussions on both the cost of heat and its selection, so those aspects of heat-input controls need not be repeated here. Instead, this chapter looks into the characteristics of the media commonly used and the equipment through which the heat is transferred. Those features most directly related to control are examined in detail: linearity, speed of response, and sensitivity to disturbance. Each heating medium and each type of reboiler has its own special needs which must be met if column performance is to be maximized. Not infrequently an engineer is surprised by a completely unexpected and apparently inexplicable response of a reboiler to control action. Enough of these characteristics are documented here to minimize the number of surprises encountered in the plant.

Reboiler Characteristics   Reboilers may be classified into three groups on the basis of the manner in which vapor is moved from the heat-transfer surface to the column:

1. Immersed
2. Natural circulation
3. Forced circulation

Each set includes subsets differing in physical construction or heat source. And each of the subsets has some outstanding physical characteristics directly affecting its control.

A reboiler tube bundle may be immersed in a "kettle" exterior to the column or internally in some designs. Figure 4.6 compares the two installations. Either may be heated with steam or liquid within the tubes. Both achieve 100 percent vaporization; i.e., there is no liquid discharged along with the vapor.

The weir in the kettle reboiler is intended to cover the tubes with liquid and seal the downcomer in the column base. This does not necessarily ensure that the tubes are always covered, however, for a sudden increase in boilup (due to a loss in pressure, for example) could vaporize liquid faster than it is returned. The column base is almost empty of liquid, and the only controllable quantity lies below the weir. Since this reservoir is small, its liquid level can be difficult to control and subject to rapid fluctuation. To gain more capacity at the point of level measurement, Lupfer [4] recommends setting the control point above the weir or removing the weir altogether.

Natural-circulation, of "thermosiphon," reboilers may be mounted either horizontally or vertically as shown in Fig. 4.7. Vertical thermosiphon reboilers are used primarily in the chemical industry—their single-pass construction and high tube-side velocity minimize fouling and facilitate cleaning. By comparison, a horizontal thermosiphon reboiler usually has the process in the shell and heating medium in the tubes. Its process-side velocity is lower, and additional liquid head is usually employed, often by trapping the liquid off the bottom tray as shown. Horizontal thermosiphon reboilers are more common in petroleum refining and are usually heated with circulating oil.

The high velocities encountered within the vertical thermosiphon reboilers are attained with as little as 5 wt % vaporization. This is an important consideration since circulation then depends on that concentration of volatile components—otherwise vaporization will be lost. Columns containing substantial concentrations of only slightly volatile components may exhibit unstable operation for this reason.

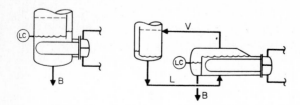

figure 4.6   *A tube bundle may be inserted either directly in the column base or in an externally mounted kettle.*

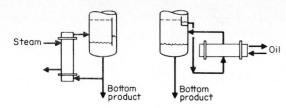

This problem can be more severe in horizontal thermosiphon reboilers where the percent vaporization is higher. Trapping the liquid from the first tray then becomes essential for the reboiler to function.

Note that the liquid circuit between column base and thermosiphon reboiler forms a "U tube." Liquid contained in a U tube has a capability of resonant oscillation like a pendulum. Its period is derived in Ref. 5 as

$$\tau_o = 2\pi\sqrt{\frac{l}{2g}} = 0.78\sqrt{l} \tag{4.5}$$

where τ_o = natural period, s
   l = total length of U tube, ft
   g = acceleration of gravity, 32.2 ft/s^2

The length of the U tube in most columns is about 15 ft, giving a 3-s period. In absence of friction, oscillation, once begun, would continue indefinitely. However, normal frictional resistance of the piping provides some damping. Nonetheless, bubble formation in the reboiler produces enough random disturbances or "noise" to induce oscillation at the resonant period. Buckley [6] describes unstable reboiler operation having a period of 3 to 4 s coincident with high heat-flux rates. A restriction in the liquid line can correct this problem by providing damping.

Forced circulation is used with vacuum distillation or when heat is obtained from the combustion of a fuel. Vacuum operation is usually intended to minimize the boiling point and therefore the rate of thermal degradation of the bottom product. However, the gravity head of liquid above the reboiler heat-transfer surface and ΔT_0 (see Fig. 2.17) both contribute to raising liquid temperature in immersed or natural-circulation reboilers. Forced circulation may be applied to either a horizontal or a vertical reboiler for a vacuum tower, but the flow in the vertical unit must be downward. At a particular flow rate, the static head and velocity head can offset one another so that relatively little net pressure drop exists through the tubes. Forced circulation also stimulates nucleation and reduces ΔT_0. Some forced-circulation systems use narrow tubes or a fixed restriction downstream of the reboiler to maintain a backpressure within the tubes. This prevents boiling in the tubes, allowing the product to flash as it leaves. Although this may be desirable from the standpoint of heat transfer, pumping costs are increased by the restriction.

Direct firing may be used to heat an oil without boiling, as shown in Fig. 2.18. But a fired heater may also be used as a reboiler, as in the fractionation of crude oil and other products having boiling points in excess of 400°F. Vaporization ranges from a few percent up to 40 to 50 percent. The flow of the liquid to the heater is nearly always forced to maintain efficient heat transfer. Reduction in flow tends to cause surface overheating within the tubes and subsequent deposition of polymers, tar, and coke. These deposits not only interfere with heat transfer but also tend to restrict flow, further aggravating the problem.

In addition, the flow is usually conducted through several parallel passes. Should the flow through one of these passes be somewhat lower than through the others, its percent vaporization will increase because essentially the same heat is being absorbed in all passes. The resulting expansion in volume increases the velocity through that pass, increasing the frictional and velocity-head loss. But if the pressure drop through all passes is the same due to manifolding, the tube generating more vapor will have its mass flow reduced even further and its percent vaporization increased more. In the absence of selective control, some tubes will be carrying high flow rates with little vaporization while others carry low flow rates with much vaporization. This condition has a detrimental long-term effect on the heater because of uneven temperature distribution. Tubes with lower flows will foul first, causing further reduction in flow. Ultimately these tubes will plug or even burn through.

Figure 4.8 shows a control system designed to provide equal flow through all passes. The highest valve signal is selected as the input to the valve-position controller. It, in turn, adjusts the set points of all flow controllers to hold that valve exactly at the fully open position. This arrangement results in the minimum attainable pressure loss through the valves. Note that the actual flow will change as tube resistance changes, but flow will always be maximum. No additional control over the total stream flow would seem necessary. If the circulating pump is oversized, flow may be reduced and horsepower saved more effectively by trimming the pump impeller than by throttling its discharge. Figure 2.16 shows a control system that equalizes temperatures instead of flows, thereby also compensating for unequal distribution of heat within the furnace.

For bottom products with a wide boiling range, percent vaporization can be indicated by temperature. However, temperature is also a function of composition and pressure. Consequently, to use it as a measurement of percent vaporization requires that temperature or composition be controlled in the column, too, with

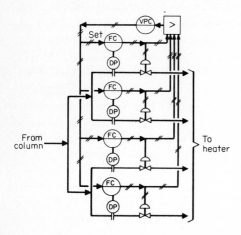

figure 4.8 *The valve-position controller sets all flows to whatever point will maintain one valve fully open.*

pressure compensation provided if necessary. Percent vaporization can also be inferred from the ratio of heater outflow to inflow using orifice meters. If both orifices are identical, the same differential pressure will be developed across each when there is no vaporization. The increased velocity brought about by expansion of liquid into vapor will increase the outflow differential with percent vaporization. Because of the two-phase outflow, an eccentric orifice must be used if the line is horizontal. Even then, calibration is not predictable and must be verified by actual field tests.

Steam and Other Vapors   Steam is not the only vapor used for heating, but it is by far the most common. When temperatures higher than can be obtained with saturated steam are required, another vapor such as one of the Dowtherms is often used. In certain low-temperature applications, a refrigerant is used; the reboiler condenses the compressed vapor and the condenser uses the liquified refrigerant for cooling. All vapors are basically the same in the manner in which they transfer heat.

Superheated steam is a relatively poor heat-transfer medium, as are most gases, from the standpoint of the rate of heat transfer per unit temperature difference. Saturated steam, on the other hand, is an excellent medium because of low resistance to heat transfer and a high latent heat of vaporization. Yet superheated steam is nearly always supplied to reboilers. Even if the steam is saturated at the supply pressure, the pressure loss across the steam control valve develops superheat. This can be seen on the temperature-entropy diagram of Fig. 2.6 by following a contour of constant enthalpy from the saturation pressure to a lower pressure. Throttling drops the pressure, keeps enthalpy the same, and has very little effect on temperature.

With the control valve on the inlet to the reboiler as shown in Fig. 4.9, the saturation pressure in the shell varies with heat load. Since the heat is being transferred between a condensing and a boiling fluid, neither changes temperature greatly in the process. If the relationship between heat-transfer rate and temperature difference is linear and the boiling point of the process fluid is constant, then steam temperature will vary directly with boilup rate. A "steam trap" or similar condensate seal is necessary to drain condensate without releasing steam. If the trap cannot remove the steam as fast as it condenses, shell pressure will rise but steam flow will fall as the shell fills with condensate.

When a steam valve is being used, boilup rate may be controlled either by steam flow or shell pressure. The steam-flow measurement is usually made upstream of the control valve where pressure is ordinarily constant, rather than downstream

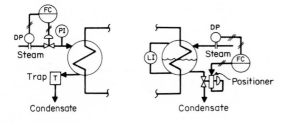

figure 4.9 *Steam flow may be controlled with a valve on either the steam or the condensate.*

where it most certainly varies. The calibration of the orifice flowmeter commonly used is sensitive to variations in pressure and temperature. Compensation may be applied for both if a mass-flow measurement is desirable. However, a heat-flow measurement is more meaningful and requires only pressure compensation. Variations in steam temperature affect density and enthalpy in opposite directions so that there is little ultimate effect on heat-flow rate.

If steam pressure rather than flow is controlled, boilup will automatically increase should the boiling point of the process fluid fall. This action compensates correctly for changes in bottom-product composition but not for changes in column pressure. Fouling of the heat-transfer surface will also affect the relationship between steam pressure and boilup. Furthermore, the rate of boilup is not linear with steam pressure nor is the steam pressure zero at zero boilup. Consequently, steam-flow control is preferred to pressure control to establish and maintain a boilup rate, prevent flooding, and control product quality.

There are certain advantages to placing the control valve on the condensate side of the reboiler. In the first place, the valve can be smaller. The density of steam at 250 lb/in^2 abs, for example, is one-hundredth that of condensate at that pressure. Because of the square-root relationship between valve size C_v and density ρ in Eq. (2.9), the condensate valve need have only one-tenth the capacity of the steam valve for the same pressure drop. But since capacity C_v varies with the square of pipe diameter, the condensate valve will be typically one-third the line size of a steam valve used for the same service. At lower pressures, the ratio is even more favorable.

Second, the steam reaching the reboiler is saturated (if the supply is saturated) and at a higher pressure than with a valve on the steam line. Hence the maximum heat-transfer rate is higher, and in addition there is no trap to limit the rate of condensate removal.

Steam flow is the rate of condensation, which in this case is controlled by adjusting the heat-transfer area exposed to condensation. As condensate flow is reduced, the level of condensate in the shell will rise, exposing less surface to the steam and more to the condensate. The condensate tends to be colder and a less efficient heat-transfer medium than steam. Therefore, a rising level results in a lower heat-transfer rate, a lower condensing rate, and a lower steam flow. Because of the flooding of the shell with condensate from which heat continues to be withdrawn, the condensate leaving tends to be subcooled. This may offer no advantage over a trap, however, since condensate discharged from a trap tends to be subcooled with respect to supply pressure when the shell pressure is low.

If the condensate valve cannot carry away the condensate as fast as the reboiler can produce it, a maximum steam flow will be reached with the shell still partially flooded. But if the reboiler cannot condense the steam as fast as the valve is directed to remove condensate, some steam will pass through the valve. The flow at which this limit is reached will decrease as the heat-transfer surface fouls and as noncondensible vapors accumulate.

The dynamic-response characteristics of the two schemes shown in Fig. 4.9 differ markedly. Manipulating the control valve in the steam line causes steam flow to change immediately. Shell pressure and hence the rate of heat transfer will lag behind a few seconds, but indicated steam flow responds directly. By contrast, the condensate valve has no direct effect on steam flow. Condensate level determines steam flow, and level takes time to change. With the valve in the steam line, the flow controller requires a wide proportional band (> 100 percent) and fast integral time (1 to 2 s) typical of most flow controllers. But when a condensate valve is used, the flow-controller settings are more representative of a level controller — a narrower proportional band (25 percent) and longer integral time (30 s). (For an explanation of how control-mode settings relate to the process, see Ref. 7.) In addition, the condensate valve should be equipped with a positioner whereas the steam valve does not need one. Reference 8 describes how a valve positioner can destabilize a flow loop but stabilize a level loop.

Although installing the control valve on the condensate line has several advantages, its slow response in changing heat-transfer rate can be detrimental to base-level control. The base of the column usually has little capacity, and the level measurement tends to be noisy and exhibits the hydraulic resonance associated with a U tube. These effects combine to make base level difficult to control stably when manipulating boilup rate. If the control valve is in the steam-condensate line, the relatively large capacity of the reboiler to store condensate is interposed between the column and the valve, and stable level control may no longer be achievable.

The steam valve is not altogether free of dynamic irregularities either. Buckley [6] notes that the sensitivity of steam flow to the steam valve increases markedly when sonic velocity is reached in the valve. This occurs when the absolute downstream pressure is less than half the absolute upstream pressure. Because downstream pressure then has no effect on flow, the influence of the reboiler capacity apparently is lost and control destabilizes. This condition is most likely to develop at low flow rates and when steam supply pressure is much higher than necessary.

Heating with Liquids   Whenever a liquid stream is used to boil a column, there arises the problem of how to measure and control the heat input. With steam heating, the rate of condensation is directly proportional to the steam flow. But as will be seen, with liquid media the relationship is very nonlinear and boilup is subject to change even with flow held constant. To demonstrate this point, the relationships governing heat transfer will be developed.

A liquid medium gives up heat flow Q proportional to its mass-flow rate W and temperature loss in passing through the reboiler:

$$Q = WC(T_1 - T_2) \tag{4.6}$$

The rate of heat transfer to a boiling liquid is a function of heat-transfer coefficient U, area A, and the residual temperature difference required for nucleate boiling ΔT_0:

$$Q = UA(\Delta T - \Delta T_0) \tag{4.7}$$

The temperature difference ΔT between the heating medium and the boiling liquid varies as the medium passes through the exchanger. Let the temperature of the heating medium at any point be T, with the temperature of the boiling liquid designated T_b. Then for an increment of heat-transfer surface dA, (4.6) and (4.7) may be written as

$$dQ = -WC\,dT = U\,dA(T - T_b - \Delta T_0) \tag{4.8}$$

Next, let (4.8) be solved for dA in terms of dT and integrated from T_1 to T_2 to yield A:

$$A = \int_0^A dA = -\frac{WC}{U} \int_{T_1}^{T_2} \frac{dT}{T - T_b - \Delta T_0}$$

$$= \frac{WC}{U} \ln \frac{T_1 - T_b - \Delta T_0}{T_2 - T_b - \Delta T_0} \tag{4.9}$$

Then outlet temperature may be related to inlet temperature and flow by converting (4.9) into an exponential form:

$$T_2 - T_b - \Delta T_0 = (T_1 - T_b - \Delta T_0)e^{-UA/WC} \tag{4.10}$$

The heat-transfer rate may be related to T_1 and W by expanding Eq. (4.6) and substituting (4.10) into it:

$$Q = WC[(T_1 - T_b - \Delta T_0) - (T_2 - T_b - \Delta T_0)]$$

$$Q = WC(T_1 - T_b - \Delta T_0)(1 - e^{-UA/WC}) \tag{4.11}$$

From (4.11) it may be seen that heat transfer is linear with temperature difference $T_1 - T_b$ but decidedly nonlinear with flow W. To assess this nonlinearity, divide (4.11) by Q_M, the maximum heat flow corresponding to an infinite flow of heating medium, i.e., when $T_2 = T_1$:

$$\frac{Q}{Q_M} = \frac{WC(T_1 - T_b - \Delta T_0)(1 - e^{-UA/WC})}{UA(T_1 - T_b - \Delta T_0)}$$

$$\frac{Q}{Q_M} = \frac{WC}{UA}(1 - e^{-UA/WC}) \tag{4.12}$$

The solution to Eq. (4.12) is plotted in Fig. 4.10. The actual operating level of WC/UA for a given reboiler may be found by solving (4.9). An equal-percentage control valve should be used to compensate this nonlinearity.

In actual practice, U is not expected to remain constant but will tend to vary in somewhat direct proportion to liquid flow W. If U varied linearly with W, then the exponent in (4.11) would be constant and Q would vary linearly with W. Heat transfer tends to vary with the 0.8 power of velocity, however. If the liquid-side film were completely controlling, then, such that U varied with $W^{0.8}$, Eq. (4.12) may be restated as

$$\frac{Q}{Q_M} = \frac{WC}{U_M A} \left\{ 1 - \exp\left[-\left(\frac{U_M A}{WC}\right)^{0.2} \left(\frac{U_M A}{W_M C}\right)^{0.8} \right] \right\} \tag{4.13}$$

where $U_M = U$ at W_M, the maximum flow.

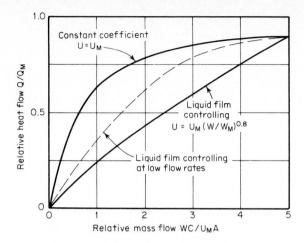

figure 4.10 *The rate of heat transfer may be sharply non-linear with the flow of liquid heating medium.*

For the case where $W_M C / U_M A = 5$, the solution to Eq. (4.13) is plotted in Fig. 4.10. Observe how the variation of film resistance with flow moderates the nonlinearity. The actual relationship for any given installation will lie somewhere between the two curves, since U is never constant nor is the liquid film controlling throughout. Typically, liquid-film resistance would be controlling at low rates but not at high rates; thus the relationship between Q and W would be relatively linear at low flow rates but would approach the limiting upper curve at high rates.

Heat transfer can be calculated directly by using temperature and flow measurements to solve Eq. (4.6). The calculation may be implemented either with analog or digital components as described in Fig. 4.11. The dynamic response of this control loop is somewhat peculiar. A step increase in liquid flow will cause a direct and linear increase in calculated heat flow. However, the increased flow will soon cause outlet temperature T_2 to rise in accordance with Eq. (4.10). As a result, the indicated heat flow will lose much of the increase it gained immediately following the change. Thus the indicated heat flow always leads the true heat flow by the residence time of the liquid in the reboiler.

Because this system has two feedback loops—one through flow and the other through outlet temperature—it is difficult to stabilize. Having the heat-flow controller set the oil-flow controller in cascade, as in Fig. 4.11, helps significantly.

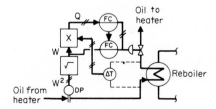

figure 4.11 *Heat input may be calculated from flow and temperature difference; cascade control of oil flow improves the stability of the heat-flow loop.*

Direct Combustion   Heating by direct combustion of fuel is common only where high temperatures are needed, as in crude-oil heating, and is practiced principally in refineries, where fuel is available from various sources. Most of these units are gas-fired, using mixtures of hydrogen and hydrocarbons vented from columns and other vessels, supplemented by natural gas and sometimes propane. One of the problems accompanying the use of mixed fuels is the varying heating value and density of the mixture.

The heat obtainable by combustion of a standard volumetric unit F of fuel gas is

$$Q = H_c F \tag{4.14}$$

where H_c is the heating value per standard volume of gas. Orifice meters are customarily used for gaseous fuels, and their differential pressure h relates to standard volumetric flow as

$$F = k \sqrt{\frac{hp}{\rho T}} \tag{4.15}$$

where k = meter coefficient
  h = orifice differential pressure
  p = absolute static pressure
  T = absolute temperature
  ρ = specific gravity of the gas

The specific gravity is simply its molecular weight divided by 29, the molecular weight of air. Equation (4.15) assumes that the fuel gases commonly used follow the ideal-gas law under metering conditions. Combining (4.14) and (4.15) yields the heat flow as a function of meter differential and specific gravity:

$$Q = k \frac{H_c}{\sqrt{\rho}} \sqrt{\frac{hp}{T}} \tag{4.16}$$

Pressure and temperature compensation may be applied where necessary. Correction for composition is the factor $H_c/\sqrt{\rho}$, known as the *Wobbe index*.

For hydrocarbon gases the Wobbe index varies uniformly with gas specific gravity as plotted in Fig. 4.12. Therefore, a specific-gravity measurement alone may be used to apply heat-flow correction. Mixtures containing hydrogen deviate substantially from the nearly linear relationship. But as Fig. 4.12 indicates, variation of the hydrogen content of a given mixture does not really need correction.

The relationship between gravity and Wobbe index breaks down completely when components such as nitrogen or carbon monoxide or dioxide enter the fuel system. Although their gravities are high, their heating values are low or nil, and their presence in the fuel cannot then be corrected by gravity alone.

Thermal calorimeters are available for measuring the heating value of a fuel gas, and some even generate the Wobbe index. But in general they respond no faster to a fuel-composition change than the heater does, and their correction is too late to be of benefit in controlling temperature. They are, in fact, a pilot model of a heater and are therefore faced with the same type of thermal lags that characterize a fired heater.

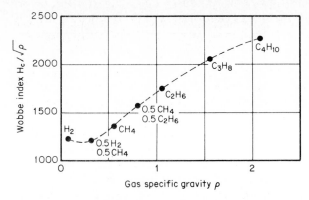

figure 4.12 *The Wobbe index varies uniformly with specific gravity for mixtures of light hydrocarbons.*

The airflow required for complete combustion varies directly with the heat flow. Therefore, if a heat-flow signal is available, it can be used to set airflow. If airflow is set in ratio to an uncompensated orifice differential, however, increasing fuel gravity will also increase the fuel-air ratio. This could mean smoky or even hazardous operation with high fuel gravity or too much excess air and inefficient combustion with low-gravity fuels.

Fuel-air ratio is most satisfactorily controlled by using an oxygen analyzer sampling the flue gas. The system shown in Fig. 4.13 provides this control. Linearizing the flow signal by square-root extraction is absolutely essential to stable control of product temperature. If the temperature controller sets differential pressure rather than flow, the gain of the temperature control loop will vary with flow because of the variable gain described in Eq. (4.2).

If the temperature controller sets h but f affects temperature, the control-loop gain increases as flow decreases. On the other hand, if the primary control loop is not temperature but column differential pressure, square-root extraction is not required because both column and flowmeter have the same nonlinear characteristic.

Burner backpressure has been used successfully as a flow signal, with the burner acting as a fixed orifice. But if the number of burners is variable or their apertures are adjustable, calibration will shift. A square-root extractor should be applied to the pressure signal when used for temperature control. Stack tem-

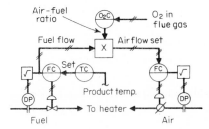

figure 4.13 *Air-fuel ratio is adjusted to hold a desired O_2 level in the flue gas.*

perature has also been used as a secondary controlled variable responding to variations in fuel quality and load. Furthermore, it is almost linear with firing rate.

Precise control of excess air is essential to conserve fuel. Combustion of methane proceeds accordingly:

$$CH_4 + 2(1 + X)O_2 + 7.52(1 + X)N_2 \longrightarrow$$
$$CO_2 + 2H_2O + 2XO_2 + 7.52(1 + X)N_2 \quad (4.17)$$

where X represents the fractional excess air and 7.52 is the number of moles of nitrogen accompanying 2 mol of oxygen in air. The products of combustion number $10.52 + 9.52X$ mol; consequently, 1 percent excess air increases the molal flow by about 1 percent and reduces internal temperatures by the same amount. For a given temperature, heat losses increase by that same amount. An oxygen analyzer on the flue gas measures the $2X$ mol of oxygen in the presence of the total moles of flue gas. The mole fraction of oxygen in the flue gas is related to excess air as

$$O_2 = \frac{X}{5.26 + 4.76X} \quad (4.18)$$

Energy loss due to excess air increases as higher flue-gas temperatures are reached. Therefore, fired heaters with high-temperature flues can more readily benefit from flue-gas analysis.

The problems of maintaining control and safe operation while fuel or air is constrained are described in Chap. 7, "Controlling within Constraints."

Controlling Preheat   Energy put into preheating the feed is less useful than if used for reboiling because the resulting vapor passes only through part of the trays. Nonetheless, preheaters are valuable for recovering lower-level heat than the reboiler can accept and for balancing the vapor loading in the column. A subcooled feed means less vapor and less liquid above the feed tray than below, which has the potential for flooding the stripping section while failing to utilize the full capacity of the enriching section. Complete vaporization of the feed is desirable in some applications, notably extractive distillation.

A common practice is to preheat feed by recovering sensible heat from the bottom product as shown in Fig. 4.14. When the product is taken from the weir chamber of a kettle reboiler, a feedback loop formed by the heat exchanger has been observed to cause cycles in liquid level of several minutes' period. A rising level causes the controller to increase product flow, thereby lowering level but also

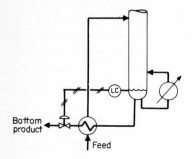

figure 4.14 *Preheating of feed with bottom product forms a negative feedback loop through the column which could destabilize base level.*

transferring more heat to the feed. The increased heat transfer reduces the flow of liquid down the column, which appears as a falling reboiler level several minutes later. The hydraulic delay through the trays is too long for the controller to accommodate, particularly in view of the small capacity of the weir chamber, so the cycles cannot be eliminated by controller adjustment. However, a temperature controller applied to the preheater is fast enough to eliminate the concurrent cycles in feed temperature and thereby is able to stabilize the level loop.

Preheating with steam follows essentially the same principles as reboiling with steam. And preheating with a liquid features the same nonlinearity as reboiling with hot oil, for example. When controlled preheating is to be achieved using a process fluid whose flow cannot be manipulated, one of the fluids must bypass the exchanger as shown in Fig. 4.15. A three-way valve is recommended since it minimizes pressure loss while allowing flow to be completely diverted in either direction. A single two-way valve bypassing the exchanger has limited rangeability in that some flow will always pass through the exchanger. Two two-way valves may be used instead of a three-way valve, but the cost and the pressure loss both tend to be higher.

Bypassing of the cold feed provides fast response of temperature to valve position. Reducing the flow through the exchanger will cause its exit temperature to rise after the fluid residence time has elapsed, however, so the net change in heat recovery may be small. As Fig. 4.12 indicates, at relatively high flow rates Q may change scarcely at all with flow. The dynamic response of this loop is very much like the heat-flow control loop where heat flow is calculated. The initial response in temperature is moderated by a later reaction in the other direction. In this case, however, the controlled variable is the true feed temperature whereas the heat-flow calculation led the true heat-flow rate.

Bypassing the heating medium does not offer the same speed of response as bypassing the feed. Its linearity is limited by the same mechanism described by Fig. 4.12. Three-way valves with equal-percentage characteristics are therefore preferred for both schemes shown in Fig. 4.15.

If the feed is simply to be heated and not vaporized, temperature control ought to suffice. From an economic standpoint, total heat recovery would be desirable, which would not require control at all. However, destabilizing reactions could occur as described earlier, particularly where the heating medium forms a feedback loop with the column feed stream.

In other cases, partial vaporization is desirable. Then temperature control may not be satisfactory since the degree of vaporization would change with feed composition. A lower-boiling mixture would require more vaporization to reach the same temperature as a higher-boiling feed. This would seem to be desirable in

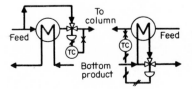

figure 4.15 *Preheat may be controlled by bypassing either fluid around the exchanger, but the response differs.*

directing more vapor upward and less liquid downward when the feed contains more-volatile components.

In cases where column pressure varies, the feed temperature measurement must be pressure-compensated if it is to be controlled. Where steam is used for preheating, simply setting steam flow in ratio to feed flow provides a constant heat gain. If the enthalpy of the feed is reasonably constant, the ratio can be set to give a certain percent vaporization. A percent-vaporization measurement may also be obtained from a ratio of the differential pressure across orifices on the inlet and outlet of the heater.

Rather than control at some arbitrary temperature or percent vaporization, a preheater can be used to balance column loading. This is most readily achieved by controlling the differential pressure across both the stripping and the enriching sections of the column. Reboiler heat input can be manipulated to control stripping section differential, and feed preheat can be adjusted to control that across the enriching section. This control system has the advantage of correcting promptly for disturbances in feed composition and enthalpy and for variations in the enthalpy of the heating medium as well. It can be applied even when the feed is not vaporized at all, since any change in feed enthalpy will affect the relative loading both above and below the feed tray.

If a maximum recovery of heat from a feed-bottom exchanger is desired, then the feed temperature cannot be controlled at a fixed point but must float. This action is achieved through the use of a valve-position controller. The input to the VPC would be the position of the valve in Fig. 4.15, while its output would set the TC. Then the temperature would be controlled stably at the highest value attainable within the range of the control valve. Operation is identical to that described for the floating pressure-control system in Fig. 2.20.

Should the feed-bottom exchanger be followed by a steam preheater, then the temperature set point should be as *low* as possible within the range of the steam valve (to conserve steam). Then the VPC would be set at minimum rather than maximum controllable position.

MANIPULATING ENERGY OUTFLOW

Energy outflow is generally much more difficult to measure and control than energy inflow. Because that energy is rejected either directly or indirectly to the environment as waste heat, its flow is sensitive to environmental conditions. Heat flow through air-cooled exchangers is most variable, depending on ambient temperature and surface conditions. Heat rejected to cooling towers is less variable, depending on wet-bulb temperature, which changes less than dry-bulb temperature. Cooling supplied by ground or surface water is still more uniform but is declining in use because of supply limitations and discharge regulations.

The factors which affect the heat flow through an exchanger are the surface area, heat-transfer coefficient, and the mean temperature difference between the fluids, as given in Eq. (4.7). However, the variables which are manipulable include only the flow rates of the streams exchanging heat. Reducing one of these flow rates is likely to alter two of the heat-controlling factors simultaneously, as the flow of oil to a reboiler affects ΔT and U at the same time in Fig. 4.10. It is important to recognize the actual mechanisms involved in controlling heat-transfer rates. Otherwise a control system that was successful in a particular installation owing to an obscure characteristic may fail when tried in another installation where that characteristic is missing. At this point, separate consideration is given to manipulating the coolant vs. the process stream.

Manipulating the Coolant   The most obvious manipulated variable for control of heat removal would be coolant flow. However, it has several limitations and is little used. Response of heat transfer is quite nonlinear, approaching the steepest curve of Fig. 4.10. At low flow rates where control is most sensitive, low coolant velocity encourages fouling of the heat-transfer surface. This is especially true of cooling water containing suspended or dissolved solids. Furthermore, the temperature rise accompanying reduced flow tends to promote scale formation in hard water.

The system shown in Fig. 2.19, which uses constant circulation of cooling water, is capable of a wider range of control without the fouling problem. Additionally, its dynamic response is superior because coolant flow is constant and high, regardless of the heat load. Both systems require the use of an equal-percentage coolant valve, owing to the nonlinearity of heat transfer with its flow.

The control of refrigerated condensers was described in Fig. 2.22. The heat-transfer surface should always be kept immersed, and control achieved by adjusting vapor flow by compressor-speed or guide-vane manipulation as shown, or by a vapor-return valve when the compressor serves several users. Heat removal by a waste-heat boiler should also be adjusted by throttling vapor flow; a liquid-level controller should manipulate coolant inflow.

Air-cooled condensers are becoming increasingly common and pose some special control problems. Airflow manipulation can be accomplished through louvers, pitch controls, or variable-speed motors. Louvers tend to be large, awkward, and easily damaged. The variable-pitch mechanism has been used for control but seems to feature deadband or hysteresis, which can cause limit cycles in pressure and liquid-level loops. The variable-speed motor is the best choice for control, but its expense may not be warranted.

Additionally, most air-cooled condensers consist of several parallel units, covering a large area. Even louvers can be quite costly, and control is usually justified for only one of the units. The other fans are switched on or off, or between low and high speeds, when the single modulating unit approaches either of the limits of its control range. Reference 9 describes an automatic system for smoothly coordinating a sequence of on-off operators with a single modulating operator.

A somewhat simpler system is shown in Fig. 4.16. The pressure-controller output modulates one fan directly, and through high- and low-pressure switches (PS), starts and stops others sequentially. As the high-pressure setting is reached, fan A is started; if this is effective in reducing pressure, the controller output will decrease, thereby reducing cooling through the louvered fan. Should the pressure not fall or should it rise again to the high-output setting, fan B will be started through a time-delay relay (TDR); a further increase in cooling demand will similarly start

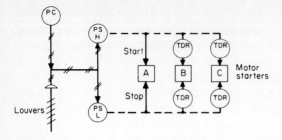

figure 4.16 *Fans are started sequentially when the controller output reaches the high setting and stopped sequentially when it reaches the low setting.*

fan C after a longer delay. A decrease in load sufficient to drop the PC output to the low setting will stop fan A first and then the others in sequence through a second set of TDRs. Because the first fan started is also the first stopped, the duty tends to be distributed; but if distribution is insufficient during certain seasons, the sequence can easily be changed by appropriate adjustments to the TDRs. Recognize, however, that the TDRs provide integrating action, and therefore need to be set at multiples of the controller's integral time.

Because air-cooled condensers must operate under a wide range of ambient conditions—hot and cold, wet and dry—it is imperative that they be capable of manipulation over a wide range. Alternatively, if a floating-pressure control system is used, the rangeability requirement is sharply reduced in that column pressure is made to float on the ambient conditions.

Flooding Submerged Condensers   More often, heat-transfer rate is changed by operating on the process side of the condenser, principally by flooding it with condensate. One well-defined example of this is the submerged condenser, physically located below the reflux accumulator, as shown in Fig. 4.17.

The rate of heat transfer is changed by allowing condensate to cover tube surface. The covered tubes have a lower heat-transfer coefficient than exposed tubes, and their mean temperature difference decreases as the condensate is subcooled. Closing the vapor bypass valve will allow the vapor in the drum to condense against the subcooled reflux and thereby lower drum pressure. This draws liquid from the condenser, exposing more tube surface and thereby increasing the rate of condensation. Opening the valve allows liquid to flow back into the condenser, covering tubes and reducing heat transfer. Changes in this direction typically take less time.

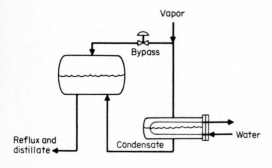

figure 4.17 *Heat transfer through the submerged condenser is adjusted by changing the level of condensate within it.*

The key to successful control using a hot-vapor-bypass valve is the sizing of the valve relative to the hydraulics of the installation. The pressure drop across the valve in Fig. 4.17 is relatively constant, being essentially the head of liquid between the two surfaces. Drum level is normally controlled so that the only change in head would be the difference between a full and empty condenser.

However, the flow through the bypass valve may change widely and is not always determinable. The valve carries all the vapor which will condense on the walls of the drum and the surface of the liquid, the latter being more significant. Reference 10 bases valve sizing on the flow of vapor that is necessary to raise all of the liquid leaving the condenser to its saturation temperature at drum pressure. This would seem excessive, however, in that liquid typically both enters and leaves the bottom of the drum, allowing a stable temperature gradient to develop within it. The reflux would therefore tend to remain subcooled, and bypass flow would be a very small percentage of the total vapor condensed.

In installations where the volume of liquid contained in the condenser is significant compared to that in the drum, there may be a pronounced interaction between the pressure- and drum-level-control loops. The pressure controller should be adjusted as tightly as possible with the drum-level loop open, and then the level controller adjusted with the pressure loop closed. This procedure gives preference to pressure control, which is the more important function.

The position of the bypass valve is an indication of condenser loading, and so it may be used to achieve floating-pressure control as in Fig. 2.20. The set point of the valve-position controller should be about 10 percent open to allow adequate margin for control. Valve sizing is not so critical then, in that the valve is always returned to the same position in the steady state.

Another type of submerged condenser is a tube bundle inserted in the base of a cooling tower. In these installations, vapor enters the bottom of the bundle, carrying condensate upward into an elevated accumulator. Functionally, it performs like that in Fig. 4.17 but can respond faster owing to the relatively low capacity of the tube bundle, part of which is occupied with vapor.

Flooding Elevated Condensers   Figure 4.18 shows a pair of vapor valves used to control heat transfer through a condenser elevated above the reflux drum. As before, control is achieved by flooding heat-transfer surface with condensate; howev-

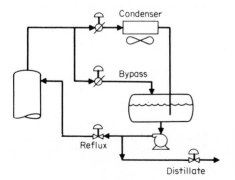

figure 4.18 *Flooding an elevated condenser requires a frictional pressure drop exceeding the static head of liquid.*

er, the liquid head is now working in the opposite direction. Flooding can be accomplished only if the condenser and its valve offer a higher pressure drop than the static head of liquid. Where the condenser throttling valve is missing, there is no way to control this pressure drop, so the vapor-bypass valve can only affect heat transfer at high loads if at all.

The throttling valve may be used to control differential pressure between the column and the accumulator, in excess of the liquid head, so that the bypass valve is effective. But if the throttling valve is present, the bypass valve becomes superfluous because heat transfer can already be changed by adjusting condenser pressure. Vapor throttling valves are not often used for several reasons. First, they are costly since the overhead vapor line is the largest in the unit. Second, throttling reduces the suction head on the reflux pump, increasing its power requirements and the tendency to cavitation. The position of the throttling valve can be controlled nearly full open to achieve floating-pressure control; this also minimizes the load on the reflux pump.

Figure 4.19 shows two preferred arrangements for controlling elevated condensers. On the left, flooding is accomplished by throttling a drain valve under the condenser, with the vapor space of the accumulator connected to the condenser inlet through an equalizing line. (The equalizing line needs to be only large enough to pass vapor at the same volumetric rate as the liquid being drained.) Control over heat transfer tends to be linear and reasonably responsive in both directions.

This concept is readily extended to flooding of the drum as well and even to eliminating the drum altogether. Heat-transfer rate can then be changed by manipulating reflux flow as shown in Fig. 4.19 (right), distillate flow, their sum, or boilup, as desired. Because they all affect the inventory of liquid in the condenser, they all influence the surface area available for heat transfer. In this system, closure of the overhead material balance and energy balance occur simultaneously. This eliminates one troublesome loop (accumulator level) from the control system.

The choice of the manipulated variable for pressure control should be based on the considerations for accumulator material-balance control outlined in Table 4.1.

Elimination of the reflux drum is effective in reducing the composition time constant and dead time. (The contribution of the drum to column dynamic response is described in connection with Figs. 3.14 and 3.15.) However, it increases the danger of cavitating the reflux pump should the condenser become overloaded

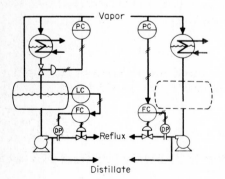

figure 4.19 *Flooding an elevated condenser can eliminate a level controller, a valve, and in some cases, even the reflux drum.*

or bound with noncondensible gas. When this possibility exists, protection against loss of liquid level must be provided, as described in Chap. 7.

Partial Condensers Partial condensation is encountered in three fundamentally different applications:

1. Where a single overhead product is taken in the vapor phase.

2. Where a small flow of noncondensible gas must be removed from a liquid product.

3. Where two overhead products are taken, the vapor product being in equilibrium with the liquid.

Only the first of these represents a two-product split. In both the second and third, the vapor stream accompanies a liquid product; these systems are discussed in detail in Chap. 8.

However, case 2 above is noteworthy in that a noncondensible gas interferes with heat transfer so that its release can be manipulated to control the rate of condensation of overhead vapor. It is therefore quite common to control column pressure by venting a small stream from the condenser. Pressure controlled in this way is both stable and responsive, as long as the flow of noncondensibles into the column is sufficient to keep the vent valve partially open. If it is not, the valve will close, and the pressure will fall below set point. The practice of using a second valve to admit a gas into the system in this eventuality is counterproductive as described in reference to the benzene-toluene column in Chap. 2. It is better to let the pressure fall to an equilibrium level, thereby achieving floating-pressure control without additional instrumentation. Alternatively, the vent valve may be sequenced with a coolant valve, so that when the first is closed by falling pressure, the second will begin to close. A means for coordinating a vent valve with condenser flooding is discussed in Chap. 7.

Figure 4.20 describes a demethanizer where the overhead product, methane and lighter components, leaves in the vapor phase. There are four valves and four controlled variables here, presenting a wide selection of control-loop arrangements. Assigning refrigerant flow to control its level in the condenser reduces the number of variable pairs to three.

This process is highly interacting. Increasing the product flow lowers column pressure and therefore reduces condensation, which in turn reduces accumulator level and reflux, and increases the ethane content of the vapor. Opening the refrigerant-return valve lowers column pressure by increasing condensation, which also raises level and takes ethane out of the product. Consequently, the

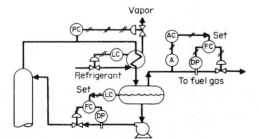

figure 4.20 *The overhead product from a demethanizer is usually withdrawn as a vapor.*

control-loop configuration shown in Fig. 4.20 would be appropriate only where the vapor product was the smallest stream, consistent with the guidelines in Table 4.1. Should the vapor be the major product, it would be best placed under pressure control to close the external material balance.

Further speculation on the assignment of the composition loop is inappropriate at this stage of the presentation. Interactions also exist between the top and bottom of the column, which are addressed in Chap. 5.

Figure 4.21 describes the overhead arrangement for a typical deethanizer. Reflux is condensed by cooling water, whereas the distillate requires chilled water. The first is a partial condenser similar to that in Fig. 4.20. The heat duty for the two condensers must be split exactly as the reflux ratio. Thus the problem of adjusting reflux ratio requires the manipulation of condenser cooling, an unwelcome complication.

Here again, considerable interaction appears among the variables. Pressure is a function of total vapor condensed and can therefore be controlled by manipulating either coolant individually or both together. The distillate condenser could be flooded for pressure control, but the reflux condenser cannot, being a partial condenser. If the distillate is the major product stream, it could then be manipulated to control pressure as suggested for the demethanizer above.

If the distillate is the smallest flow, it could be manipulated to control its own composition, as shown in Fig. 4.21. Then any change in its flow will affect the self-regulating level in the flooded condenser, causing column pressure to vary. The pressure controller must then readjust cooling at the reflux condenser to keep the total heat load in balance.

The inability to flood a partial condenser forces the manipulation of coolant, which is not as satisfactory from the standpoint of linearity, speed of response, or rangeability. Therefore the release of noncondensible gas (as in the demethanizer) or flooding of the distillate condenser (as in the deethanizer) will provide superior pressure control. Unfortunately, the failure to manipulate cooling at the reflux condenser results in reflux that is subcooled and therefore pressure that is higher than needed, increasing energy requirements unnecessarily. Additionally, superior composition control may be obtained if the distillate is manipulated for that purpose. Again, the assignment of the loop is pursuant to interaction considerations, and is examined again in Chap. 8.

Columns operated under vacuum also feature partial condensers in that minute leaks from the atmosphere introduce noncondensibles which must be removed.

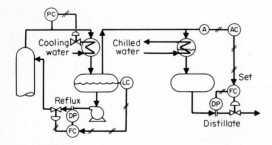

figure 4.21 *In a deethanizer, the condensing duty is usually shared between cooling water and refrigeration.*

Where water is the overhead product, a direct-contact barometric condenser may be used, with the overhead product leaving in the company of the cooling water. While this is common practice with evaporators, the opportunity seldom appears in distillation. However, condensation by direct contact is nonetheless resorted to in vacuum distillation where minimum pressure loss is mandatory. The objective is achieved by subcooling the distillate and recirculating through a spray condenser as shown in Fig. 4.22.

In a vacuum column, noncondensible gases must be continuously removed to maintain the desired absolute pressure. As in other partially condensing systems, another variable besides column pressure must be used to manipulate heat removal. Figure 4.22 shows a reflux-temperature controller making that adjustment.

INVENTORY CONTROLS

Although control over inventories of vapor and liquid might seem to occupy a rather mundane place among the loops assigned to a distillation column, it is essential in achieving dynamic responsiveness for the other loops. Unless inventories are controlled precisely, there can be no steady state for the process; by the same token, the faster the inventory-control function, the faster the process will reach a steady state. Consequently, this chapter concludes with specific recommendations for implementing the inventory loops.

Pressure Control   Although column pressure is an indication of vapor inventory, it also has a specific relationship to other column variables such as relative volatility, boiling point, and flooding limit. In this respect, it differs from liquid inventory variables, whose actual value may have no influence whatever on the performance of the column. A falling liquid level may simply indicate an unsteady state, whereas a falling pressure will cause boiling to increase rapidly, even to the point of flooding the column. As a result, pressure control is generally more demanding than level control.

Temperature is used more than any other measurement to infer composition. But column temperatures are affected equally by pressure. Controlling column pressure at the same location as the temperature measurement will improve the correlation between temperature and composition. Pressure must be controlled tightly for the temperature controller to perform its function.

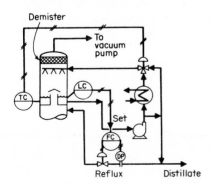

figure 4.22   *Condensing by direct contact with subcooled spray is sometimes used in vacuum columns.*

For many mixtures, relative volatility increases as pressure and temperature are reduced, as described in reference to Fig. 3.7. The pressure set point should, in these cases, be lowered as far as possible, while remaining in the control range, to take advantage of the energy savings attainable.

An estimate of these savings may be made using the separation model given in Eq. (3.58):

$$S^{1/nE} = \frac{\overline{\alpha}}{\sqrt{1 + D/Lz}} \tag{4.19}$$

If separation and tray efficiency are constant, then changing average relative volatility from $\overline{\alpha}_1$ to $\overline{\alpha}_2$ will allow a corresponding change in reflux:

$$\frac{1 + D/L_2 z}{1 + D/L_1 z} = \left(\frac{\overline{\alpha}_1}{\overline{\alpha}_2}\right)^2 \tag{4.20}$$

Solving for the new reflux ratio,

$$\frac{L_2}{D} = \left[\left(z + \frac{D}{L_1}\right)\left(\frac{\overline{\alpha}_2}{\overline{\alpha}_1}\right)^2 - z\right]^{-1} \tag{4.21}$$

Then using the overhead material balance,

$$V = L + D$$

the new overhead vapor flow can be calculated:

$$\frac{V_2}{V_1} = \frac{1 + L_2/D}{1 + L_1/D} \tag{4.22}$$

Finally, the heat flow corresponding to the vapor flow can be determined:

$$\frac{Q_2}{Q_1} = \frac{V_2 \Delta H_{D,2}}{V_1 \Delta H_{D,1}} \tag{4.23}$$

This is a necessary step, because latent heat ΔH_D also changes with pressure, partially offsetting the savings promised by the improvement in relative volatility.

example 4.1

Consider a butane splitter separating a 50-50 mixture of normal and isobutane into products each 95 percent pure. At a reflux temperature of 140°F, the column pressure is about 125 lb/in²abs and the relative volatility of the overhead mixture is 1.29. The latent heat of vaporization is 121 Btu/lb. At the same pressure, reboiler temperature is 163°F and the relative volatility there is 1.26. The reflux ratio is 7.03.

If the reflux temperature can be reduced to 100°F, the column pressure could fall to 72 lb/in²abs, giving $\alpha = 1.35$ and $\Delta H_D = 135$ Btu/lb. The reboiler temperature would fall to 123°F and at that point, $\alpha = 1.32$. Calculate the energy savings possible by operating under these conditions.

$$\overline{\alpha}_1 = \frac{1}{2}(1.29 + 1.26) = 1.275$$

$$\overline{\alpha}_2 = \frac{1}{2}(1.35 + 1.32) = 1.335$$

$$\frac{L_2}{D} = \frac{1}{(0.5 + 1/7.03)(1.335/1.275)^2 - 0.5} = 4.90$$

$$\frac{V_2}{V_1} = \frac{1 + 4.90}{1 + 7.03} = 0.735$$

$$\frac{Q_2}{Q_1} = 0.735 \frac{135}{121} = 0.820$$

Energy savings would be 18 percent.

Following this procedure, Table 4.2 was prepared to compare the expected energy savings with column pressure and reflux temperature along with the coolant temperature needed to bring it about. It is assumed that the condenser is air-cooled, requiring a 50°F temperature difference to transfer the design heat load. As ambient temperature falls, the heat load can be reduced which proportionally reduces the temperature difference across the condenser. The top row represents typical summer conditions while the bottom row represents winter conditions in the northern United States.

The center line of Fig. 4.23 is the energy-to-feed ratio from Table 4.2 plotted against percent of maximum pressure; the relationship is seen to be linear. If the same column is used to make higher-purity products (upper line), energy consumption is naturally higher, but the percentage conserved by reducing column pressure is also increased. For this reason, there is no general factor that can be applied to estimate energy savings possible through adjustment of column pressure — each column must be evaluated individually.

The savings given in Table 4.2 and Fig. 4.23 are perhaps the best that can be expected with seasonal variations, there being a host of limitations which apply:

1. Ambient-temperature variations in many parts of the world are smaller than those shown in Table 4.2.

2. Water-cooled condensers have a narrower range of temperature variation.

3. Accumulation of noncondensibles may prevent pressure from being reduced significantly.

4. Tray efficiencies may change unfavorably with falling pressure.

5. Columns operating at subatmospheric pressure or subambient temperature show little potential for energy savings.

Despite all of these reservations, the energy savings achievable by operating at minimum pressure are substantial in many installations and may require very

TABLE 4.2 Energy Required to Separate Butanes as a Function of Pressure

Reflux temperature, °F	Pressure, lb/in²abs	Q/F, %	Ambient temperature, °F
140	125	100	90
120	95	90.1	75
100	72	82.0	59
80	53	75.3	42
60	38	70.4	25

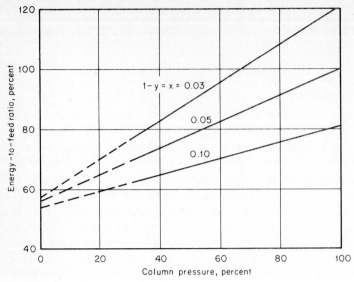

figure 4.23 *The potential for energy savings in a given column increases with separation factor.*

little investment. Some operators simply adjust the pressure set point seasonally, taking care to adjust any temperature set points accordingly. Yet this procedure cannot take advantage of day-night variations or weather changes. Table 4.2 indicates a 10 percent energy saving accompanying a drop in ambient temperature from 90 to 75°F, a common day-night change in many locations. To follow ambient variations closely, a continuously floating-pressure set point is required and is discussed below.

Another potential advantage of reduced-pressure operation is increased capacity. Because most columns are energy-limited, a decrease in the required energy-to-feed ratio may allow an increase in feed rate. Whether this is available depends on how the flooding limit of a column changes with pressure. If the Q/F ratio changes with pressure at a rate greater than the Q limit, then feed rate can be increased as pressure is reduced. The author has shown this to be the case for columns separating isopentane and lighter components and rejecting heat directly into the environment [11]. Flooding limits are examined in detail in Chap. 7.

Floating-Pressure Control   Operating a totally condensing column with no pressure control whatsoever is an attractive concept. The familiar control valves used to throttle, bypass, or flood the condenser would simply be eliminated. Although this would be ideal in the steady state, transient upsets to the heating or cooling system could cause undesirable reactions. Consider, for example, a column with an air-cooled condenser exposed to a sudden rainstorm. The wetted surface of the condenser becomes capable of transferring more heat, causing a sharp reduction in column pressure. The column itself is not cooled by the rain, but some of its sensible heat is quickly converted to latent heat by the fall in pressure. The vapor-rate increase caused by the transient reduction in sensible heat can be enough to flood the column. Its least effect would be to move substantial quantities of high-boiling components up the column and into the reflux drum. Protection must be provided against this type of upset.

While some sort of control needs to be provided to counter the short-term upsets, the pressure should be minimum in the long term. This can be achieved by judiciously adjusting the set point of the pressure controller. However, the adjustment must be slow and continuous rather than stepwise. Operators tend to adjust set points in steps, however small; and each step introduces the type of upset the pressure controller is intended to eliminate.

The solution is to use a valve-position controller (VPC) to adjust the pressure set point as shown in Fig. 2.20. It is intended to hold the condenser control valve in its fully open or fully closed position, depending on whether the valve bypasses, throttles, or floods the condenser. In any case, the condenser should be fully loaded in the long term.

The distinction between short-term and long-term control is made in the mode settings of the two controllers. The pressure controller has proportional and integral action adjusted to provide prompt recovery from upsets. The pressure set point will then be followed as closely as possible, as in any pressure-control loop. The valve-position controller, however, should have integral action alone. Then a sudden change in valve position will not elicit a proportional change in the pressure set point. Instead, the set point will move at a rate proportional to the deviation of valve position from its limit. The actual rate is adjustable through the integral time, which can be set up to 60 min in contrast to perhaps 2 min for the pressure controller. The effects of control action on valve position and column pressure are depicted in Fig. 2.21.

Since weather conditions usually deteriorate more rapidly than they improve, protection ought not be necessary for conditions causing a rise in pressure. Consequently, the valve-position set point could be safely held at 10 percent closed (or open as the case may be), which provides control over falling pressure only.

The scheme shown in Fig. 2.20 allows the operator to hold pressure constant at will by placing the pressure controller in its locally set mode. Either locally set pressure control or manual operation will take control away from the valve-position controller. Opening of the valve-position loop would then cause the VPC to reduce its output continuously in an effort to restore valve position to 10 percent. The operator might then find it difficult to return to the remotely set mode since the remote set point would be well downscale. This windup of the VPC can be prevented by using the pressure measurement as feedback. As long as there is no deviation between the pressure remote set point and measurement, the VPC will integrate normally. But when a pressure deviation develops, integrating action will eventually come to rest with the VPC deviation equalling the pressure deviation. To place the system in valve-position control, the pressure controller must be transferred from the locally to the remotely set mode while in manual to avoid bumping the valve. This configuration presents the operator with only the conventional pressure controller at the panel. The VPC can be blind (no display). It may be mounted away from the panel and requires no auto-manual transfer station.

If no other action accompanies the lowering of column pressure, no energy will be saved. Instead, the increased relative volatility will tend to enhance the purities of both products. If both are under product-quality control, their controllers will then reduce reflux and/or boilup in an effort to return compositions to their respective set points. In the meantime, the combination of falling pressure with continued high boilup may bring the column close to flooding.

The entire process of pressure change may be expedited, compositions maintained at set points, and the threat of flooding eliminated by changing boilup and/or reflux directly as a function of pressure. Figure 4.23 shows how the boilup-to-feed ratio should change with pressure for a butane splitter; very similar curves have been derived for a depropanizer and a pentane splitter [12]. Since reflux ratio is linear with V/F ratio, similar relationships can be readily prepared for L/D as a function of pressure.

The set of lines given in Fig. 4.23 could be approximated by the function

$$\frac{Q'}{F} = 0.56 + mp' \tag{4.24}$$

where the primes identify terms expressed in normalized (i.e., percentage) values; m would be the slope of the line adjusted as required to achieve the desired separation factor. A control system implementing Eq. (4.24) is shown in Fig. 4.24. Column pressure changes heat input directly to achieve immediate energy savings and avoid upsets to product compositions. If a change in composition were required, however, the controller (AC) would change the demand for heat by readjusting slope m, thereby recalibrating the pressure function for the new condition.

The dynamic response of this system is noteworthy in that setting of heat input directly proportional to pressure forms a positive-feedback loop. If there were no pressure controller, the system would be unstable because falling pressure would reduce heat input causing it to fall further. However, the pressure controller acting on the condenser valve forms a negative-feedback loop having a higher gain than m, which ensures steady-state stability. Dynamic stability is achieved by limiting the rate of change of pressure by appropriate selection of integral time in the valve-position controller (VPC) of Fig. 2.20.

The use of floating-pressure control solves some problems but creates others. Among its advantages is the elimination of reflux subcooling — pressure is continuously floating on coolant temperature. Second, rangeability of the pressure-control loop is improved — the control valve is always operating in a favorable por-

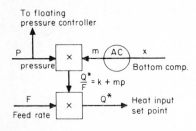

figure 4.24 *Column pressure can be used to manipulate heat input, promoting faster response without upsetting compositions.*

tion of its range. This allows the use of coolant manipulation which is ordinarily discouraged. For example, cooling water could be throttled for pressure control; since its flow would always be high, fouling associated with low velocity and high temperature would be avoided. Similarly, manipulation of speed of a single fan in a bank of several may be adequate; the VPC would always return the output of the pressure controller to the same fan speed.

The problems posed by floating column pressure are the variability of column temperatures and flooding limits. The former may be compensated for variable pressure using methods described in Chap. 6; the latter is discussed in detail in Chap. 7.

In some pressure-control loops, the position of the valve does not represent the degree of condenser loading, and therefore the system of Fig. 2.20 cannot be applied. If, for example, noncondensibles are vented for pressure control, the position of the vent valve is more a function of the rate at which noncondensibles are entering the column. To use floating-pressure control on this (partial) condenser requires the coordination of reflux temperature and pressure. Because the column is in a strict sense a multiple-product unit, this situation is treated in Chap. 8.

Another situation requiring special consideration is the manipulation of reflux and/or distillate for pressure control as in Fig. 4.19 (right). The position of the valve is determined by the flow required for material-balance closure and is unrelated to the condition of the condenser in the steady state. Condenser loading is a function of the level of condensate inside it. Then adjustment of pressure set point to maximize condenser loading requires a condensate-level controller as shown in Fig. 4.25. Level should be maintained near the bottom of the condenser to expose as much heat-transfer surface as possible, but some must remain covered to allow responsive pressure control. The level controller would be adjusted in a manner similar to the valve-position controller described earlier.

Accumulator-Level Control Control of accumulator level is important not because the level itself has any bearing on the performance of the column, but only in the responsive coordination of external and internal material balances. If the level were perfectly controlled, then any change imposed in distillate flow, however minute, would be transferred directly into a change in reflux or boilup, whichever were manipulated by the level controller. The same dynamic response on overhead composition would then be attained whether reflux, boilup, or distillate were chosen to control it.

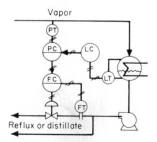

figure 4.25 *When pressure is controlled by reflux or distillate, a condenser-level controller is needed to float the pressure set point.*

However, holding liquid level constant by means of a high-gain feedback controller is neither possible nor desirable. In every control loop there is a maximum
controller gain which cannot be exceeded without bringing about undamped oscillations. For the typical level-control loop, this limit is relatively high: a proportional
band in the 5 to 10 percent range is possible, but this is still far from zero percent.

As a result, the manipulated variable lags behind any disturbance in the other
streams into or out of the accumulator. This is illustrated by the response of reflux
manipulated to control accumulator level in Fig. 4.26. When a step increase in
distillate flow is imposed, the first action is for accumulator level a to begin falling.
A proportional controller will then reduce reflux proportionately to the falling level,
reaching a steady state when it has been decreased by an amount exactly equal to
the increase in distillate flow. The trajectory follows a first-order lag whose time
constant is

$$\tau_a = \frac{v}{L_{\max}} \frac{P}{100} \tag{4.25}$$

where v is the volume of the accumulator represented by the span of the level
transmitter, $L_{\max}$ is the full-scale volumetric flow of reflux, and P is the proportional
band of the level controller expressed in percent. This time constant is typically in
the range of 1 to 5 min.

Adding integral action restores the liquid level to its original steady-state value by
temporarily changing reflux more than distillate changed. However, integral action
is inherently slower than proportional action so that it does not add the needed
dynamic responsiveness. While derivative action or a tighter proportional band
could help in this regard, the noise and resonance that are present in the level
measurement are thereby amplified excessively, causing wide fluctuations in the
manipulated flow.

A preferred solution to the dynamic response problem is the addition of a
feedforward-control loop, first used by van Kampen (13). In Fig. 4.27, distillate flow
is fed forward to change reflux flow without having to wait for the level controller
to respond. In theory this makes perfect level control attainable without a high-gain
feedback controller. In practice, however, the flow rates of the two streams cannot
be measured with perfect accuracy, so the level controller needs to bias the calculation:

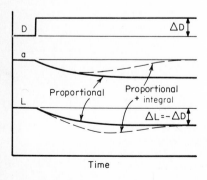

figure 4.26 *A feedback controller
manipulating reflux flow will respond to a change in distillate
flow with a lag.*

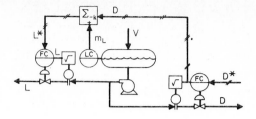

figure 4.27 *Distillate flow is fed forward to change reflux before accumulator level is upset.*

$$L^* = m_L - kD \qquad (4.26)$$

where m_L is the output of the accumulator level controller and k is an adjustable gain. Linear flow signals must be used when adding or subtracting.

If k is set at unity, then changes in distillate flow will be imposed equally on reflux, tending to keep liquid level constant. This action essentially eliminates the dynamics of the reflux accumulator from the quality-control loop. In actual practice, van Kampen found it beneficial to increase k above unity so that variations in distillate are amplified. This causes an over-correction which converts the accumulator from a lag to a lead. The response of reflux to distillate is summarized in Fig. 4.28 for values of k of 0, 1, and 2.

For the case of $k = 0$, the response of reflux to distillate under proportional level control was determined to be a first-order lag of time constant described by (4.25). Where $k = 2$, the response is that of a first-order lead-lag function, having a lead time constant twice that of the lag, with the lag still described by (4.25). The lead-lag ratio is, in fact, coefficient k. At $k = 1.0$, lead equals lag and the net effect is the elimination of any dynamic contribution. Values of k less than 1.0 give partial cancelation of the lag by a smaller lead. Using the feedforward loop, van Kampen was able to reduce the period of oscillation of his distillate-quality loop from 5 h to 30 min.

Base-Level Control   In most installations, base level is controlled by bottom-product flow. This loop arrangement is considered stable and responsive except for kettle reboilers, particularly where a feed-bottoms heat exchanger is used. This problem is discussed in relation to Fig. 4.14.

However, if bottom-product flow is extremely small compared with other stream flows, then it cannot control base level well and will undergo wide fluctuations in

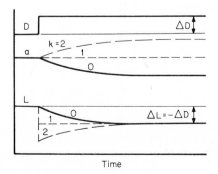

figure 4.28 *The accumulator can be converted from a lag to a lead by appropriate adjustment of k.*

the attempt. In situations such as this, it may be wise to use it for composition control, in which case base level is to be controlled by heat input to the reboiler. This loop arrangement has been used extensively by at least one petroleum company [4].

As overhead composition response to distillate flow depended on tight control of accumulator level, so bottom composition response to bottom-product flow depends on tight control of base level. In most columns, the volume of product stored in the base is much smaller than that in the accumulator. However, turbulence associated with boiling produces a much higher noise level, requiring a wider proportional band for the level controller. As a result, the time lags of the accumulator and base-level loops in response to changes in product flows tend to be quite similar. Although feedforward could be added to the base-level loop, it is rarely done, possibly because of the faster response of compositions to changes in boilup than reflux.

The principle obstacle to controlling base level with heat input is the possibility of inverse response. In cases where it is severe, an alternate loop arrangement must be found. Buckley et al. [3] were forced to use reflux to control base level because boilup produced an 11-min inversion time, and bottom-product flow was too small.

Where noise or resonance in the level signal is high enough to interfere with stable control, nonlinear filtering has been found helpful. A zone of low proportional gain is provided for small deviations about set point so that low-amplitude fluctuations are not amplified. True disturbances, however, will be sufficient to drive the level measurement out of the low-gain zone and allow the controller to take corrective action. This solution is preferable to dynamic filtering, which retards response to signals of all magnitudes.

REFERENCES

1. Anderson, R. H., G. Garrett, and M. Van Winkle: "Efficiency Comparison of Valve and Sieve Trays in Distillation Columns," *Ind. Eng. Chem. Process Des. Dev.,* vol. 15, no. 1, 1976.
2. Lupfer, D. E., and M. L. Johnson: "Automatic Control of Distillation Columns to Achieve Optimum Operation," *ISA Trans.,* April 1964.
3. Buckley, P. S., R. K. Cox, and D. L. Rollins: "Inverse Response in a Distillation Column," *Chem. Eng. Prog.,* June 1975.
4. Lupfer, D. E.: "Distillation Column Control for Utility Economy," presented at 53d Annual GPA Convention, Denver, Col., March 25–27, 1974.
5. Shinskey, F. G.: "Process Control Systems," 2d ed., McGraw-Hill, New York, 1979, pp. 66, 67.
6. Buckley, P. S.: "Material Balance Control in Distillation Columns," presented at AIChE Workshop on Industrial Process Control, Tampa, Fla., November 11–13, 1974.
7. Reference 5, p. 80.
8. Reference 5, pp. 144, 145.
9. Reference 5, pp. 149, 150.
10. Anaya D., Alejandra: "Sizing Hot Vapors Bypass Valve," *Chem. Eng. (N.Y.),* August 25, 1980.
11. Shinskey, F. G.: "Energy-Conserving Control Systems for Distillation Units," *Chem. Eng. Prog.,* May 1976.
12. Shinskey, F. G.: "Control Systems Can Save Energy," *Chem. Eng. Prog.,* May 1978.
13. Van Kampen, J. A.: "Automatic Control by Chromatographs of the Product Quality of a Distillation Column," presented at Convention in Advances in Automatic Control, Nottingham, England, April 1965.

CHAPTER FIVE

Control-Loop Interactions

Any process with more than a single control loop has the possibility of interaction between its loops. And interaction produces some strange results. When there are two valves to be manipulated by two controllers, there are two ways of connecting those single loops, one pairing more effective than the other. If one pairing produces very little interaction, i.e., the loops operate as if they were nearly independent of each other, the opposite pairing will be almost totally ineffective. It is also possible that both valves have similar influence over both controlled variables, in which case either pairing will produce essentially the same results. But because of their similar response, interaction will then be maximum, and operation of the unit will be destabilized when both controllers are in automatic.

The situation of two interacting loops is readily described using the relative-gain concept presented in the following pages. Choices are limited and directions are clear. But distillation columns have as many as four to six manipulated and controlled variables. The number of possible configurations of single control loops increases factorially with the number of variables. While there are only two possible single-loop pairs of two controlled and manipulated variables, there are 6 possible pairs of three loops, 24 possible pairs of four loops, and 120 possible configurations of five single loops.

But even this does not exhaust all the possibilities; if we include all possible ratios of the 5 manipulated variables, their number is augmented by 10. Then the number of possible control-loop configurations is the permutation of 15 variables taken 5 at a time, or 360, 360. This is the principal reason why the design of control systems for distillation is so difficult. Furthermore, the best pairing for one column may not be the best for another since interaction depends heavily on relative flow rates, compositions, etc. Through the use of the relative-gain concepts, the degree of interaction can be estimated as a guide to the designer in structuring his control system even before the column itself has been sized.

The discussions on interaction assume an equal number of controlled variables and valves. There are, of course, two other possibilities: there could be an excess of either. The case of an excess of controlled variables is presented in Chap. 7. The active valves are shared among the controlled variables by simple selection or restructuring as needed to satisfy the most pressing objectives. Where there is an extra valve, it offers the possibility of adjustment to obtain some optimum economic performance of the unit. This subject is covered in Chap. 11.

THE RELATIVE-GAIN CONCEPT

The expression of control-loop interaction presented here was first proposed by E. H. Bristol in 1966 [1]. The concept was then used by the author in *Process-Control Systems* (1967), where the interaction measure first appeared under the name "relative gain" [2]. It was explained later in more detail in the first edition of this book (1977). The usefulness of the concept led to wider applications and more complete definition [3, 4]. But owing to the essentially multivariable character of distillation, the relative-gain concept has its greatest potential here. Furthermore, the distillation process has received more attention than any other multivariable process, both in industry and university, at home and abroad.

Open-Loop Gains   The effectiveness of a feedback controller and the optimum values of its mode settings are a function of the open-loop characteristics of the process to which it is applied. For processes in which variables interact with each other, the process as it appears to a given controller may exhibit different characteristics, depending on the condition of other controllers acting on other variables in the same process.

In the simplest case, the controlled process could be represented by the block diagram shown in Fig. 5.1. Here it is assumed that both manipulated variables affect both controlled variables through linear combinations of steady-state gains K and dynamic gain vectors g. (This is not true of interactions in distillation columns where combinations tend to be nonlinear and coefficients variable. However, the block diagram of Fig. 5.1 is useful to introduce the concept and simplify the mathematical operations.)

Consider the appearance of the process as seen by controller 1. If controller 2 is *not* in automatic (so that m_2 is constant), then the response of c_1 to m_1 is simply

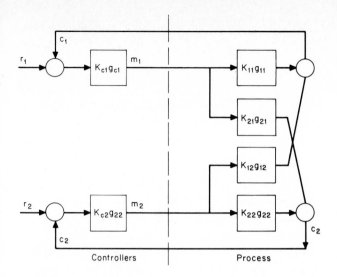

figure 5.1 *Linear representation of two interacting control loops.*

$$\frac{\partial c_1}{\partial m_1}\bigg|_{m_2} = K_{11}g_{11} \tag{5.1}$$

However, any change in m_1 is imposed on loop 2 through element $K_{21}g_{21}$. If controller 2 is then placed in automatic, it will react to this disturbance by changing m_2 and thereby producing a secondary change in c_1 through $K_{12}g_{12}$.

This secondary response brought about by the interaction between the two loops is complicated by the presence of loop 2. Controlled variable 2 responds to both manipulated variables:

$$dc_2 = K_{22}\,g_{22}\,dm_2 + K_{21}g_{21}\,dm_1 \tag{5.2}$$

Then controller 2 reacts with negative feedback:

$$dm_2 = -K_{c2}\,g_{c2}\,dc_2 \tag{5.3}$$

Combining these expressions gives the response of m_2 to m_1:

$$dm_2 = -dm_1\frac{K_{21}g_{21}}{K_{22}g_{22} + 1/K_{c2}g_{c2}} \tag{5.4}$$

Then c_1 responds to both manipulated variables:

$$dc_1 = K_{11}g_{11}\,dm_1 + K_{12}g_{12}\,dm_2 \tag{5.5}$$

Substituting for dm_2 in Eq. (5.5) gives the open-loop gain for loop 1 when controller 2 is in automatic:

$$\frac{dc_1}{dm_1} = K_{11}g_{11} - \frac{K_{12}g_{12}K_{21}g_{21}}{K_{22}g_{22} + 1/K_{c2}g_{c2}} \tag{5.6}$$

Equation (5.6) shows that both the steady-state gain and dynamic response of loop 1 are altered as a function of the parameters chosen for controller 2. In practice, controller 2 could present a variety of responses, but the extremes are logically examined first:

1. If controller 2 has a very low dynamic gain in the frequency region where loop 1 is principally active, then Eq. (5.6) approaches the case of constant m_2 described by Eq. (5.1).

2. If controller 2 has a very high dynamic gain in that region owing to integral action and the fast response of loop 2, $1/K_{c2}g_2$ will approach zero. Then c_2 will be essentially constant, in which case the open-loop gain for loop 1 appears as

$$\left.\frac{\partial c_1}{\partial m_1}\right|_{c_2} = K_{11}g_{11} - \frac{K_{12}g_{12}K_{21}g_{21}}{K_{22}g_{22}} \tag{5.7}$$

The *relative gain* is defined as the ratio of these two extremes, i.e., the gain of loop 1 without control of loop 2 to its gain when loop 2 is controlled perfectly:

$$\lambda_{11} \equiv \left.\frac{\partial c_1}{\partial m_1}\right|_{m_2} \bigg/ \left.\frac{\partial c_1}{\partial m_1}\right|_{c_2} \tag{5.8}$$

For the case of the process described thus far,

$$\lambda_{11} = \left(1 - \frac{K_{12}g_{12}K_{21}g_{21}}{K_{11}g_{11}K_{22}g_{22}}\right)^{-1} \tag{5.9}$$

Relative-Gain Arrays (RGA)   Each controlled variable in an interacting process is subject to influence by each manipulated variable. In a system comprising n pairs of variables, then, there are n^2 open-loop gains and therefore n^2 relative gains. The relative gain for a selected pair of variables c_i and m_j is defined:

$$\lambda_{ij} \equiv \left.\frac{\partial c_i}{\partial m_j}\right|_m \bigg/ \left.\frac{\partial c_i}{\partial m_j}\right|_c \tag{5.10}$$

The relative gains for a set of manipulated and controlled variables are arrayed in a square matrix as shown below:

$$\Lambda = \begin{array}{c} \\ c_1 \\ c_2 \\ \\ c_i \\ \\ \end{array} \begin{array}{ccccc} m_1 & m_2 & \dots & m_j & \dots \\ \left[\begin{array}{ccccc} \lambda_{11} & \lambda_{12} & \dots & \lambda_{1j} & \dots \\ \lambda_{21} & \lambda_{22} & \dots & \lambda_{2j} & \dots \\ \multicolumn{5}{c}{\dots\dots\dots\dots\dots\dots} \\ \lambda_{i1} & \lambda_{i2} & \dots & \lambda_{ij} & \dots \\ \multicolumn{5}{c}{\dots\dots\dots\dots\dots\dots} \end{array}\right] & \begin{array}{c} 1.0 \\ 1.0 \\ \\ 1.0 \\ \\ \end{array} \\ \begin{array}{ccccc} 1.0 & 1.0 & \dots & 1.0 & \dots \end{array} \end{array} \tag{5.11}$$

The array has the very helpful property of having the numbers in every row and column summing to 1.0. This is particularly valuable in a two-loop system in that only one relative-gain term need be estimated. Another useful property of the concept is that it is unaffected by scaling or, in fact, nonlinearities. Because the process gain is, in effect, divided by itself under equivalent conditions, these factors disappear.

To investigate the significance of the relative-gain numbers, it is helpful to rearrange the block diagram of Fig. 5.1 to represent the case where loop 1 is open but loop 2 is controlled perfectly as described by Eq. (5.7). This arrangement is shown in Fig. 5.2 and represents the denominator in the relative-gain ratio.

If $K_{11}g_{11}$ is zero, λ_{11} will also be zero, and loop 1 will not respond to control action if loop 2 is open. However, if loop 2 is closed as in Fig. 5.2, loop 1 can also be closed through it. Therefore, a relative gain of zero does not rule out control but only conditions it on the closure of other loops. This property becomes important in controlling product compositions, as demonstrated later in this chapter.

If the product of the signs of K_{21}, K_{22}, and K_{12} in Fig. 5.2 is *opposite* to that of K_{11}, then the negative sign appearing at the summing junction will cause that loop to have the same sign as K_{11}. Then *both* loops give *negative* feedback, and the relative gain as calculated by Eq. (5.9) will fall in the range of 0 to 1.0.

If *either* K_{21} or K_{12} is zero, the second loop has no effect on the first. Then numerator and denominator in the relative-gain expression are equal and their ratio is 1.0. In general, relative gains near 1.0 indicate minimal interaction; these loops are the best choices for closure.

If the product of the signs of K_{21}, K_{22}, and K_{12} is the *same* as K_{11}, the negative sign at the summing junction will cause the lower path to oppose the upper. Then if controller 1 is set for negative feedback through K_{11}, it will produce *positive* feedback through the interaction path. This can provide two different results:

1. If $K_{11} > K_{21}K_{12}/K_{22}$, then the negative-feedback loop will dominate in the steady state and the system will be stable. This is indicated by $\lambda_{11} > 1.0$.

2. If $K_{11} < K_{21}K_{12}/K_{22}$, the positive-feedback loop will dominate in the steady state, indicated by a *negative* relative gain.

In the former case where $\lambda_{11} > 1.0$, interaction reduces the open-loop gain of loop 1, as evidenced by the denominator being lower than the numerator. This reduces the effect that the controller has over the process when the other controller is in automatic.

In the second case where λ_{11} is negative, numerator and denominator are of opposite sign. Then the net feedback through loop 1 changes between positive and negative when loop 2 is opened or closed. If controller 1 is adjusted with loop 2 open, it will have the incorrect action when loop 2 is closed. If it is adjusted when loop 2 is closed, its action will be wrong when loop 2 is opened — whether controller 2 is placed in manual or m_2 reaches a physical limit. Neither of these situations is tolerable, and therefore pairing variables having negative relative gains is to be avoided. These properties are summarized in Table 5.1.

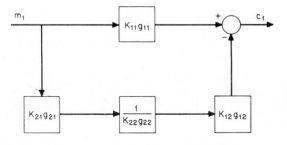

figure 5.2 *A rearrangement of Fig. 5.1 for the case of loop 1 open and loop 2 controlled perfectly.*

TABLE 5.1   Significance of Relative Gains

λ_{ij}	Significance
<0	Conditionally stable—do not close loop
0	Control depends on other loop(s)
0–1	Interaction extends period and raises gain
1.0	No interaction with other loops
>1	Interaction reduces control effectiveness
∞	Loops are completely dependent

The last category in Table 5.1 is the relative gain approaching infinity. In this case, the denominator in Eq. (5.10) is zero, indicating failure of c_i to respond to m_j when other loops are closed. In effect, c_i is controlled by other manipulated variables than m_j and therefore is not capable of being controlled independently. This is the case of a superfluous control loop. As an example, consider the attempt to control both temperature and pressure in a saturated-steam boiler by manipulating feedwater flow and firing rate. Because temperature and pressure are dependent on each other, any attempt to change one independently will be completely undone by the other loop.

The 2 × 2 array, owing to the property of summation to unity, is symmetrical and unambiguous. If any of the elements is 1.0, the others must be 0 and 1.0; choice of loop configuration is then obvious. If any element is between 0 and 1.0, so are the others, and the logical choice is to close those loops whose relative gains are closer to 1.0. If a number greater than 1.0 appears, there will also be a negative number in the same row and in the same column. Because of the conditional stability of the loops with negative numbers, the only tolerable choice is to pair those with positive numbers. Whenever an infinity appears, a negative infinity appears in the same row and column; then only one control loop is functional.

When the process contains more than two pairs of interacting variables, the resulting RGA may be unsymmetrical and there may be no clear choice for loop selection. Consider, for example, the following possibility:

$$\Lambda = \begin{matrix} & m_1 & m_2 & m_3 \\ c_1 & 1 & 2 & -2 \\ c_2 & 0 & 0 & 1 \\ c_3 & 0 & -1 & 2 \end{matrix}$$

From a cursory evaluation, it appears that m_1 must be used to control c_1 in that it has no effect on the others. By the same token, c_2 only seems to be affected by m_3. But if these two choices are made, one is left with m_2 to control c_3, which pairing has a relative gain of -1. If such a process does indeed exist, a viable control-system structure cannot be determined from the information above. To avoid the -1 pairing, one of the zeros must be selected, which may be operable, depending on dynamic response.

Dynamic Effects   The open-loop gains described by Eqs. (5.1) and (5.7) contain dynamic elements as does the relative gain (5.9) calculated from them. The dynamic elements **g** have both gain and phase lag, which vary as a function of frequency or period. Consequently, the numbers that appear in the RGA are frequency-dependent.

To facilitate calculations, most practitioners evaluate only the steady-state RGA obtained by eliminating the dynamic elements. For the 2 × 2 system:

$$\overline{\lambda}_{11} = \left(1 - \frac{K_{12}K_{21}}{K_{11}K_{22}}\right)^{-1} \tag{5.12}$$

Whereas this gives much information for a small investment in time and effort, it by no means defines the problem completely. What appears to be a good loop selection based on steady-state gains may turn out to have poor dynamic response, and occasionally a loop having a $\overline{\lambda}$ of zero will be quite controllable owing to its dynamic responsiveness.

But instead of making the laborious calculations needed to evaluate an RGA at all frequencies of interest, it is more expedient to make qualitative determinations of dynamic responsiveness. Additionally, evaluating a dynamic RGA using Eq. (5.9) will *not* give a completely true picture of dynamic interaction; ideally, Eq. (5.6) should be used as the denominator instead of (5.7) which assumes perfect control of c_2. Finally, dynamic process models are rarely available whereas quite accurate steady-state models usually are, having been necessary to design the process.

To appreciate the role of dynamic elements in interaction, it is helpful to examine the effects that a fast and a slow loop have on each other. Such an example is given in Fig. 5.3 where the total flow and composition of a mixture are to be controlled by manipulating the two streams. Flow begins to change as soon as a valve is moved, having only the inertial lag of the fluid, which is typically less than a second. However, the response of the analyzer is delayed by the time required to transport the mixture to the detector as well as the response of the detector itself. Consequently, the analyzer is at least an order of magnitude slower than the flow measurement.

Assume that the flow controller (FC) is connected to one valve and the analyzer controller (AC) to the other. A change in flow set point would cause the flow controller to move its valve directly, delivering the entire change requested. Because only one valve is moved by the FC, the flow ratio would be upset, causing a change in composition. As this change is detected by the analyzer, the AC will *gradually* reposition its valve to return composition to its set point. This gradual valve adjustment will affect flow, but only slightly, since the flow controller is fast enough to reposition its valve to keep up with the upset. In effect, the composition

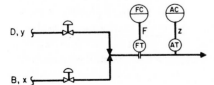

figure 5.3   *The flow loop is typically much faster than the composition loop.*

loop is too slow to appreciably upset the flow loop, although the flow loop sends major load changes into the composition loop.

Note also that the flow controller moves only the one valve, and therefore its mode settings are independent of the presence of the composition loop. By contrast, the AC moves both valves — one directly and the other indirectly as the FC keeps total flow constant. Hence the process gain seen by the AC is changed by the closure of the flow loop. But because of the speed of the flow loop, the gain change is essentially that of the steady state; the dynamic response of the composition loop is not noticeably affected by the faster flow loop.

In Fig. 5.1, let c_1 represent composition and c_2 represent flow. Then g_{21} and g_{22} would be identical short time lags, and g_{11} and g_{12} would be identical longer lags. A different effect is achieved where g_{11} and g_{22} are shorter lags than g_{21} and g_{21}. This disparity would *reduce* the dynamic interaction between the loops, possibly to the point of extinction.

A more interesting but also more difficult case is that of similar dynamic elements; then interaction can have a pronounced effect on controller tuning. This can be envisioned by examining the diagram of Fig. 5.2. If all dynamic elements are equal, then signals through the two paths do not arrive at the summing junction at the same time. If both paths form negative feedback loops (i.e., λ_{11} falls between 0 and 1.0), the interaction path increases loop gain but with a delayed response owing to the three dynamic elements in series. Reference 5 shows through dynamic analysis that two loops with identical dynamics and $\overline{\lambda} = 0.5$ will experience a substantial increase in period as a result of interaction, requiring an increase in both proportional-band and integral settings.

If $\overline{\lambda}_{11}$ exceeds 1.0, however, the feedback loop formed by interaction is positive but not dominant. Because positive-feedback loops have 180° less phase lag than comparable negative-feedback loops, interaction does not affect the period of the loops [5]. Still, the controllers require readjustment for stabilization.

Figure 5.4 compares the set-point response of interacting loops having identical dynamic elements, after the controllers have been readjusted to accommodate the interaction. The response curves for relative gains above unity (superior) and below unity (inferior) are fundamentally different. The longer period of oscillation for the loop having $\overline{\lambda} = 0.5$ is evident; yet it returns to set point faster. By contrast, the loop having $\overline{\lambda} = 2.0$ retains its original period of oscillation but is characterized by a slow, asymptotic return to set point.

Recognize also that $\overline{\lambda}_{11} = 0.5$ represents the worst case for interaction in the 0 to 1.0 range. (If $\overline{\lambda}_{11}$ were less than 0.5, the loop assignments would be exchanged because $\overline{\lambda}_{12}$ would be greater than 0.5.) However, $\overline{\lambda}_{11}$ of 2.0 represents a mild degree of interaction in the range above 1.0. Values in the 5 to 10 range are not unusual, and the prospect of exchanging loop assignments is ruled out because the opposite pairing has a negative relative gain. Loops with values of $\overline{\lambda}$ exceeding 2.0 turn away from the set point even earlier than the lower curve in Fig. 5.4 and take even longer to reach a steady state.

The dynamic response of interacting loops having negative relative gains is noteworthy. Consider that the open-loop gain of loop 1 in Fig. 5.2 will change sign as loop 2 is closed; i.e., the lower path has the higher steady-state gain, but the lower

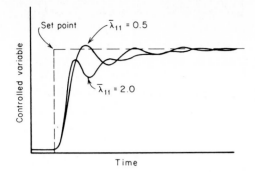

figure 5.4 *The set point response of interacting loops with identical dynamic elements shows a relative gain of 0.5 to give faster return to set point than that of 2.0, although at a longer period.*

path also has more time lags. As a result, the upper path (through $K_{11}g_{11}$) gives a response in one direction, followed by a stronger response in the other direction through the lower path. The combination of the two produces inverse response as shown in Fig. 5.5.

In the upper left-hand corner, the open-loop response of loop 1 is shown with loop 2 open. In the lower left-hand corner, the closure of loop 2 is seen to reverse the direction of c_1 but only after some time has elapsed. The result is an inversion time marked where c_1 crosses its original value. As mentioned earlier, this produces the same effect as dead time in a closed loop.

If loop 2 were left open, loop 1 could be closed and produce the set-point response shown at upper right. However, closure of loop 2 requires the reversal of control action and considerable detuning necessary to accommodate the slow dynamic response shown at lower right.

Another point to be made is that *both* loops have negative relative gains, although only *one* requires its controller action to be reversed—the second controller placed

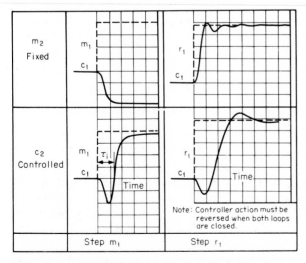

figure 5.5 *Negative values of relative gain result in inverse response, greatly extending the period of the affected control loop.*

in automatic. This is also the controller that must be detuned owing to inverse response. This discussion may seem academic in light of the previous warnings against pairing variables with negative relative gains; however, the ability to recognize their presence as contributing to inverse response may prove helpful in diagnosing strange behavior patterns.

Cases of severe interaction with positive relative gains can be mitigated by detuning one of the controllers. By increasing its proportional band and integral time, one can simulate the case of unequal dynamic elements. The tightly tuned loop will be capable of upsetting the loosely tuned loop, but the reverse will not be true. The selection of which controller to detune should be based on economic or safety criteria.

While there have been some studies of dynamic relative gains evaluated at various frequencies [6], at this writing it is largely an unexplored area without definite conclusions. Therefore, the emphasis from this point on will be on the evaluation and use of steady-state relative gains.

Methods for Evaluating the RGA   Fortunately there are several viable methods for filling in the numbers in a steady-state RGA. The first simply follows the general definition given in Eq. (5.10). Controlled variable c_i is expressed in terms of the set of manipulated variables and differentiated with respect to m_j: this gives the numerator. Next, the mathematical model is rearranged by substituting for all manipulated variables except m_j, in terms of the other controlled variables. The resulting expression is differentiated again with respect to m_j: this provides the denominator. This method has the advantage of expressing relative gains algebraically, in terms of either manipulated or controlled variables. The result is a general solution whose variability is self-evident. The following example illustrates the method.

example 5.1

In Fig. 5.3, stream D containing y fraction of a particular component is blended with stream B containing x fraction of that component to form a blend of flow F and fraction z. The total and component balances are already familiar:

$$F = D + B \tag{5.13}$$

$$Fz = Dy + Bx \tag{5.14}$$

The steady-state gain of F with respect to D, with B constant, is the derivative of (5.13):

$$\left. \frac{\partial F}{\partial D} \right|_B = 1 \tag{5.15}$$

To obtain the steady-state gain of F with respect to D, with B manipulated to control z, requires that B in (5.13) be replaced by z from (5.14). Combining the two balances yields

$$F = D \frac{y - x}{z - x} \tag{5.16}$$

whose derivative is

$$\left. \frac{\partial F}{\partial D} \right|_z = \frac{y - x}{z - x} \tag{5.17}$$

Then the relative gain of F to D may be found:

$$\overline{\lambda}_{FD} = \frac{\left.\dfrac{\partial F}{\partial D}\right|_B}{\left.\dfrac{\partial F}{\partial D}\right|_z} = \frac{z - x}{y - x} \tag{5.18}$$

When all terms in the array are entered, it appears as

$$\Lambda = \begin{array}{c} \\ F \\ \\ z \end{array}\begin{array}{|cc} \overset{D}{\dfrac{z - x}{y - x}} & \overset{B}{\dfrac{y - z}{y - x}} \\ \dfrac{y - z}{y - x} & \dfrac{z - x}{y - x} \end{array} \tag{5.19}$$

As z approaches y, Λ approaches

$$\begin{array}{c} \\ F \\ z \end{array}\begin{array}{|cc} \overset{D}{1} & \overset{B}{0} \\ 0 & 1 \end{array}$$

indicating that flow is almost totally dependent on D and composition on B. But this is reasonable: if z is nearly equal to y, B must be a relatively small flow and would therefore have little influence over F. On the other hand, as z approaches x, the elements reverse and B assumes the primary influence over flow.

One lesson to be learned from this example is that the relative gains are sensitive to feed compositions and set points. Therefore, a control system that will satisfy one process may not satisfy another—each deserves individual evaluation.

If an analytical model is not available, a numerical model may be incremented under both conditions to obtain $\Delta c_i / \Delta m_j$. However, this approach is less accurate than true derivatives would give in proportion to the size of the increments. Furthermore, the results are not general but apply only to a specific point. The same procedure may be used in actual plant tests, but again, with the same qualifications.

Observe, that only two pieces of information are needed to evaluate a relative gain using either of the above methods. Observe also that a 2×2 system requires the calculation of only one relative gain in that the others are either equal or complementary; an $n \times n$ system requires a minimum of $(n - 1)^2$ relative gains for calculation.

A third approach, involving matrix manipulation, may be applied to the 2×2 process shown in Fig. 5.1. In essence, the process was described by two linear equations:

$$c_1 = K_{11}m_1 + K_{12}m_2 \tag{5.20}$$

$$c_2 = K_{21}m_1 + K_{22}m_2 \tag{5.21}$$

This relationship can be expressed for any number of variables using vector-matrix notation:

$$\mathbf{c} = \mathbf{K} \cdot \mathbf{m} \tag{5.22}$$

where matrix K for the 2×2 system is

$$K = \begin{vmatrix} K_{11} & K_{12} \\ K_{21} & K_{22} \end{vmatrix} \tag{5.23}$$

Another means of representing the same process would be to express the manipulated variables in terms of the controlled variables:

$$m_1 = H_{11}c_1 + H_{12}c_2 \tag{5.24}$$

$$m_2 = H_{21}c_1 + H_{22}c_2 \tag{5.25}$$

or, in vector-matrix notation,

$$\mathbf{m} = \mathbf{H} \cdot \mathbf{c} \tag{5.26}$$

Notice that Eq. (5.26) is the inverse of (5.22), so that

$$\mathbf{H} = \mathbf{K}^{-1} \tag{5.27}$$

This is a useful relationship, because as

$$K_{ij} = \left. \frac{\partial c_i}{\partial m_j} \right|_m$$

so

$$H_{ji} = \left. \frac{\partial m_i}{\partial c_i} \right|_c$$

Therefore,

$$\overline{\lambda}_{ij} = K_{ij}H_{ji} \tag{5.28}$$

This allows the complete evaluation of an RGA from an open-loop gain matrix K, by first inverting it to give H and then multiplying each K_{ij} by the corresponding H_{ji}. (Observe that the elements in the H matrix need transposition because m_j is expressed in terms of c_i.)

For the simplest case, the 2×2 system, the result is expressed simply in Eq. (5.12). However, note that this solution requires *four* gains to fill out Λ, although only two are needed according to Eqs. (5.10) and (5.28). Arrays larger than 2×2 do not yield to a simple formula such as Eq. (5.12) but must be solved by differentiating numerator and denominator in Eq. (5.10) or by the matrix inversion procedure given above.

A useful corollary reverses the above procedure, begins with matrix H, and develops K by inversion; the individual elements of the RGA are then calculated by Eq. (5.28). In some cases open-loop gains H are more available than gains K. For the 2×2 system, the formula for relative gain in terms of gains H is identical to that for gains K:

$$\overline{\lambda}_{11} = \left(1 - \frac{H_{21}H_{12}}{H_{11}H_{22}} \right)^{-1} \tag{5.29}$$

There is a fourth procedure which may be very helpful, depending on the form of available data. It begins by substituting derivatives into gains $\mathbf{K}$ in Eq. (5.12):

$$\bar{\lambda}_{11} = \left(1 - \frac{\partial c_2}{\partial m_1}\bigg|_{m_2} \frac{\partial c_1}{\partial m_2}\bigg|_{m_1} \bigg/ \frac{\partial c_1}{\partial m_1}\bigg|_{m_2} \frac{\partial c_2}{\partial m_2}\bigg|_{m_1}\right)^{-1} \tag{5.30}$$

The manipulated variables may be canceled, leaving:

$$\bar{\lambda}_{11} = \left(1 - \frac{\partial c_1}{\partial c_2}\bigg|_{m_1} \bigg/ \frac{\partial c_1}{\partial c_2}\bigg|_{m_2}\right)^{-1} \tag{5.31}$$

This expression will become quite useful when evaluating interactions between composition loops in a distillation column.

Equation (5.31) also has a corollary, developed by substituting differentials into the $\mathbf{H}$ gains of Eq. (5.29):

$$\bar{\lambda}_{11} = \left(1 - \frac{\partial m_2}{\partial c_1}\bigg|_{c_2} \frac{\partial m_1}{\partial c_2}\bigg|_{c_1} \bigg/ \frac{\partial m_1}{\partial c_1}\bigg|_{c_2} \frac{\partial m_2}{\partial c_2}\bigg|_{c_1}\right)^{-1} \tag{5.32}$$

This time, controlled variables are canceled, giving

$$\bar{\lambda}_{11} = \left(1 - \frac{\partial m_2}{\partial m_1}\bigg|_{c_2} \bigg/ \frac{\partial m_2}{\partial m_1}\bigg|_{c_1}\right)^{-1} \tag{5.33}$$

The residual gains in (5.33) become useful in decoupling systems which are described in Chap. 6. For example, $(\partial m_2/\partial m_1)_{c_2}$ is the change in m_2 needed to be coordinated with a change in m_1 if c_2 is not to be upset.

COMPOSITION-LOOP INTERACTIONS

The most serious interactions experienced in controlling distillation columns are between the product-composition loops. Compositions are most difficult to control, and the loops tend to have dynamic elements that are both slow and similar. In the past, control of a single composition was usually considered adequate, and hence the interaction problem did not exist. But with the sharp increase in energy costs, the economic return attainable by controlling both products at minimum specifications has become too attractive to ignore. Consequently, many, if not most columns, can benefit by dual-composition control, and interaction has become a factor of considerable importance.

The 5 × 5 System   One approach to addressing the problem of interaction in a distillation column would be to evaluate an RGA for all the variables taken together. Figure 5.6 locates the five manipulated and five controlled variables for a typical column. They could then be arranged in an RGA such as that shown in Fig. 5.7.

The relative gains for manipulation of distillate and bottom-product flow rates can be determined by inspection. Assume that a steady state exists, with all valves in fixed positions. Then the distillate valve is incremented: the resulting change in flow will affect only accumulator level. Therefore the D column must contain all zeros except for a 1.0 at level a. By the same token, incrementing the bottom valve

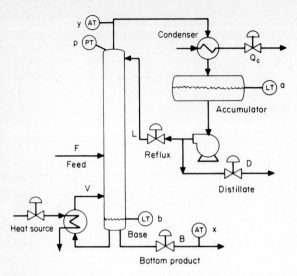

figure 5.6 *The typical distillation column has five pairs of variables.*

will only change base level: the B column then contains all zeros except for a 1.0 at level b.

This does not present a true picture of the control-loop configurations which are viable for a distillation column. There are many instances where D has been successfully used to control y, and others where B has successfully controlled x. Part of the problem is the attempt to apply a steady-state RGA to describe a process that has no steady state — at least it has no steady state while its liquid-level loops are open. To construct a true steady-state RGA, it is necessary to close the level loops. If some of the loops are closed and others are open, then the numerator in the relative-gain equation requires redefinition. For example, one might be interested in the interaction between the composition loops when reflux is controlling accu-

$$\lambda_{yD} = \frac{\left.\dfrac{\partial y}{\partial D}\right|_{V,\,L,\,B,\,Q_C}}{\left.\dfrac{\partial y}{\partial D}\right|_{x,\,a,\,b,\,p}}$$

	D	V	L	B	Q_C
Comp. y	O			O	
Comp. x	O			O	
Level a	1			O	
Level b	O			1	
Press. p	O			O	

c ⟵ columns, m ⟵ rows

figure 5.7 *The 5 × 5 RGA discourages the use of product flow rates to control compositions.*

mulator level, bottom-product flow is controlling base level, and heat removal is controlling condenser cooling. The 5 × 5 RGA would be essentially reduced to the 2 × 2 array shown in Fig. 5.8.

This order reduction is justified if the loops arbitrarily assigned for control are at least an order of magnitude faster than the loops remaining to be evaluated. This is a valid assumption for the distillation column. The relative gain for the composition subset in Fig. 5.8 is redefined as

$$\overline{\lambda}_{yD}(\Lambda_{DV}) = \left.\frac{\partial y}{\partial D}\right|_{V,a,b,p} \Big/ \left.\frac{\partial y}{\partial D}\right|_{x,a,b,p} \tag{5.34}$$

The numerator is evaluated for constant liquid levels and pressure, and the denominator for all variables controlled, as is customary. The relative gain $\overline{\lambda}_{yD}$ thus calculated is stipulated to apply only to the subset (Λ_{DV}), that is, where D and V are selected for composition control. If some other variable such as reflux were to replace boilup in the subset, $\overline{\lambda}_{yD}$ would have a numerically different value, now belonging to the subset (Λ_{DL}).

All Possible 2 × 2 Subsets At this point it is necessary to determine how many xy subsets require evaluation to describe completely the 5 × 5 process. This is actually the number of combinations of five variables taken two at a time—the total is 10. The number can be reduced however, if condenser cooling is not used to control composition. This seems to be a universal rule. The combination of the four remaining variables taken two at a time leaves only six xy subsets.

One additional point to be made is that D and B affect compositions in exactly the same way in the steady state—the external material balance requires that $dD = -dB$. Then $\overline{\lambda}_{yB}$ (Λ_{BL}) will have exactly the same value as $\overline{\lambda}_{yD}$ (Λ_{DL}). There remains then only three *independent* manipulated variables: D, V, and L, leaving only three *independent* xy subsets: DV, DL, and LV.

However, the number of choices for manipulated variables can be expanded by including all ratios of D, V, L, and B. This is also a combinational problem: again, four variables taken two at a time gives six ratios. Yet one of these, D/B, is not

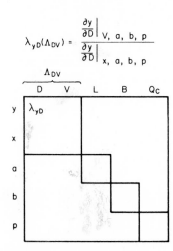

figure 5.8 *The 5 × 5 array can be reduced to 2 × 2 by placing three of the loops under control.*

independent of D and therefore can be excluded. Another observation is that because $V = L + D$, a ratio of any two of these variables determines L/D, as indicated by Eqs. (3.14) and (3.15), and therefore determines separation factor S through Eq. (3.58). Ratios L/B and V/B affect compositions still differently, bringing the total number of independent manipulated variables to six: S, D/F, L/F, V/F, L/B, and V/B. Figure 5.9 compares the xy relationships for the six independent variables.

This diagram is extremely useful in estimating the range of relative gains which can be expected for each selected pair of manipulated variables. To do this, Eq. (5.31) is restated in terms of x and y:

$$\overline{\lambda}_{y1}(\Lambda_{12}) = \left(1 - \frac{\partial y}{\partial x}\Big|_1 \Big/ \frac{\partial y}{\partial x}\Big|_2\right)^{-1} \tag{5.35}$$

The selected manipulated variables, here identified simply as 1 and 2, could be any two of the six. Note that the slope of the line representing constant D/F in Fig. 5.9 is opposite in sign to the other curves. Then when D (or B) is selected to control composition, its opposition in slope to that of any other manipulated variables cancels the negative sign in Eq. (5.35), so that $\overline{\lambda}_{yD}$ for all subsets falls between zero and unity.

If, however, two manipulated variables are chosen whose xy slopes have the same sign, the negative sign in Eq. (5.35) remains, and the resulting relative gains will fall *outside* the 0 to 1.0 range. The closer the slopes are to each other, the smaller the denominator becomes and the higher will be the absolute value of the relative gain — identical slopes mean complete dependence.

In general, the curves can be expected to fall as illustrated in Fig. 5.9. The curve for constant S is always symmetrical about the diagonal in a binary system. The curve for constant V/B shows the greatest departure and would be expected to be the first to change slopes. If it happened to be vertical at the operating point, $\overline{\lambda}_{yV/B}$ would be zero, regardless of the choice of the other manipulated variable. Hence $\overline{\lambda}_{xV/B}$ would be unity, and nothing could upset bottom composition as long as V/B

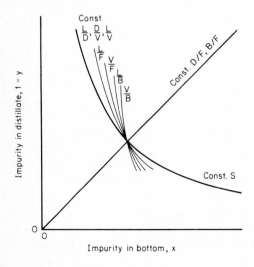

figure 5.9   *These are the six independent operating lines.*

were held constant; this was already discussed for V/F in connection with Fig. 3.11.

The constant L/F and V/F curves tend to fall closer together as L/D increases, which means as L approaches V. Then relative gains for the LV subset tend to increase in magnitude with the L/D ratio.

Evaluating the *DS* Subset   There are in the two-product distillation column, 15 independent 2×2 subsets to evaluate, each identifiable by a single relative gain. Therefore, for brevity the coding of the relative gain terms will be reduced in the following manner: Let $\bar{\lambda}_{y1}(\Lambda_{12})$ be identified simply as Λ_{12}; then $\bar{\lambda}_{x2}(\Lambda_{12})$ is also described by Λ_{12}, and $\bar{\lambda}_{y2}$ or $\bar{\lambda}_{x1}$ is identified as Λ_{21}. That is, the first subscript for Λ is designated as that assigned to composition y in the evaluation. Recognize also that $\Lambda_{21} = 1 - \Lambda_{12}$.

The model used to identify the operating point of a column in Chap. 3 consists of a material balance and a separation equation. Therefore the subset that can be evaluated most directly is that combination Λ_{DS}. The method used follows Eq. (5.35). First, the xy relationship is expressed in terms of D:

$$\frac{D}{F} = \frac{z - x}{y - x}$$

It is then differentiated:

$$\frac{\partial y}{\partial x}\bigg|_{D/F} = 1 - \frac{1}{D/F} = -\frac{y - z}{z - x} \tag{5.36}$$

Next, the xy relationship is described in terms of S:

$$S = \frac{y(1 - x)}{x(1 - y)}$$

and differentiated:

$$\frac{\partial y}{\partial x}\bigg|_{s} = \frac{y^2}{Sx^2} = \frac{y(1 - y)}{x(1 - x)} \tag{5.37}$$

Combining the two derivatives by Eq. (5.35) yields:

$$\Lambda_{DS} = \left(1 + \frac{y - z}{z - x} \bigg/ \frac{y(1 - y)}{x(1 - x)}\right)^{-1}$$

$$\Lambda_{DS} = \left(1 + \frac{(y - z)x(1 - x)}{(z - x)y(1 - y)}\right)^{-1} \tag{5.38}$$

The most influential terms in Λ_{DS} are the product impurities x and $1 - y$ as they are the smallest numbers. Should x be much smaller than $1 - y$, Λ_{DS} approaches unity—this condition corresponds to an operating point on the upper left portion of the separation curve in Fig. 5.9. Conversely, if $1 - y$ is much smaller than x, Λ_{DS} will approach zero, corresponding to an operating point toward the lower right side of the separation curve. This observation leads to a general rule that the material balance should be manipulated to control the composition of the less-pure product, and the separation factor manipulated to control that of the more-pure product.

Before proceeding to other subsets, let us evaluate Λ_{DS} for multicomponent two-product columns. Because more variables are involved, the derivation is necessarily more complex. We are principally interested in the key impurities y_H and x_L.

Differentiation of the material-balance equation in any component produces the results shown in (5.36). For example,

$$\left.\frac{\partial y_H}{\partial x_H}\right|_{D/F} = 1 - \frac{1}{D/F} = -\frac{y_H - z_H}{z_H - x_H} \tag{5.39}$$

To convert ∂y_H into ∂x_L we use the component balance in the bottom product:

$$x_H + x_L + x_{HH} = 1 \tag{5.40}$$

However, x_{HH} depends only on z_{LL} and D/F; therefore if D/F is constant, $\partial x_L = -\partial x_H$. Then

$$\left.\frac{\partial y_H}{\partial x_L}\right|_{D/F} = \left.\frac{\partial y_H}{\partial x_H}\right|_{D/F} \cdot \left.\frac{\partial x_H}{\partial x_L}\right|_{D/F} = \frac{y_H - z_H}{z_H - x_H} \tag{5.41}$$

The slope of the constant-separation curve is more difficult to derive because x_{HH} and y_{LL} are varied by D/F. The solution begins with the multicomponent separation equation:

$$S = \frac{y_L x_H}{x_L y_H} \tag{5.42}$$

Then it may be solved for y_H and differentiated by parts:

$$\left.\frac{\partial y_H}{\partial x_L}\right|_S = -\frac{y_L x_H}{x_L^2 S} + \frac{y_L}{x_L S}\left.\frac{\partial x_H}{\partial x_L}\right|_S + \frac{x_H}{x_L S}\left.\frac{\partial y_H}{\partial x_L}\right|_S \left.\frac{\partial y_L}{\partial y_H}\right|_S \tag{5.43}$$

By substituting (5.42) for S and solving,

$$\left.\frac{\partial y_H}{\partial x_L}\right|_S = -\left(\frac{1}{x_L} - \frac{1}{x_H}\left.\frac{\partial x_H}{\partial x_L}\right|_S\right) \bigg/ \left(\frac{1}{y_H} - \frac{1}{y_L}\left.\frac{\partial y_L}{\partial y_H}\right|_S\right) \tag{5.44}$$

If x_{HH} and y_{LL} were constant, the partials in numerator and denominator of (5.44) would each be -1. However, they are not, so Eq. (5.40) must be differentiated:

$$\left.\frac{\partial x_H}{\partial x_L}\right|_S = -1 - \left.\frac{\partial x_{HH}}{\partial x_L}\right|_S \tag{5.45}$$

The same procedure is followed for distillate compositions:

$$\left.\frac{\partial y_L}{\partial y_H}\right|_S = -1 - \left.\frac{\partial y_{LL}}{\partial y_H}\right|_S \tag{5.46}$$

Next, restating Eq. (5.44) in terms of (5.45) and (5.46) yields

$$\left.\frac{\partial y_H}{\partial x_L}\right|_S = -\frac{\dfrac{1}{x_L} + \dfrac{1}{x_H}\left(1 + \left.\dfrac{\partial x_{HH}}{\partial x_L}\right|_S\right)}{\dfrac{1}{y_H} + \dfrac{1}{y_L}\left(1 + \left.\dfrac{\partial y_{LL}}{\partial y_H}\right|_S\right)} \tag{5.47}$$

Because both slopes have changed sign as a result of operating on key impurities, Λ_{DS} remains in the 0 to 1.0 range:

$$
\Lambda_{DS} = \left(1 + \frac{y_H - z_H}{z_H - x_H} \left[\frac{\dfrac{1}{x_L} + \dfrac{1}{x_H}\left(1 + \left.\dfrac{\partial x_{HH}}{\partial x_L}\right|_S\right)}{\dfrac{1}{y_H} + \dfrac{1}{y_L}\left(1 + \left.\dfrac{\partial y_{LL}}{\partial y_H}\right|_S\right)} \right] \right)^{-1}
\tag{5.48}
$$

Unfortunately, there seems to be no direct solution for the partial derivatives remaining in Eq. (5.48). However, they may be found using an estimated value of Λ_{DS} in the following manner:

$$
\left.\frac{\partial x_{HH}}{\partial x_L}\right|_S = \frac{dx_{HH}}{d(D/F)} \Bigg/ \left.\frac{\partial x_L}{\partial(D/F)}\right|_S
\tag{5.49}
$$

$$
\left.\frac{\partial x_L}{\partial(D/F)}\right|_S = (1 - \Lambda_{DS})\left.\frac{\partial x_L}{\partial(D/F)}\right|_{y_H}
\tag{5.50}
$$

Differentiating Eq. (3.8) gives

$$
\frac{dx_{HH}}{d(D/F)} = \frac{z_{HH}}{(1 - D/F)^2} = \frac{x_{HH}}{1 - D/F}
\tag{5.51}
$$

Next x_L is found in terms of D/F, with Eq. (3.7) substituted for y_L:

$$
x_L = \frac{(1 - y_H)D/F - z_L - z_{LL}}{D/F - 1}
$$

Differentiating gives

$$
\left.\frac{\partial x_L}{\partial(D/F)}\right|_{y_H} = -\frac{1 - y_H - x_L}{1 - D/F}
\tag{5.52}
$$

Then combining (5.49) through (5.52) gives a solution in terms of Λ_{DS}:

$$
\left.\frac{\partial x_{HH}}{\partial x_L}\right|_S = -\frac{x_{HH}}{(1 - \Lambda_{DS})(1 - y_H - x_L)}
\tag{5.53}
$$

Following a similar procedure, we can derive

$$
\left.\frac{\partial y_{LL}}{\partial y_H}\right|_S = -\frac{y_{LL}}{\Lambda_{DS}(1 - y_H - x_L)}
\tag{5.54}
$$

To solve for Λ_{DS}, then, it is first necessary to evaluate Eq. (5.48) using zero for the partial derivatives. Then the resulting value of Λ_{DS} is substituted into (5.53) and (5.54) to evaluate the partial derivatives. These values are then entered into Eq. (5.48) to give a second approximation of Λ_{DS}.

example 5.2

Calculate Λ_{DS} for the multicomponent column described in Example 3.4.

	z	y	x
LL	0.0846	0.333	—
L	0.176	0.640	0.020
H	0.162	0.020	0.210
HH	0.577	—	0.769

First approximation:

$$\Lambda_{DS} = \left[1 + \frac{0.020 - 0.162}{0.020 - 0.210} \frac{1/0.020 + 1/0.210}{1/0.020 + 1/0.640} \right]^{-1} = 0.558$$

$$\left. \frac{\partial x_{HH}}{\partial x_L} \right|_S = -\frac{0.769}{(1 - 0.558)(1 - 0.020 - 0.020)} = -1.81$$

$$\left. \frac{\partial y_{LL}}{\partial y_H} \right|_S = -\frac{0.333}{0.558(1 - 0.020 - 0.020)} = -0.622$$

Second approximation:

$$\Lambda_{DS} = \left[1 + \frac{0.020 - 0.162}{0.020 - 0.210} \frac{1/0.020 + (1 - 1.81)/0.210}{1/0.020 + (1 - 0.622)/0.640} \right]^{-1} = 0.595$$

A third approximation using $\Lambda_{DS} = 0.595$ gives $\Lambda_{DS} = 0.599$.

Evaluating the Remaining Slopes   Rather than derive all 15 relative gain equations, it is more expeditious to evaluate the six slopes of Fig. 5.9 from which they may be directly calculated. This also simplifies the selection process because we will be interested primarily in only two of the curves: those with the highest and lowest slopes.

Two of the slopes have already been derived in (5.36) and (5.37). Because these two will be used to find the rest, they are given special designations:

$$\left. \frac{\partial y}{\partial x} \right|_{D/F} = \delta \tag{5.55}$$

$$\left. \frac{\partial y}{\partial x} \right|_S = \sigma \tag{5.56}$$

The remaining slopes may all be evaluated as vector sums of these two, as described in Fig. 5.10. Let m represent any of the four remaining manipulated variables.

The method begins by introducing a departure from the operating point by an amount $d(D/F)$. If separation were constant, this would produce changes ∂y_1 and ∂x_1 which are related through σ. However, at constant m, separation is affected by $d(D/F)$, causing additional changes ∂y_2 and ∂x_2 in proportion to the departure from the separation curve, as determined by $\partial S/\partial(D/F)$ at constant m. The resulting slope is the ratio of the sum of these components:

$$\left. \frac{\partial y}{\partial x} \right|_m = \frac{\partial y_1 + \partial y_2}{\partial x_1 + \partial x_2}$$

Dividing top and bottom by ∂x_1 gives

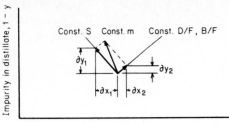

Const. S   Const. m   Const. D/F, B/F

∂y_1

∂y_2

∂x_1 ∂x_2

Impurity in distillate, 1 − y

Impurity in bottom, x

figure 5.10 *The change in y with respect to x at constant m is the sum of two vectors.*

$$\left.\frac{\partial y}{\partial x}\right|_m = \frac{\partial y_1/\partial x_1 + (\partial y_2/\partial x_2)(\partial x_2/\partial x_1)}{1 + \partial x_2/\partial x_1} = \frac{\sigma + \delta\,\partial x_2/\partial x_1}{1 + \partial x_2/\partial x_1} \tag{5.57}$$

Next it is necessary to evaluate the remaining differentials:

$$\partial x_1 = \left.\frac{\partial x}{\partial(D/F)}\right|_S d(D/F)$$

and

$$\partial x_2 = \left.\frac{\partial x}{\partial S}\right|_{D/F} dS = \left.\frac{\partial x}{\partial S}\right|_{D/F} \left.\frac{\partial S}{\partial(D/F)}\right|_m d(D/F)$$

When the differentials are divided, $d(D/F)$ disappears:

$$\frac{\partial x_2}{\partial x_1} = \left.\frac{\partial x}{\partial S}\right|_{D/F} \left.\frac{\partial S}{\partial(D/F)}\right|_m \bigg/ \left.\frac{\partial x}{\partial(D/F)}\right|_S \tag{5.58}$$

The partials of x can be determined using the relative gain already derived:

$$\left.\frac{\partial x}{\partial S}\right|_{D/F} = \left.\frac{\partial x}{\partial S}\right|_y \Lambda_{DS} = -\frac{x(1-x)}{S}\Lambda_{DS} \tag{5.59}$$

$$\left.\frac{\partial x}{\partial(D/F)}\right|_S = \left.\frac{\partial x}{\partial(D/F)}\right|_y (1 - \Lambda_{DS}) = -\frac{y-x}{1-D/F}(1 - \Lambda_{DS}) \tag{5.60}$$

The change in S with D/F at constant m is related to its effect on reflux ratio:

$$\left.\frac{\partial S}{\partial(D/F)}\right|_m = \left.\frac{\partial(D/L)}{\partial(D/F)}\right|_m \frac{dS}{d(D/L)} \tag{5.61}$$

The last term is the derivative of the separation model (3.58):

$$\frac{dS}{d(D/L)} = -\frac{nES}{2(z + D/L)} \tag{5.62}$$

When Eqs. (5.58) to (5.62) are combined, the result is

$$\frac{\partial x_2}{\partial x_1} = -\epsilon\left(\frac{L}{F}\right)\left.\frac{\partial(D/L)}{\partial(D/F)}\right|_m \tag{5.63}$$

where ϵ represents a lump of terms that will appear again and again:

$$\epsilon = \frac{nEy(1-y)}{2(1+zL/D)(y-x)} \tag{5.64}$$

Next it is necessary to derive the last term in (5.63) for each of the four variables designated as m. In the case of reflux

$$\frac{D}{L} = \frac{D/F}{L/F}$$

$$\left.\frac{\partial(D/L)}{\partial(D/F)}\right|_{L/F} = \frac{1}{L/F} \tag{5.65}$$

For boilup

$$\frac{D}{L} = \frac{D/F}{V/F - D/F}$$

$$\left.\frac{\partial(D/L)}{\partial(D/F)}\right|_{V/F} = \frac{1+D/L}{L/F} \tag{5.66}$$

For the L/B ratio

$$\frac{D}{L} = \frac{D/F}{(L/B)(1-D/F)}$$

$$\left.\frac{\partial(D/L)}{\partial(D/F)}\right|_{L/B} = \frac{1}{(L/F)(1-D/F)} \tag{5.67}$$

and for the V/B ratio

$$\frac{D}{L} = \frac{D/F}{(V/B)(1-D/F) - D/F}$$

$$\left.\frac{\partial(D/L)}{\partial(D/F)}\right|_{V/B} = \frac{1}{L/F}\left(\frac{1+D/L}{1-D/F}\right) \tag{5.68}$$

When each of the last four numbered equations is inserted into (5.63) and the result factored into (5.57), the four slopes are obtained:

$$\left.\frac{\partial y}{\partial x}\right|_{L/F} = \frac{\sigma - \delta\epsilon}{1 - \epsilon} \tag{5.69}$$

$$\left.\frac{\partial y}{\partial x}\right|_{V/F} = \frac{\sigma - \delta\epsilon(1+D/L)}{1 - \epsilon(1+D/L)} \tag{5.70}$$

$$\left.\frac{\partial y}{\partial x}\right|_{L/B} = \frac{\sigma - \epsilon(\delta - 1)}{1 - \epsilon(\delta - 1)/\delta} \tag{5.71}$$

$$\left.\frac{\partial y}{\partial x}\right|_{V/B} = \frac{\sigma - \epsilon(1+D/L)(\delta - 1)}{1 - \epsilon(1+D/L)(\delta - 1)/\delta} \tag{5.72}$$

Programs for solving for all six slopes in BASIC and on an HP-11C calculator are given in Appendix A. Having obtained the slopes, selected values are inserted into Eq. (5.35) to determine the corresponding relative gain. This operation is included in the programs. To simplify notation, the symbol μ will henceforth be used to designate $\partial y/\partial x$.

The slopes above are arranged in order of increasing value, with σ being lowest of all and δ being negative; Fig. 5.9 describes the progression, although the signs of all slopes appear reversed owing to the use of $1 - y$ instead of y as the vertical scale.

As the denominator in a slope equation passes through zero, the slope will pass through infinity, changing sign. The most likely curve to reverse its slope in this way is that of constant V/B, since its value is highest of all. The reversal will take place when

$$\epsilon(1 + D/L)(\delta - 1)/\delta > 1$$

or by substitution

$$\frac{nEy(1 - y)(1 + D/L)}{2(1 + zL/D)(y - z)} > 1$$

This is most likely to happen with columns of low distillate purity, low reflux ratio, and a proportionately large number of trays.

example 5.3

Evaluate all six slopes for the column described in Fig. 3.11 and Example 3.5.

$$\delta = -1.00$$

$$\sigma = 1.00$$

$$\mu_{L/F} = 2.31$$

$$\mu_{V/F} = 3.92$$

$$\mu_{L/B} = 8.60$$

$$\mu_{V/B} = -11.67$$

Next, reduce the number of theoretical trays from 30 to 20 (this corresponds to a mixture of higher relative volatility) and repeat the calculations.

$$\mu_{L/F} = 1.72$$

$$\mu_{V/F} = 2.31$$

$$\mu_{L/B} = 3.24$$

$$\mu_{V/B} = 8.60$$

(Slopes δ and σ do not change.)

In practice, the four slopes containing ϵ are nearly always positive so that relative gains in combination with each other and with σ are greater than unity or negative; in combination with δ, relative gains are always between 0 and 1.0.

For multicomponent separations, δ is evaluated as in Eq. (5.41); note that it is still B/D, but that the sign is positive owing to the use of impurity y_H where purity y appeared in binary calculations. Then σ is calculated from Eq. (5.47) using the subsequent trial-and-error procedure if high accuracy is required. The remaining parameter ϵ is calculated as in Eq. (5.64), with $z_L + z_{LL}$ substituted for z.

LOOP-SELECTION CRITERIA

The steady-state RGA forms the basis for selecting the optimum control-system structure, but it is not the only consideration. While relative gains close to unity are the most desirable, superior numbers produce different results from inferior numbers and require evaluation using different criteria. Dynamic responsiveness is also important. Additionally, some variables can be manipulated with greater accuracy than others, and some may not be available for manipulation at all. Fortunately, several choices are usually available for any particular subset having a desirable relative gain. The first order of business is to display all the possible choices in a manner that will facilitate selection and not obscure any viable candidate.

Comparing Relative Gains   Accordingly, the worksheet shown in Fig. 5.11 has been prepared to aid in selecting the optimum loop configuration. The vertical column at left lists the viable candidates for controlling bottom composition. Across the top are listed the viable candidates for controlling distillate composition. Note that B, V, and their ratio are omitted from this list: they will always affect x sooner than they affect y and therefore will tend to produce unfavorable dynamic interaction. (It should be noted, however, that this limitation applies only when controlling *both* x and y.)

The worksheet also contains shaded areas representing impossible or unacceptable combinations. For example, D cannot be used to control both x and y at the same time, nor can D and B be used together. The entire lower right corner is missing because separation cannot be used to control both compositions simultaneously. The right side has a shaded area because a ratio involving D would be dynamically unfavorable to control y while x was controlled with D; however controlling x with either B or V under these conditions may be satisfactory.

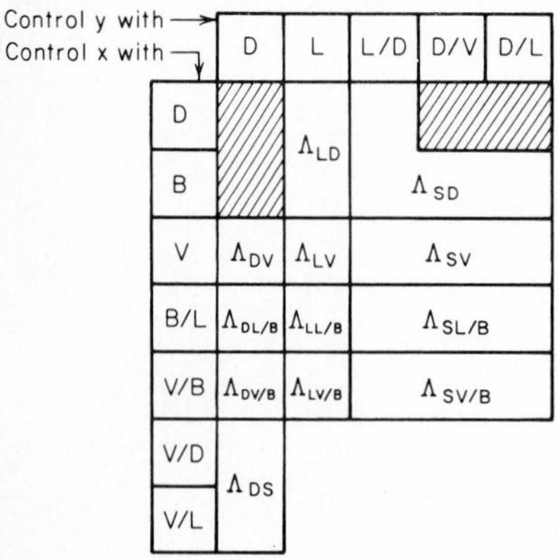

Control y with → Control x with ↓	D	L	L/D	D/V	D/L
D	//////	////// Λ_{LD}		//////	//////
B	//////		Λ_{SD}		
V	Λ_{DV}	Λ_{LV}	Λ_{SV}		
B/L	$\Lambda_{DL/B}$	$\Lambda_{LL/B}$	$\Lambda_{SL/B}$		
V/B	$\Lambda_{DV/B}$	$\Lambda_{LV/B}$	$\Lambda_{SV/B}$		
V/D	Λ_{DS}				
V/L					

B, V & ratios of B & V omitted because affect x faster

figure 5.11   *The worksheet is intended to include all viable pairs for composition control.*

To use the worksheet, the pertinent relative gains are calculated and entered into it.

example 5.4

Complete the relative-gain worksheet for the following column: $x = 0.10$, $y = 0.994$, $z = 0.80$, $nE = 50$, $L/D = 4.75$. Following the same procedure as in example 5.3, we calculate:

$$\delta = -0.2771 \qquad \epsilon = 0.347$$
$$\sigma = 0.0663$$
$$\mu_{L/F} = 0.0786$$
$$\mu_{V/F} = 0.0813$$
$$\mu_{L/B} = 0.1317$$
$$\mu_{V/B} = 0.1488$$

From these, the relative gains are calculated and entered into Fig. 5.12.

It is not always necessary to calculate all 12 relative gains shown in the worksheet. We generally are interested only in the greatest and least slopes. In Example 5.4, they are σ and $\mu_{V/B}$ which, when combined, produce the most favorable relative gain in Fig. 5.12. Furthermore, the resulting $\Lambda_{SV/B}$ turns out to be the most favorable for the majority of two-product columns, principally because $\mu_{V/B}$ tends to be the highest slope. However, one may encounter a constraint which prevents V/B from being manipulated to control composition. Then the second highest slope should be used.

Table 5.2 shows how relative gains are affected by compositions and column parameters. The first three separations are all symmetrical, i.e., product purities are equal and feed is an equimolar mixture; as a result, Λ_{DS} is 0.50 for them all. The higher number of stages and reflux ratio for column 2 denote a lower relative volatility but do not significantly affect the relative gains. Higher purities in column 3 raise all superior relative gains.

Control y with → Control x with ↓	D	L	L/D	D/V	D/L
D		0.779			
B			0.807		
V	0.227	29.9	5.39		
B/L	0.322	2.48	2.01		
V/B	0.349	2.12	1.80		
V/D	0.193				
V/L					

figure 5.12 *For the column in Example 5.4, the best relative gain is SV/B at 1.80.*

TABLE 5.2   Relative Gains for Typical Binary Separations

	COLUMN							
	1	2	3	4	5	6	7	8
COMPOSITIONS								
x	0.05	0.05	0.01	0.01	0.01	0.001	0.001	0.01
y	0.95	0.95	0.99	0.99	0.99	0.99	0.99	0.999
z	0.50	0.50	0.50	0.80	0.20	0.50	0.80	0.50
PARAMETERS								
$n\ E$	20	50	50	50	50	100	100	100
L/D	2.0	5.0	5.0	5.0	5.0	10.0	10.0	10.0
ϵ	0.264	0.377	0.072	0.051	0.126	0.083	0.056	0.0084
SLOPES								
δ	-1.00	-1.00	-1.00	-0.24	-4.16	-0.98	-0.24	-1.02
σ	1.00	1.00	1.00	1.00	1.00	9.91	9.91	0.101
$\mu_{L/F}$	1.72	2.21	1.16	1.07	1.75	10.9	10.5	0.110
$\mu_{V/F}$	2.31	2.65	1.19	1.08	1.92	11.0	10.6	0.111
$\mu_{L/B}$	3.24	7.13	1.34	1.44	1.96	12.1	14.0	0.120
$\mu_{V/B}$	8.60	20.0	1.42	1.56	2.19	12.4	14.7	0.122
RELATIVE GAINS								
Λ_{DS}	0.50	0.50	0.50	0.81	0.19	0.91	**0.98**	0.09
Λ_{SD}	0.50	0.50	0.50	0.19	0.81	0.09	0.02	**0.91**
Λ_{DV}	0.70	0.73	0.54	0.82	0.32	0.92	**0.98**	0.10
$\Lambda_{DV/B}$	0.90	0.95	0.59	0.87	0.35	**0.93**	**0.98**	0.11
$\rightarrow \Lambda_{SV}$	**1.76**	**1.61**	6.27	13.5	**2.09**	10.0	15.9	10.6
Λ_{LV}	3.89	6.00	34.9	76.9	10.9	100	166	116
$\Lambda_{LL/B}$	2.13	**1.45**	7.36	3.87	9.21	9.98	3.97	12.6
$\Lambda_{LV/B}$	**1.25**	**1.12**	5.39	3.14	4.89	8.34	3.53	10.7
$\rightarrow \Lambda_{SL/B}$	**1.45**	**1.16**	3.96	3.29	**2.04**	5.49	3.40	6.31
$\rightarrow \Lambda_{SV/B}$	**1.13**	**1.05**	**3.39**	**2.77**	**1.84**	5.00	**3.09**	5.82

Columns 4 and 5 illustrate changes in feed composition, which have a pronounced effect on relative gains containing D and also on Λ_{SV}. Columns 6, 7, and 8 have unsymmetrical product purities that are quite high, requiring more trays and a higher reflux ratio. As a result, the selection of viable relative gains in reduced, and material-balance systems become more favorable.

The viable relative gains are all indicated in bold type. Note that the SV/B configuration is the best selection for almost all columns. By contrast, the SV system is favorable only for three, and material-balance systems for three.

Dynamic Considerations   Each worksheet will contain two relative gains bracketing unity — one inferior and one superior number. (In Fig. 5.12, the two are Λ_{SD} at 0.807 and $\Lambda_{SV/B}$ at 1.80.) Figure 5.4 gives a general indication of how the dynamic response differs for inferior and superior relative gains. However, it is possible to construct a more complete picture of the dynamic response of a distillation column using only steady-state relative gains [7].

Consider the response of bottom composition to a step increase in boilup under conditions of constant reflux. Composition x will fall monotonically, as in the lowest curve of Fig. 5.13, because the increase in V will carry a proportional increase in light material out with the distillate. Since reflux is fixed, $\Delta D = \Delta V$. If reflux is then used to control y, the upset to y caused by the increase in boilup will cause L to be

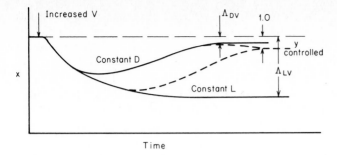

figure 5.13 *The response curve for bottom composition depends on the configuration of the loops at the top of the column.*

increased. This action returns much of the light material to the bottom of the column, raising x along the lower broken line labeled "y controlled." The ratio of the change in x produced with constant L to that produced with y controlled is the relative gain Λ_{LV}:

$$\overline{\lambda}_{xV}(\Lambda_{LV}) = \left.\frac{\partial x}{\partial V}\right|_L \bigg/ \left.\frac{\partial x}{\partial V}\right|_y = \Lambda_{LV}$$

If D is held constant, however, the effect of increasing V on bottom composition is sharply reduced. The increase ΔV produces an equal increase ΔL, which brings more of the light material back down the column than with y controlled. The slight reduction in x achieved in the steady state is due entirely to a proportional change in distillate composition because distillate flow is fixed. If y is then controlled by manipulating D, the controller will respond to this change in composition by increasing D, thereby removing additional light material from the bottom. The trajectory taken by x in this case reverses direction twice: once owing to the increase in reflux reaching the bottom, and again when it decreases slightly following the action of the y controller. This double reversal is essentially the same as inverse response, introducing a substantial delay into the bottom-composition loop. This will ordinarily be the case when material-balance control is used because all subsets containing D produce inferior numbers. In Fig. 5.13, $\Lambda_{LV} = 5$ and $\Lambda_{DV} = 0.5$.

The ideal trajectory for x following a change in boilup would be to move its final value (corresponding to a controlled y) as quickly as possible and remain there. This will be achieved most closely when the relative gain is 1.0. But given the choice of two relative gains on either side of 1.0, the superior number will provide the faster dynamic response for the bottom-composition loop. This will ordinarily be achieved when separation is used to control top composition and the V/B ratio is used to control x, because $\Lambda_{SV/B}$ is normally the lowest of the superior numbers.

The faster dynamic response of the SV configuration compared with the DV configuration was confirmed in a dynamic column simulation by T. J. McAvoy [8]. Furthermore, this simulation demonstrated the vulnerability of the DV configuration to a failure of the distillate-composition loop. With that loop out of service, the material balance became locked, and the column could not tolerate an upset in feed composition: the bottom-composition controller drove the heat-input valve to saturation without restoring control. However, when *separation* was used to control y, bottom composition could still be controlled with the top loop open because changes in boilup were able to shift the material balance as needed. (C. J. Ryskamp's similar observations on production columns prompted McAvoy's simulations.)

The curves of Fig. 5.4 show a relative gain of 0.5 to be slightly superior to a relative gain of 2.0 in set-point response when all dynamic elements are equal. However in McAvoy's simulation, the system manipulating D/V for top-composition control clearly out-performed that manipulating D, although its relative gain was 11.7 compared to 0.52 for the DV subset.

Conflicts and Compromises   Two additional considerations are the accuracy with which a variable can be manipulated and its sensitivity to disturbance from external inputs such as weather changes. This subject is discussed at some length in Chap. 4.

To summarize here, the smallest flow should be manipulated to control composition as it is capable of regulating the material balance to the highest absolute accuracy. If there is very little difference among the various flow rates, then this consideration is not important, and relative-gain and dynamic considerations dominate. But in those separations where one flow is clearly much smaller than the others, using it to control a composition will enhance stability and protection against environmental disturbances. This has been demonstrated again and again in operating columns. An especially well-documented case is the conversion of an ethylbenzene-xylene column, having a reflux ratio of 70 from reflux to distillate control of top composition [9]. Composition control changed from impossible to so easy that an on-off controller was used.

Therefore, in columns having high reflux ratios, D rather than L is preferred to control composition. However if $B \ll D$, then the material balance is more accurately maintained by manipulating B for composition control.

These accuracy considerations can present conflicts with relative gains. For example, if B is by far the smallest stream and $\Lambda_{DV/B}$ is close to unity, a dilemma exists. Although $\Lambda_{BV/B} = \Lambda_{DV/B}$, both B and V cannot be used to control compositions as that would leave base level without a manipulated variable. Furthermore, because $\Lambda_{BL/B}$ would also be close to unity, B could not be used to control x and L to control y; yet the opposite configuration gives poor dynamic response. In this case, it may be necessary to use the SV/B configuration as a compromise.

A conflict is less likely to be encountered when D is the smallest flow, because accumulator level can be controlled satisfactorily by boilup (whereas base level is not well controlled by reflux). Then, in the case that Λ_{DS} should approach zero, Λ_{SD} will approach unity, allowing D to control x and reflux ratio to control y. This arrangement is presented in more detail in the next section.

Accuracy considerations also should prevail when stipulating the form of the flow ratios. Note, for example, that the top row of variables in Fig. 5.11 contains both D/L and L/D. The smallest ratio will tend to yield the best composition control. Therefore, if $D < L$, then D should be manipulated in ratio to L for composition control while L is manipulated for level or pressure control. The D/V ratio is even smaller than D/L and is therefore preferred even more. However, should $D > L$, then L should be manipulated in ratio to D, which would be under level or pressure control. Aternatively, if D were used to control x because of a favorable value of Λ_{SD} and $D < B$, then L would be manipulated in ratio to D for control of y.

Similar relationships appear at the bottom of the column. Variables used to control x in Fig. 5.11 include ratios B/L and V/B. In the event that B is placed under base-level control because it is larger than D, then V would be set in ratio to it for control of x. However, if $B < D$, then it could be manipulated proportionally to reflux or total overhead vapor to control x.

Logical Combinations   At this point, it may be worthwhile to consolidate all the above considerations into a method for selecting the most-effective combinations or at least eliminating the least-effective ones. To reduce the number of possibilities, this presentation is restricted to the "normal" case, i.e., where only δ is negative.

For these columns, the lowest superior relative gain is $\Lambda_{SV/B}$. The highest inferior relative gain will be either $\Lambda_{DV/B}$ or Λ_{SD}, depending primarily on whether x or $1 - y$ is the smallest impurity.

Of the four manipulated flow rates, vapor flow is ordinarily the largest. This does not make it useless in controlling composition, as evidenced by the favorable relative gains of subsets containing V. However, the other variable used in the same subset should be evaluated for accuracy of manipulation.

Figure 5.14 is a matrix where intersections between the smallest manipulable flow and the most-favorable relative gain are evaluated. If D is the smallest flow and $\Lambda_{SV/B}$ the most-favorable gain (in the 1 to 5 range), D/V should be manipulated to control y, and V/B to control x. If $\Lambda_{DV/B}$ is more favorable (0.9 to 1.0 when $\Lambda_{SV/B} > 5$), then D should be used to control y. The third case applies when neither of the above relative gains are favorable. Then Λ_{SD} may be in the 0.9 to 1.0 range, in which case L/D can be used to control y, and D to control x; accumulator level must then be controlled by heat input. These three configurations are described in detail in the next section of this chapter.

When B is the smallest flow, conflicts arise, as noted earlier. If Λ_{SD} is favorable (which is also Λ_{SB}), then reflux ratio should be manipulated to control y, and B used to control x. If Λ_{SD} is unfavorable, B should be set in ratio to L to control x; typically this relative gain is only slightly above $\Lambda_{SV/B}$, as indicated in Fig. 5.12.

Smallest flow

Most favorable Λ	D	B	L
SV/B	D/V V/B	L/D B/L	L/V V/B
DV/B	D V/B		D V/B
SD (SB)	L/D D	L/D B	L/V B

figure 5.14 *Optimum selection of manipulated variables to control y (upper) and x (lower).*

Reflux is rarely the smallest of the flowing streams, yet the possibility has been included in Fig. 5.14. In this case, L/V would tend to be smaller than L/D and therefore is suggested for controlling y in both subsets containing S. In the last, B is preferred to D for controlling x if $B < D$. If $B > D$ or if inverse response prevents V from being used to control base level, then L/D and D should be used to control x and y, respectively, with accumulator level controlled by V.

Because of the wide variations in column characteristics encountered, the above recommendations should be taken as general. In many cases there will be a close similarity between gains such as Λ_{SV} and $\Lambda_{SV/B}$ or Λ_{SD} and Λ_{LD}. Then a suboptimum pairing might be preferred if it simplifies the control system or avoids a particular constraint.

OPTIMUM SYSTEM STRUCTURES

Based on the last discussion, it is possible to identify four principal system structures which, by combining the most-favorable relative gain with manipulation of the smallest flow, would be considered optimum for certain classes of columns. Not all columns can be satisfied by one of the four, yet they represent general solutions that will be found applicable to most. They are identified in Fig. 5.14 as the SV/B, DV/B, SD, and SB configurations.

The *SV* and *SV/B* Configurations   In recent experience, the SV/B configuration seems to be applicable over the broadest range. It has the advantage of a faster dynamic response for the bottom loop and yet has the lowest of the superior relative gains. Therefore, it is probably the best choice whenever $\Lambda_{SV/B}$ falls between 1 and 5, almost without regard to other considerations. The SV configuration is a simpler version that has seen more use to date [10].

Control of y can be achieved by manipulating any of the three overhead ratios D/L, D/V, or L/V. As noted above, D/V is ordinarily the best choice from the standpoint of accuracy. However, dynamic response of the top-composition loop is also affected by the choice and is an important consideration. This can be demonstrated by comparing D/L and D/V manipulations.

Consider the configuration in Fig. 5.15 in which reflux flow is used to control accumulator level and distillate is set in ratio to it. An increase in boilup will raise accumulator level, thereby increasing both reflux and distillate while maintaining their ratio constant. This action determines slope σ and will be the same regardless of the choice of variables used to establish the separation factor. However, note that the top composition controller manipulates *only* distillate flow. As mentioned earlier, distillate has no direct effect on composition, and therefore response is delayed until the changing level in the accumulator brings about a subsequent adjustment to reflux flow.

If the system is reconfigured to manipulate the D/V ratio for composition control, this delay is eliminated at the cost of an increase in complexity. Figure 5.16 shows that the output of the level controller is now designated V and is sent to both reflux- and distillate-flow controllers. Multiplication by the output of the composition controller generates the distillate-flow set point; consequently the controller output is D/V. Subtraction of measured distillate flow generates the reflux set point.

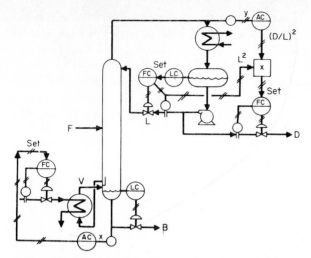

figure 5.15   *Manipulation of D/L is uncomplicated but lacks dynamic responsiveness; here V is manipulated to control x, giving an SV configuration.*

This system responds to a change in boilup the same as that of Fig. 5.15—both reflux and distillate change in a constant ratio. However, the top-composition controller now moves both flows and in opposite directions. An increase in D/V causes D to increase and L to decrease by exactly the same amount since their sum (V) is constant. This is the system which Ryskamp introduced so successfully to control a debutanizer and butane splitter as described in Ref. 10; he used a flooded

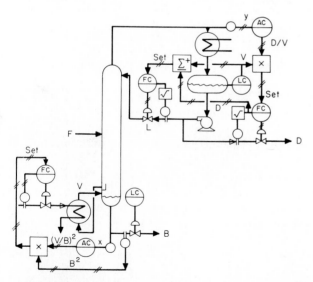

figure 5.16   *Manipulation of D/V allows the composition controller to move distillate and reflux simultaneously; here V/B is used to control bottom composition.*

accumulator and condenser so that the column-pressure controller was used to set the sum of reflux and distillate. (A pressure controller is not shown in Figs. 5.15 or 5.16 for simplicity; it may be assumed to manipulate condenser cooling.)

Figure 5.16 contains three more blocks than Fig. 5.15. The subtractor is needed to convert changes in distillate to proportional changes in reflux; this is the same function introduced in Fig. 4.27. Additionally, square-root extractors are required to linearize the two flow signals if head-type meters are used. They were not required in Fig. 5.15 because the two flows could be set in ratio without linearization — in effect, the composition controller set $(D/L)^2$. However, whenever flow rates are added or subtracted the signals must be linear.

While it is important to keep control systems simple so they are easy to understand, operate, and maintain, the improved response provided by the system of Fig. 5.16 seems worth the added complication. Albert Einstein said, "Things should be kept as simple as possible, but no simpler." It should be noted that if an on-line indication of top composition is not available, its controller would be missing, and the simpler system of Fig. 5.15 would be adequate. Reflux ratio would then be set manually, based on design conditions or from the results of laboratory analysis.

The bottom-composition loop of Fig. 5.16 is also more advanced than that of Fig. 5.15. By setting heat input in ratio to B, the configuration is changed from SV to SV/B, with a resulting improvement in relative gain. Now both product-flow rates are coordinated with boilup, resulting in what should be the most effective and universally applicable of configurations for two-product columns. From the standpoint of interaction, its relative gain is considerably closer to unity than that of the SV system whose performance has been well-established. It would also seem to be insensitive to feed-rate disturbances as noted in Chap. 6 under "Feedforward Control." Its principal (and perhaps only) limitation would seem to be a sensitivity to inverse response in that heat input is manipulated along with bottoms flow to control base level; however, because they are *both* manipulated, inverse response would not present the same problem encountered when heat input alone is used to control level.

The *DV* and *DV/B* Configurations   The DV/B configuration is recommended when $\Lambda_{DV/B}$ is the highest of the inferior relative gains, and $\Lambda_{SV/B}$ is too high to be considered useful. An example is an ethylene-ethane fractionator, where, owing to a combination of high bottom-product purity and high reflux ratio, $\Lambda_{DV/B}$ is around 0.95 and $\Lambda_{SV/B}$ near 5.0.

Bottom composition is controlled by V/B ratio as was the case for the SV/B configuration. But here, the top-composition controller manipulates distillate set point directly, without any intermediate device. The speed of response of the top-composition loop can be enhanced by changing reflux flow at the same time, through the system described in Fig. 4.27. This would apply whether reflux were under control of accumulator level or of column pressure in the case of a flooded accumulator and condenser.

The DV configuration has been used successfully for many years on columns that have been impossible to control with reflux and boilup. It has been called the *direct material-balance* system [11] because a product flow is manipulated to control one of the compositions. Its principal disadvantage is sluggish response, especially of the bottom-composition loop as noted in Fig. 5.13.

The *DV* configuration has been used successfully to control columns with high reflux ratio, such as an ethylbenzene-xylene column [9], a benzene-toluene column [12], a propylene-propane column [13], an ethanol-water column [14], a pentane-isopentane column [15], an ethylbenzene-styrene column [16], and the ethylene-ethane fractionator mentioned previously. Incorporation of B into the bottom-composition loop upgrades it to DV/B, with an improvement in relative gain.

These systems are generally not recommended for columns having low reflux ratios, but the relative-gain calculations will bear this out. Relative-gain Λ_{DS} is a function of compositions alone and is unaffected by reflux ratio. However, as reflux ratio increases, Λ_{DV} becomes closer to Λ_{DS}, and $\Lambda_{SV/B}$ generally rises. For columns with low reflux ratios or low bottom purity, $\Lambda_{SV/B}$ tends toward unity and therefore represents the optimum configuration.

The *SD* Configuration   When the top product is substantially purer than the bottom product, Λ_{DS} approaches zero. Then its complement Λ_{SD} approaches unity, exceeding $\Lambda_{DV/B}$. If the column has an unacceptably high value of $\Lambda_{SV/B}$ and D is smaller than B, then the *SD* configuration would be the best choice. As with the DV/B configuration, this would apply principally to columns of high reflux ratio — it is a direct material-balance system. Therefore it complements the DV/B system where product purities are reversed.

The *SD* configuration is described in Fig. 5.17. Bottom-product composition is controlled by manipulating distillate flow, with reflux set in ratio. Dynamic response is achieved by controlling accumulator level with boilup. Then any change

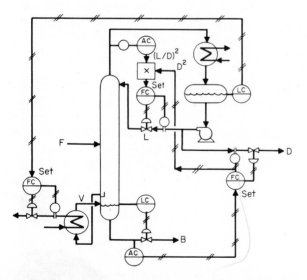

figure 5.17 *For columns with a small flow of high-purity distillate, this configuration may be optimum.*

in distillate flow will cause level to change and boilup to react proportionately, which will start bottom composition to respond.

The top-composition controller adjusts reflux alone without affecting the external material balance. Here the dynamic coupling to bottom composition through accumulator-level control of boilup may be undesirable. This may call for detuning of the top-composition controller.

The SD configuration may be found useful in controlling the separation of a high-purity propylene product from the effluent of a catalytic-cracking unit. It is not a commonly used configuration.

The SB and SB/L Configurations   The remaining optimum system structures apply to the situation where B happens to be smaller than D. If $\Lambda_{SV/B}$ is unacceptably high with Λ_{DS} approaching zero so that Λ_{SD} would approach unity, the SB configuration would be preferred.

Here, bottom composition would be controlled by its own flow, as shown in Fig. 5.18. Then base level must be controlled by boilup, which can be a problematic loop. Yet when properly functioning, it responds quickly to changes in bottom-product flow, causing boilup to affect composition promptly. The principal limitation to this configuration is the occurrence of inverse response of base level to boilup, an increasingly common characteristic, owing to the prevalence of valve trays. This is not an easy problem to solve, and alternate configurations manipulating larger flows for composition control tend to be much more sensitive to disturbances. Buckley [17] used reflux to control base level, which in itself is a very slow loop and also retards the response of the bottom-composition loop.

The accumulator-level controller closes the external material balance by manipulating distillate flow, with reflux set in ratio. Top-composition control is achieved by adjusting the L/D ratio. The overhead system could be simplified by controlling composition with reflux flow rather than reflux ratio. This would then be the LB configuration, whose relative gain Λ_{LD} is ordinarily less than Λ_{SD}. Whether the addition of the multiplier is warranted depends on the difference between the two

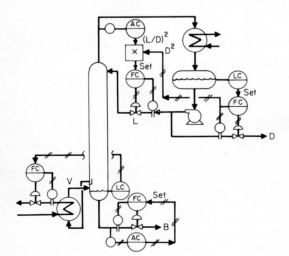

figure 5.18 *This system is intended for columns separating a small amount of heavy contaminant from a high-purity distillate.*

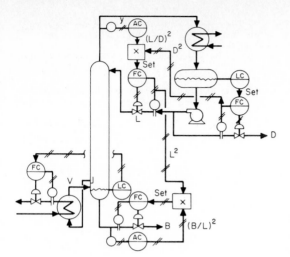

figure 5.19 *Setting B in ratio to L upgrades the previous system to SB/L, with a superior relative gain.*

relative gains — it may be insignificant. In a certain depropanizer, the difference was only between 0.919 and 0.906, probably not worth the added complication. Dale Lupfer and his associates were proponents of the *LB* configuration, as described in Ref. 18.

If $\Lambda_{SV/B}$ is favorable and *B* is small, then the configuration shown in Fig. 5.19 is applicable. Here *L/D* is used to control *y*, and *B/L* to control *x*, as recommended in the top row of Fig. 5.14. The only difference from the configuration of Fig. 5.18 is the setting of *B* in ratio to *L* instead of manipulating *B* alone. Yet this changes the relative gain from an inferior number to the second lowest of the superior numbers. This is the system which would best fit the column described in Example 5.4, where $B/D = 0.277$ and $\Lambda_{SL/B} = 2.01$. The *SB* relative gain at 0.807 and the *LB* at 0.799 would probably be much less satisfactory.

REFERENCES

1. Bristol, E. H.: "On a New Measure of Interaction for Multivariable Process Control," *IEEE Trans. Autom. Control,* January 1966.
2. Shinskey, F. G.: *Process-Control Systems,* McGraw-Hill, New York, 1967, pp. 188–202.
3. Shinskey, F. G.: *Process-Control Systems,* 2d ed., McGraw-Hill, New York, 1979, pp. 196–221.
4. Shinskey, F. G.: *Controlling Multivariable Processes,* Instrument Society of America, Research Triangle Park, N. C., 1981, pp. 97–113.
5. Reference 3, pp. 206, 207.
6. McAvoy, T. J.: "Some Results on Dynamic Interaction Analysis of Complex Control Systems, *Ind. Eng. Chem. Process Des. Dev.,* March 1982.
7. Shinskey, F. G.: "Predict Distillation Column Response Using Relative Gains," *Hydrocarbon Proc.,* May 1981.
8. McAvoy, T. J.: "Dynamic Simulation of a Nonlinear Dual Composition Control Scheme," presented at the 2d World Congress of Chemical Engineering, Montreal, October 4–9, 1981.
9. McNeill, G. A., and J. D. Sacks: "High Performance Column Control," *Chem. Eng. Prog.,* March 1969.
10. Ryskamp, C. J.: "New Strategy Improves Dual Composition Control," *Hydrocarbon Proc.,* June 1980.
11. Bojnowski, J. J., R. M. Groghan, and R. M. Hoffman: "Direct and Indirect Material Balance Control," *Chem. Eng. Prog.,* September 1976.
12. Boyd, D. M.: "Fractionating Column Control," *Chem. Eng. Prog.,* June 1975.

13. Van Kampen, J. A.: "Automatic Control by Chromatographs of the Product Quality of a Distillation Column," presented at Convention on Advances in Automatic Control, Nottingham, England, April 1965.
14. Shinskey, F. G.: "Controlling Distillation Processes for Fuel-Grade Alcohol," *In Tech,* December 1981.
15. MacMullan, E. C., and F. G. Shinskey: "Feedforward Control of a Superfractionator," *Contr. Eng.,* March 1964.
16. Shinskey, F. G.: "The Material-Balance Concept in Distillation Control," *Oil & Gas J.,* July 14, 1969.
17. Buckley, P. S., R. K. Cox, and D. L. Rollins: "Inverse Response in a Distillation Column," *Chem. Eng. Prog.,* June 1975.
18. Lupfer, D. E.: "Distillation Column Control for Utility Economy," presented at 53d GPA Annual Convention, Denver, Col. March 25–27, 1974.

CHAPTER SIX

Composition Control

In Chap. 5, the optimum basic structure for the control of two-product columns is determined using relative-gain analysis. This structure represents the skeleton which can be built upon by the addition of auxiliary regulating functions. While these additional functions can enhance or improve control effectiveness, they are not a substitute for an optimum structure and have limited capability when applied to an inferior structure.

Decoupling is especially notable in this regard. Its intent is to reduce the interaction that naturally exists in a system. But a system that is optimally configured will exhibit a minimum amount of interaction and should not require decoupling. By constrast, a less-effective structure typically cannot be raised to the same level of performance, even at the cost and complexity of a decoupling system.

COMPOSITION MEASUREMENTS

Before composition can be successfully controlled, it must be measured. This section deals with various methods to measure composition and presents, in a general way, the features and disadvantages of each.

The first to be considered is the composition analyzer applied to the product streams themselves. Where a method of analysis is unavailable or unwarranted, a measurement of temperature, with or without pressure compensation, is often used to infer composition. Proper location of the measuring point is critical to success, and so the composition profile of the column becomes important. This also helps identify any nonlinear properties which the control system may encounter.

Composition Analyzers   The principal function of a distillation column is to deliver products meeting certain composition specifications; therefore, composition analyzers are ideally an integral part of these control systems. Nonetheless, composition is difficult to measure, and on-line analyzers tend to be complex, expensive, and demanding of maintenance. Furthermore, most analytical sensors cannot be mounted on the column or inserted in a flowing stream. They must be mounted in a protected location and have a prepared sample brought to them. All these considerations discourage the use of analyzers in control, except where the economic return they make possible exceeds the cost and effort they require.

The most common analyzer found on distillation columns is the gas chromatograph. It separates components by adsorption on a packed bed and by successive desorption by a stream of carrier gas such as helium. Because the adsorption-desorption process operates on the basis of differences in vapor pressure, components that are difficult to separate by distillation may be difficult to analyze by chromatography. Azeotropes and close-boiling isomers fit this category. But because the process is similar to distillation, it lends itself to most fractionators encountered in industry.

The chromatograph has the distinct advantage of being able to measure the concentration of several components in a mixture. This allows the control of some mathematical function of two or more components in a product. For example, the purity of the propane product leaving the depropanizer in Fig. 1.2 could be controlled using the sum of ethane and butane concentrations. Similarly, its ethane content could be regulated by controlling the ethane-to-propane ratio in the stream leaving the bottom of the deethanizer.

The principal disadvantage of a chromatograph would be the dead time encountered in transporting a sample to the detector and the intermittent (sampling) nature of its operation. The dead time appears in various places. First, a sample must be delivered from the process to the analyzer, which might be located some distance apart. While this distance should be minimized, there are limitations associated with the electrical classification of the analyzer and the hazard rating of the area. Analyzers are usually mounted in a shed to provide protection from the elements and ease of maintenance. To minimize transportation time, a vapor sample is preferred to a liquid, although it may not always be available. Reducing the sample pressure helps by reducing density and eliminating the possibility of fractionation by partial condensation. Velocity should be maximized by continuous

circulation of sample to the analyzer and back to the process — an absolute necessity for liquid samples. The pressure drop needed can be supplied by a product
pump or taken across a control valve as shown in Fig. 6.1.

Dead time also exists within the analyzer itself. Periodically, a small volume of
sample is injected into the column by a valve and followed by a continuous flow
of carrier gas. Dead time elapses in removing the adsorbed component from the
column and transporting it to the detector where its presence is measured. However, different components arrive at the detector at different times, depending on
their relative volatility. All the information is customarily stored and transmitted at
the same time. Consequently, the dead time for all components is the same: the
time between sample injection and transmittal of the analysis.

There is also an *apparent* delay associated with the sampling nature of the
analyzer. Composition is measured and results reported only on a periodic basis.
The information is already old (by the dead time cited above) when reported, but
it becomes progressively older until the results of the next analysis arrive. The
composition is displayed to the operator and to the controller as a series of steps
separated by equal intervals of time, known as the *sample interval* Δt. A portion of
such a record is shown in Fig. 6.2.

When the sampling function is placed in a control loop with a dominant time
constant, such as the composition time constant for a distillation column, the signal
is smoothed as it passes through the loop. Thus, control action — if continuous — is
applied to the *average* value of composition, which is the reported value delayed by
half the sample interval.

The dead time between sample injection and report is somewhat less than the
sample interval because the analyzer cycle also includes flushing of the column to
remove unwanted components. Although the dead time of the sample in the analyzer and the sampling interval are concurrent events, they *both* contribute to
control-loop dead time.

Occasionally, an analyzer is shared between two streams, such as the overhead
and bottom of a distillation column. This doubles the sample interval without
affecting the analyzer dead time. Although this practice saves capital cost

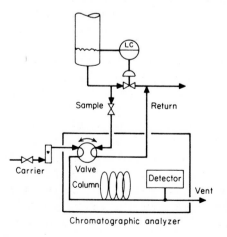

figure 6.1   *The volume of sample
trapped within the injection valve
is transported to the detector in
fractions by the carrier gas.*

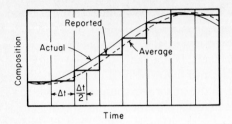

figure 6.2 *The sampling nature of a chromatograph introduces an apparent dead time of half the sample interval.*

and maintenance, the resulting increase in dead time of both control loops can deteriorate performance significantly.

The stepwise nature of the signal from a chromatographic analyzer prevents the use of derivative in the control algorithm. Ryskamp [1] has applied two first-order lags to smooth the leading edge of the steps sufficiently to allow the use of derivative. Another alternative is the sampled-data control algorithim, which is executed but once, at the receipt of new information. It not only enables the use of derivative but also overcomes the effective delay caused by the sample interval. It is described with "Feedback Control Algorithms" later in this chapter.

Other composition analyzers occasionally used include infrared and ultraviolet spectrometers for such components as CO, CO_2, and SO_2; infrared liquid-composition analyzers are also available now. Refractive index is used to measure components like styrene in ethyl-benzene. Distillation analyzers are used for complex mixtures such as crude oil and its fractions; there are also flashpoint analyzers for kerosene and jet-fuel products. Most require a sample to be withdrawn and transported to a location convenient for installation and maintenance. Where an in-line detector is available, it should be used.

Additional physical-property measurements find application to specific mixtures. Among them are density, viscosity, dielectric constant, electrical conductivity, and thermal conductivity. These measurements tend to be simpler than the more-specific analyzers, lest costly, and easier to install in-line. However, they also are more susceptible to interference from foreign components. Each application requires a detailed evaluation before success can be expected.

Temperature Measurement   The most common measurement used for composition is the boiling point. In an ideal binary mixture, boiling point and composition are related by means of the vapor pressures of the components. Consider an ideal mixture of components A and B. By the law of partial pressures,

$$y_A = \frac{p_A}{p} \qquad y_B = \frac{p_B}{p} \tag{6.1}$$

where y = mole fraction of designated components in the vapor
  p_A, p_B = their partial pressures
  p = total pressure
Raoult's law states

$$p_A = x_A p_A^\circ \qquad p_B = x_B p_B^\circ \tag{6.2}$$

where x is the mole fraction of the designated component in the liquid and $p°$ is its vapor pressure. In a binary system, $y_A + y_B = 1$ and $x_A + x_B = 1$. When these relationships are combined with (6.1) and (6.2), a solution for x_A in terms of vapor pressures and total pressure is obtained:

$$x_A = \frac{p - p_B°}{p_A° - p_B°} \qquad (6.3)$$

Since the vapor pressures of the pure components are unique functions of temperature, a relationship exists between composition, temperature, and total pressure per Eq. (6.3). The curves calculated for the benzene-toluene system in Fig. 6.3 attest that the relationship is nearly linear.

For systems which depart significantly from ideality, and for ternary or more complex systems, the relationship among composition, temperature, and pressure can be determined using equilibrium constants. The procedure for calculating the vapor pressure of a given ternary mixture at a given temperature was demonstrated in Example 3.2. This procedure must be modified by selecting a pressure and estimating the temperature which develops that pressure for a given mixture. A solution can only be found by trial and error. An approximate temperature is selected, K values for the components are determined for that temperature and pressure, and a corresponding vapor composition is calculated as $y_i = Kx_i$. The temperature is adjusted until $\Sigma y_i = 1.0$.

There are three basic limitations in using temperature to infer composition:
1. Variations in off-key components will cause errors.
2. Sensitivity of the measurement is too low for many applications.
3. Variation in pressure will cause an error.

The sensitivity of the measurement to off-key components creates problems in separating natural-gas liquids. Temperature is typically used to control the amount of propane leaving in the bottom product from a depropanizer in the company of butanes and gasoline components. Should a change in feed quality increase the average molecular weight and therefore the boiling point of the heavier components, the temperature controller will tend to let more propane pass to keep the boiling point constant.

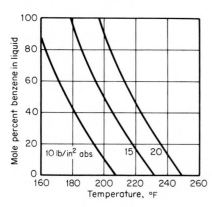

figure 6.3 *The relationship between boiling point and composition for the benzene-toluene mixture is nearly linear.*

In many applications, however, boiling point or vapor pressure is the desired controlled variable. This is true of many of the products of a petroleum refinery, which are complex mixtures of components. Some analyzers used, particularly in fractionating crude oil, are forms of boiling-point devices — they report the initial boiling point, end point, flashpoint, or vapor pressure.

Lack of sensitivity is a drawback in using temperature measurements as a guide to separating isomers. The difference in boiling points between isobutane and n-butane at 60 lb/in^2 gage is only 23°F. A variation of 1 mol % isobutane in the n-butane bottom product would change the indicated boiling point only 0.25°F. If the problem could be solved by devising a more sensitive thermometer, it might be a simple matter. But the same deviation of 0.25°F could also be caused by a pressure variation of only 0.25 lb/in^2 or by a change in the concentration of off-key isopentane of only 0.6 mol %.

In an effort to improve the sensitivity of a temperature measurement to the concentration of a key component, the point of measurement is usually located several trays from the end of the column. This practice has the additional advantage of improving the speed of response to a feed-rate or composition upset and thereby accelerating control action. However, the measurement then becomes less representative of product quality and begins to be sensitive to feed composition. More is said on locating temperature measurements under the heading "Column Profiles."

Sensitivity to pressure variations can be minimized by controlling pressure at the point of temperature measurement. This presumes that pressure can be controlled tightly and in fact seems to be the principal reason for emphasis on tight pressure control in the past. However, if pressure control is lost because of a condenser limitation, the temperature controller allows off-specification product to be discharged. This is also a problem for floating-pressure-control systems.

A simple and generally applicable technique is to compensate the temperature measurement mathematically for variations in pressure. The objective is to reference a temperature measurement made at a variable pressure to a base pressure. Then both the controller and the operator will recognize the boiling point at the reference pressure. The compensated temperature T_b is related to measured temperature T, pressure p, and reference pressure p_b as

$$T_b = T - \left.\frac{\partial T}{\partial p}\right|_x (p - p_b) \tag{6.4}$$

The partial derivative is the inverse slope of the vapor-pressure curve at a normal product composition.

For the benzene-toluene mixtures appearing in Fig. 6.3, $\partial T/\partial p$ at 15 lb/in^2 abs is 3.8°F/lb/in^2 at 100 percent benzene and 4.2°F/lb/in^2 at 100 percent toluene. The correction factor changes more with pressure than with composition, decreasing from 5°F/lb/in^2 at 10 lb/in^2 abs to 3.1°F/lb/in^2 at 20 lb/in^2 abs for pure benzene. If compensation is required over this broad a range, a second-order approximation should be used:

$$T_b = T - a\,(p - p_b) + b\,(p - p_b)^2 \tag{6.5}$$

Coefficients a and b are selected to fit the temperature-pressure curve across the operating range. For pure benzene, values of $4.1°F/lb/in^2$ for a and $0.1°F/(lb/in^2)^2$ for b match the boiling-point curve at 10, 15, and 20 lb/in^2 abs. Pressure compensation is essential when the boiling point is to be controlled under floating-pressure operation. Rademaker et al. [2] report that for hydrocarbons, the product $p(\partial T/\partial p)_x$ is "of the order of 40°C." This property is very useful when compensating a temperature measurement in a refinery column where mixtures having many components are separated on the basis of boiling point.

Luyben [3] and Boyd [4] describe the use of differential-temperature and double-differential-temperature measurements to obtain a more exact representation of product quality. Pressure variations have relatively little effect on these systems since all temperatures are influenced to essentially the same degree. In attempting to hold the column temperature profile constant, they provide better regulation over product quality than control of a single temperature point but must be matched to the characteristics of the column for maximum effectiveness. This subject is also examined under "Column Profiles."

Vapor-Pressure Measurement   A device which combines the pressure compensation and high sensitivity with the simplicity and response of a temperature measurement is the Differential-Vapor-Pressure Cell Transmitter (DVP) manufactured by The Foxboro Company [5]; it is shown schematically in Fig. 6.4. A temperature bulb filled with a reference fluid (liquid and vapor) is connected to the low-pressure side of the sensor, while the high-pressure side is connected directly to the column at the same elevation as the temperature bulb.

If the vapor pressure exerted by the mixture in the column equals that exerted by the reference fluid at the same temperature, then the DVP cell produces a midscale (zero) signal. The presence of more-volatile components in the column will raise the output, whereas components of lower volatility will reduce it. The device gives exact pressure compensation, although only at zero differential pressure. Therefore the reference fluid should be selected to match the vapor pressure of the mixture in the column *at the point of measurement.*

The DVP measurement suffers from some of the same limitations as a temperature measurement in that it should be inserted not at the terminal point of a column but a few trays inside it. This is necessary to avoid errors caused by subcooled reflux and dissolved gases at the top and partial vaporization of high boilers at the bottom. These locations also improve the sensitivity of the measurement, since composiiton

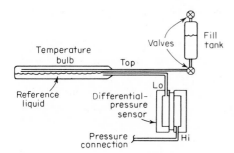

figure 6.4 *The Foxboro DVP cell transmitter compares the vapor pressure of a reference fluid against the pressure in the column at the same temperature.*

changes are greater at points farther away from the products where impurities are minimal.

A common application of the DVP cell is an ethanol-water column. A cell containing distilled water as a reference fluid is inserted near the bottom of the column to control ethanol content there. Ethanol concentration in the bottom product is in the parts-per-million range and would not be detectable. But owing to its high relative volatility at low concentrations, only a few trays are able to raise the ethanol concentration to a measurable level. Furthermore, the high sensitivity of the DVP cell makes this less of a problem than with an uncompensated temperature measurement; as little as 0.15 wt % ethanol will exert 1-in H_2O differential vapor pressure, measurable over a 20-in span [6].

A second DVP cell is used near the top of the ethanol-water column. This can be filled with a mixture of ethanol and water representing the product desired (different mixtures are used for beverage columns as opposed to fuel-alcohol columns). The reference bulb may be filled at the factory, or in situ, using a sample from the column itself. However, great care must be exercised to remove all air from the bulb, owing to its high vapor pressure.

The reference fluid must also be perfectly stable. Any changes in composition which alter the vapor-pressure curve over time will affect the calibration of the device. Some distilled products are unsuitable in this regard. Acetic acid, for example, contains acid impurities which attack stainless steel, releasing hydrogen — this raises vapor pressure intolerably. In these cases, an alternative, stable fluid having the same vapor-pressure curve as the product should be substituted. In the case of acetic acid, methyl cellosolve is used.

Proper installation is essential to success. The temperature bulb has a top and a bottom and must be installed accordingly. This assures that the capillary tube connecting the bulb to the cell body is always filled with liquid. If it should contain vapor, the vapor may condense and reflux, affecting the measurement. The pressure and temperature connections should also be made at the same elevation to avoid a difference in liquid head. Similarly, the pressure connection should be short and direct to avoid trapping liquid. When properly installed and calibrated, a DVP cell is a very useful composition transmitter, filling the gap between less-sensitive temperature measurements and higher-cost, slow-responding analyzers.

Another application of vapor-pressure control is in products used for motor fuels. For this service, the vapor pressure at 100°F is required, based on anticipated ambient storage conditions; it is commonly known as the *Reid vapor pressure*. Analyzers are available to determine the Reid vapor pressure of a product withdrawn from a distillation column, and the signal can be used in the same way as any temperature or composition measurement. The DVP cell output does not correlate well with Reid vapor pressure and is not recommended as a substitute for it. The problem lies in the very large change in pressure from distillation conditions at 300 to 400°F and Reid vapor pressure at 100°F. Vapor pressure tends to double for every 50°F rise in temperature. Changes in the combinations of high and low boilers in the mixture seem to affect vapor pressure differently under the two conditions.

Column Profiles Analyzers measure the composition of the products themselves and are therefore able to report on their true properties. Whereas this involves considerable expense and some delay, it frees the engineer from a concern with the profile of compositions within the column. But when a temperature or vapor-pressure sensor needs to be inserted in the column, the composition profile becomes quite important. The measuring devices should be located where they will be sufficiently sensitive to changes in product quality yet remain representative of that quality. Furthermore, attempts to achieve pressure compensation or enhanced sensitivity using multiple measurements compound the problem.

Ryskamp [7] observed that a column whose trays are sharing the separation load equally will have a profile that appears as a straight line when composition is plotted on a probability scale against trays on a linear scale. The probability scale expands from 50 percent symmetrically in both directions. The distance between 1 and 2 percent is the same as that between 40 and 50 percent, 50 and 60 percent, and 98 and 99 percent, as shown in Fig. 6.5.

If the profile plotted on these coordinates is nonlinear, then separation is less than it could be because some trays are not being fully utilized. A distortion appearing at the feed tray indicates that the feed does not match the tray contents in composition, temperature, or both. A curvature reveals that the reflux ratio is inconsistent with the number of trays provided. For example, as minimum reflux ratio is approached, a pinch point develops near the feed tray so that little separation is achieved per tray in that region. Conversely, at very high reflux ratios, separation per tray may exceed that attained elsewhere.

Separation factor and material balance affect the profile differently. An increase in separation factor resulting from an adjustment to reflux ratio decreases the slope of the profile. Shifting the material balance, however, moves the profile up or down the column without changing its slope. Both changes are illustrated in Fig. 6.5.

Although the probability plot is useful in pointing out inefficiencies in column operation, a linear composition scale is needed to properly locate measurement points and identify nonlinear behavior. When the straight lines of Fig. 6.5 are

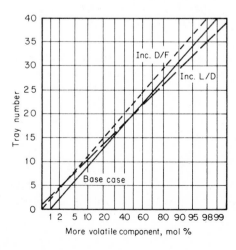

figure 6.5 *Composition plotted on a probability scale against tray number gives a straight line when separation is uniformly distributed among the trays.*

plotted on linear coordinates in Fig. 6.6, curvatures appear at each end of the column. The curves also converge at the ends. This reveals that the sensitivity of the composition of the products is generally much lower than that of compositions elsewhere in the column. A significant increase in sensitivity is attainable, in Fig. 6.6, about five trays from each end of the column. Beyond ten trays from each end, sensitivity no longer increases as the linear portion of the profile is entered.

Figures 6.5 and 6.6 are idealized in that no discontinuity appears at the feed point — unlike almost all real columns. Furthermore, any change in operating conditions that affects the composition at the feed tray will introduce a discontinuity if it were absent originally. If the column described by the figures had a feed composition of 50 percent introduced at tray 20, then no discontinuity would develop on a change in reflux ratio, which causes the profile to rotate around tray 20. However, changing the external material balance affects compositions everywhere and therefore always distorts the profile.

In practice, the profiles are more important toward the ends of the column. Temperature or vapor-pressure measurements are usually made near the ends so that they are representative of product compositions yet sufficiently removed to increase sensitivity and to avoid errors caused by subcooled reflux and partial vaporization in the reboiler.

Figure 6.3 shows temperature to be nearly linear with composition over its entire range. Therefore the temperature profile for a column will have a shape almost identical to its composition profile and will be affected in the same way by manipulation of L/D and D/F.

Locating a measurement at a point other than where product is withdrawn risks lack of representation. While the temperature at the point of measurement may be controlled precisely, the product composition could still vary with feed composition and throughput. For example, note that two of the lines in Fig. 6.5 cross at about

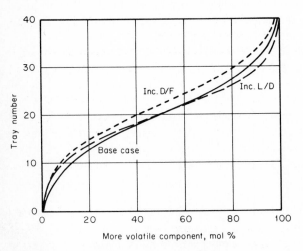

figure 6.6 *When plotted on linear coordinates, the ideal profile curves sharply at both ends of the column.*

tray 7. Both these conditions then satisfy a controller operating on the tempera-ture at that tray, although each results in a different product composition. While Fig. 6.5 shows this to be brought about by a certain combination of D/F and L/D, it could be caused by a shift in feed composition as well. One approach to solving this problem is to control the profile in more than one place.

The optimum location for temperature control points must provide the best com-bination of sensitivity, representation of product compositions, freedom from effects of feed composition and reflux enthalpy, and dynamic responsiveness. The choice will change from column to column as a function of boiling-point difference, num-ber of trays, and variability of feed composition. Reference 8 considers all these factors and gives guidelines for selecting the control trays.

Knowledge of the composition or temperature profile and its variability has also led to the use of multiple measurements to control compositions. The most obvious advantage of using two temperature measurements is to provide pressure compen-sation. One bulb would be located midway between the feed and the product where sensitivity is maximum; the other would be located close to the end of the column in an area of low sensitivity, yet removed from the effect of reflux subcooling. Changes in column pressure would tend to affect both temperatures equally and their difference would be insensitive to that source of upset.

The author used this method to control the flow of hydrocarbon reflux to an ethanol dehydration column [6]. It functioned to maintain a relatively small tem-perature rise over a large number of trays to ensure a constant hydrocarbon inven-tory throughout that section of the column. The system is described in more detail in Chap. 10. One danger in using a differential-temperature signal is the possibility of ambiguity. In the ethanol dehydrator, for example, a low differential temperature could also be attained without any reflux at all or when the column was cold. Therefore, users of differential-temperature controllers should be wary of the possibility of multiple steady states, especially during start-up or following severe disturbances.

Luyben [3] used the difference between two temperature differences across the bottom section of a column to manipulate heat input. Boyd [4] subtracted a ΔT mea-sured above the feed tray of a benzene-toluene column from another ΔT measured below the feed tray, manipulating benzene flow to control it. He was able to control composition of the benzene product in the parts-per-million range because the ΔT difference changed as much as 40°F for a toluene concentration change of only 3 parts in 10,000. However successful systems like this might be, they require a detailed knowledge of composition and temperature profiles over the foreseeable range of operating conditions, supplemented by significant experimental or simu-lation effort, and unfortunately, the experience gained on one column may not be transferable to another.

Nonlinear Characteristics   The nonlinearity apparent toward the ends of the curves in Fig. 6.6 affects the gain of the composition control loops. So do discontin-uities at the feed tray if measurements are made in that vicinity. Consider that 100 percent purity represents the absolute limit to product composition. Response of composition to *any* manipulated variable must therefore be nonlinear as the key impurity approaches zero concentration. This can be seen in the composition changes caused by increasing D/F from the base case in Fig. 6.6. The change in composition at the top of the column is much greater than at the bottom because the concentration of the key impurity is moving *away* from zero rather than toward it.

The variability of process gain is best determined by differentiation. For example, the gain of y with respect to D/F can be readily evaluated under conditions of constant separation:

$$\left.\frac{\partial y}{\partial(D/F)}\right|_s = \Lambda_{DS}\left.\frac{\partial y}{\partial(D/F)}\right|_x \tag{6.6}$$

The last term is a simple derivative of the material-balance equation:

$$\left.\frac{\partial y}{\partial(D/F)}\right|_x = -\frac{z-x}{(D/F)^2} = -\frac{(y-x)^2}{z-x} \tag{6.7}$$

The terms in Eq. (6.7) do not change significantly as x approaches zero or as y approaches unity and therefore do not contribute to the variability experienced in the gain of the composition loop.

However, Λ_{DS} contains the significant terms:

$$\Lambda_{DS} = \left[1 + \frac{(y-z)x(1-x)}{(z-x)y(1-y)}\right]^{-1}$$

The ratio of the key impurities is the determining factor.

As $\dfrac{x}{1-y} \to 0$, $\Lambda_{DS} \to 1$

As $\dfrac{x}{1-y} \gg 1$, $\Lambda_{DS} \to \dfrac{1-y}{x}$

Therefore, the gain of the top-composition loop approaches zero as its key impurity $1-y$ approaches zero. This produced the type of closed-loop response described in Fig. 1.3.

Examining the alternate loop pairing,

$$\left.\frac{\partial x}{\partial(D/F)}\right|_s = (1 - \Lambda_{DS})\left.\frac{\partial x}{\partial(D/F)}\right|_y \tag{6.8}$$

The last term is

$$\left.\frac{\partial x}{\partial(D/F)}\right|_y = -\frac{(y-x)^2}{y-z} \tag{6.9}$$

Again, Eq. (6.9) shows no wide variations so that $(1 - \Lambda_{DS})$ contains the significant terms. Here the condition contributing to gain variability is the reverse of what was observed earlier:

$$\text{As } \frac{x}{1-y} \to 0, (1 - \Lambda_{DS}) \to \frac{x}{1-y}$$

$$\text{As } \frac{x}{1-y} \gg 1, (1 - \Lambda_{DS}) \to 1$$

In summary, the composition of the purer product shows the larger variations in gain when manipulating the material balance. (In practice, this pairing is already undesirable because the variability is coincident with relative gain approaching zero.)

The gain of compositions with respect to the other manipulated variable, e.g., L/D, requires evaluation:

$$\left.\frac{\partial x}{\partial (L/D)}\right|_{D/F} = \Lambda_{DS}\left.\frac{\partial x}{\partial S}\right|_y \frac{dS}{d(L/D)} \tag{6.10}$$

The second term now contains significant factors:

$$\left.\frac{\partial x}{\partial S}\right|_y = -\frac{x^2(1-y)}{y}$$

Based on Eq. (5.62),

$$\frac{dS}{d(L/D)} = \frac{nES}{2(1 + zL/D)(L/D)}$$

Combining gives

$$\left.\frac{\partial x}{\partial (L/D)}\right|_{D/F} = -\Lambda_{DS}\frac{nE\,x(1-x)}{2(1 + zL/D)(L/D)} \tag{6.11}$$

This indicates that loop gain varies directly with key impurity x.

Another configuration worth examining is the SV subset. Assume manipulation of D/V:

$$\left.\frac{\partial y}{\partial (D/V)}\right|_v = \Lambda_{SV}\left.\frac{\partial y}{\partial S}\right|_x \frac{dS}{d(D/V)} \tag{6.12}$$

The second term again contains significant factors:

$$\left.\frac{\partial y}{\partial S}\right|_x = \frac{(1-y)^2 x}{1-x}$$

Again, based on (5.62),

$$\frac{dS}{d(D/V)} = -\frac{nES(1 + D/L)^2}{2(z + D/L)}$$

Combining the above,

$$\left.\frac{\partial y}{\partial (D/V)}\right|_v = -\Lambda_{SV}\frac{nE(1 + D/L)^2 y(1-y)}{2(z + D/L)} \tag{6.13}$$

Loop gain is determined by impurity $1 - y$. The other loop of the SV subset has again

$$\left.\frac{\partial x}{\partial (V/F)}\right|_s = \left.\frac{\partial x}{\partial (D/F)}\right|_s \left.\frac{\partial D}{\partial V}\right|_s \qquad (6.14)$$

The last term is simply D/V.

To illustrate the gain variations which can be expected in composition loops, consider a column having 40 theoretical trays separating a mixture having a relative volatility of 1.4. As a base case, $x = 0.05$ and $y = 0.95$. All relative gains and open-loop gains have been calculated for the base case and for independent variation of x to 0.01 and y to 0.99; the results appear in Table 6.1.

The bold figures in the table indicate the two gains which are to be compared to the corresponding change in the related controlled variable. For example, in the DS configuration, the gain of the y, D loop changes from -0.90 to -0.316 as $1 - y$ moves from 0.05 to 0.01; the gain is nearly proportional to concentration, as predicted by the corresponding variation in Λ_{DS}. Also, the gain of the x, L/D loop changes from -0.0344 to -0.0073 as x moves from 0.05 to 0.01; the gain is nearly proportional to x, as predicted by Eq. (6.11). The SD configuration is a mirror image of the DS configuration.

TABLE 6.1 Open-Loop Gains as a Function of Product Compositions

Variable	I	II	III
	($nE = 40$, $\alpha = 1.40$)		
x	0.05	0.01	0.05
y	0.95	0.95	0.99
Configuration DS			
Λ_{DS}	0.5	0.839	0.161
$\left.\dfrac{\partial y}{\partial (D/F)}\right|_s$	**−0.90**	−1.51	**−0.316**
$\left.\dfrac{\partial x}{\partial (L/D)}\right|_{D/F}$	**−0.0344**	**−0.0073**	−0.0067
Configuration SD			
Λ_{SD}	0.5	0.161	0.839
$\left.\dfrac{\partial y}{\partial (L/D)}\right|_{D/F}$	**0.0344**	0.0067	**0.0073**
$\left.\dfrac{\partial x}{\partial (D/F)}\right|_s$	**−0.90**	**−0.316**	−1.51
Configuration SV			
Λ_{SV}	1.72	2.93	3.38
$\left.\dfrac{\partial y}{\partial (D/V)}\right|_{V/F}$	**−3.39**	−5.68	**−1.37**
$\left.\dfrac{\partial x}{\partial (V/F)}\right|_s$	**−0.166**	**−0.046**	−0.222

In the SV configuration, the gain of the y, D/V loop changes from -3.39 to -1.37 as $1 - y$ moves from 0.05 to 0.01; the gain is less than proportional to $1 - y$. The gain of the x, V loop changes from -0.166 to -0.046 for the same change in x, again roughly proportional to composition.

The realization to be drawn from the above discussion is that the gain of a typical composition loop tends to vary directly with the concentration of the key impurity, regardless of the choice of manipulated variable. If compensation is not provided in the control system, damping will be variable and oscillation nonsinusoidal as typified by Fig. 1.3.

Even without compensation, the gain variation need not destabilize the loop. The average gain developed by cycling about a set point seems to be lower than if composition is constant at set point, as indicated by closed-loop simulation. Therefore, if damping is provided by proper controller adjustment to bring the composition to set point, there should be no danger of losing damping under upset conditions. If the set point is changed significantly, however, the controller gain must be readjusted accordingly. Gain compensation is described in the next section.

FEEDBACK CONTROL ALGORITHMS

Nearly all the work of regulating process variables is accomplished by feedback controllers. Consequently, their performance is central to the task of maintaining compositions at set point. As noted above, special features may occasionally be required such as compensation for variable process gain. While there will be no attempt at describing the functioning of a wide variety of algorithms in detail here, the nature of the PID algorithm will be explored, along with those features required to control distillation columns. For the reader who seeks more general or basic information on feedback control, an earlier work by the author [9] is recommended.

The PID Algorithm   The basic feedback control algorithm combines proportional action (P), integration of the deviation from set point (I), and derivative action (D) applied to the controlled variable. Derivative is usually reserved for temperature and composition loops, therefore PI controllers are far more common than PID controllers.

The simple PI algorithm is described by the following differential equation:

$$m = \frac{100}{P} \left(e + \frac{1}{I} \int e \, dt \right) \tag{6.15}$$

where m = output of controller
e = deviation of controlled variable from set point
P = proportional band, percent
I = integral time constant, same units of time as t

Proportional action is the principal stabilizing function and determines the damping of the loop. If the controller output always operated about the same value, all upsets being transitory, proportional action alone would be sufficient for regulation. However, most disturbances have a more-lasting character, so the controller is required to change its output to a new steady-state value to rebalance the loop and return measurement to set point. Without integral action, this is impossible because an output change Δm must be accompanied by a change in deviation Δe.

Integrating action will continue to move output m as long as deviation e is non-zero. Hence, it is capable of eliminating any deviation given enough time. For the integrating mode, timing is critical — its time constant I must be sufficiently long to allow response to return from the process through its dead time. If I is set shorter than the process dead time, integration takes place faster than the process can respond to it, resulting in loss of loop stability identified by a slow cycle. The optimum value of I is about twice the process dead time when derivative is used in the controller and three times when it is not. Table 6.2 summarizes the optimum controller settings for countering step load changes with PI and PID controllers.

With only proportional action, the period of oscillation of a control loop is approximately $4\tau_d$ (dead time). The retarding effect of integral action, and the accelerating effect of derivative action on the period can be seen in the table. The proportional band needed for stability is the product of the steady-state process gain K_p and the ratio of dead time to the dominant time constant τ_1. In the case of a composition loop, this would be τ_x described in Fig. 3.15.

The two criteria examined in Table 6.2 are integrated error (IE) and integrated absolute error (IAE). Integrated error was described by Eq. (1.14). By contrast, IAE cannot be calculated from the controller settings but must be determined by closed-loop tests. If, following an upset, the controlled variable returns to set point without crossing it, IE and IAE will have identical values. However, if it does overshoot, some of the negative deviation will cancel some of the positive deviation so that IE will be lower than IAE, which penalizes deviations without regard to sign.

In general, a minimum IAE gives more acceptable performance than a minimum IE, which could be satisfied in the presence of perpetual cycling about set point. However, in those instances where a product is being accumulated in a tank or other large vessel, plus and minus variations in quality do cancel, and a minimum IE would be preferred.

In adjusting a controller, the first step is to determine loop dead time or its period under proportional control, which is then divided by 4 to obtain the effective dead time. Then the appropriate value of I (and D if used) should be introduced and the period again observed. If the period exceeds the value indicated in Table 6.2, I (and D) should be increased until the desired period is found; if it is too low, I (and D) should be decreased. Finally, the proportional band should be readjusted to provide the desired degree of damping.

TABLE 6.2 Optimum Controller Settings for Step Load Changes

Controller	Period	P	I	D	Criterion
PI	$5.7\tau_d$	$111 K_p \tau_d / \tau_1$	$3.3\tau_d$	—	IAE
	6.9	134	2.5	—	IE
PID	3.4	90	2.0	$0.5\tau_d$	IAE
	4.5	143	0.7	0.7	IE

An integrator always requires a constant of integration, which is the accumulated integrations of past deviations or the initial condition of the output when placed in automatic. In an open-loop situation where the process cannot respond to output of the controller, any deviation will cause the integrator to drive continuously into saturation. This is known as a *windup* condition and will persist even if the deviation should return suddenly to zero. Because windup is a recurring problem whenever constraints are encountered, it is given special consideration in Chap. 7.

Derivative action is available by the addition of that function to the controlled variable:

$$c' = c + D\frac{dc}{dt} \tag{6.16}$$

where c' is the controlled variable modified by derivative action of time constant D expressed in the units of time t. In its ideal form, derivative gain increases directly with the frequency of the input signal, which makes it hypersensitive to noise. Stability is provided by limiting the gain of the derivative function to the region of 10 to 20 by appropriate filtering. Still, some signals such as flow and liquid level are too noisy to be able to use derivative control.

The principal function of derivative is to reduce the period of the loop by canceling the phase lag accompanying integration. This also allows a reduction of the proportional band. As a result, the integrated error sustained by a *PID* controller is only one-third to one-half that sustained by a *PI* controller encountering the same upset on the same process.

Compensating for Variable Gain   The variable gain common to composition control loops on distillation columns can be compensated by modifying the control algorithm. If the control algorithm is *adaptable*, it will allow its mode settings to be adjusted from a remote source. This property is common to most of the digital control algorithms; because mode settings are entered at the keyboard, they are also calculable from other information.

Because these composition loops show gain changing directly with the concentration of the key impurity, compensation can take the form of the proportional band changing directly with the same variable. For example:

$$P = K_P y_H \tag{6.17}$$

where K_P is an adjustable coefficient and y_H is the heavy key in the distillate product.

If an adaptable controller is unavailable, gain compensation can be provided by a calculation at the input of the controller, performed as shown in Fig. 6.7. The gain of the divider varies with the controlled variable y_H because it appears in the denominator along with set point y_H^*. Its gain also changes with the set point. At zero deviation, the output of the divider is $1/2$, requiring a set point of $1/2$ for the composition controller. While this circuit does not produce a gain that varies directly with y_H, it moves in the proper direction. Furthermore, all the gain changes

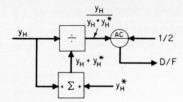

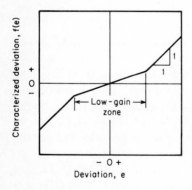

figure 6.7 *This system compensates reasonably well for variations in process gain proportional to key impurity y_H.*

noted in Table 6.1 are *less* than the composition changes of 5:1, reducing the need for exact compensation.

As the measuring device is moved from the end of the column, response tends to become more linear, as indicated by the column profile in Fig. 6.6. Therefore, gain compensation should not be required for most temperature control loops. There are exceptions, however, especially where profiles show extreme gradients. In these cases, there also tend to be substantial gradients around the feed point.

When this happens, a temperature bulb located midway in either the top or bottom section of the column may give linear response only under relatively uniform operation. Large upsets causing substantial shifts in the column profile in either direction may cause the temperatuare measurement to enter a low-gain region of the profile. If the temperature controller has been properly adjusted for stable operation in the high-gain region, loop gain may fall too low for a reasonable rate of recovery from the severe upset.

A symmetrically characterized nonlinear controller designed for pH control has been found useful for correcting this nonlinearity. It contains the characterizing function shown in Fig. 6.8. The width of the low-gain zone should be adjusted to match the linear portion of the temperature profile so that the gain of the controller increases when the nonlinear portion is approached in either direction. The gain within the zone is also adjustable in relation to unity gain outside.

Sampled-Data Control   A digital control algorithm is processed periodically at intervals of time known as the sample interval. The algorithm is therefore updated at each calculation but its output is held constant between calculations. This process adds dead time to the loop just as did the sampling nature of the chromatographic analyzer described in Fig. 6.2. In most cases then, the controller's

figure 6.8 *This nonlinear function is useful in controlling temperature in columns with sharp profiles.*

sample interval is kept small relative to the process dead time to avoid deteriorating the response of the loop.

When the process already has a sampling element in it, such as the chromatographic analyzer, loop performance can be improved if the control algorithm is synchronized with the device. Because the loop is open between samples, there is no point in continuing to run the control algorithm. The sampled-data algorithm is best keyed to the analyzer report, making one calculation only on receipt of new information.

The *PI* sampled-data algorithm operates incrementally, calculating each new output (n) based on the last output $(n - 1)$ and changes observed since that time:

$$m_n = m_{n-1} + \frac{100}{P} (e_n - e_{n-1} + K_I e_n) \tag{6.18}$$

Synchronizing the controller to the analyzer eliminates the delay contributed by the sampling operation. Also, integral action is keyed to the sample interval Δt through K_I, which equals $\Delta t/I$. Then should Δt change for some reason, integral time also changes; K_I is typically set in the range of 0.1 to 0.5.

The same algorithm can be used to close the loop when Δt is variable, as is the case when laboratory analyses are used for control. New information can be entered and the algorithm executed once manually. In the absence of information, the algorithm is not executed so it cannot wind up.

Derivative action can also be applied incrementally to the controlled variable:

$$c'_n = c_n + K_D(c_n - c_{n-1}) \tag{6.19}$$

where the derivative gain K_D equals $D/\Delta t$. Because the derivative gain is adjustable, as opposed to the fixed value of 10 to 20 in conventional algorithms, sampled derivative is useful on the stepwise signal from a chromatographic analyzer. Typically, K_D would be set in the range of 1 to 5.

FEEDFORWARD CONTROL

Feedback has the fundamental limitation of being unable to initiate corrective action until the controlled variable begins to deviate from set point. And in loops having slow dynamic response, as composition loops do, the resulting deviation can become sizeable before the corrective action takes effect. This disadvantage can be completely overcome by the use of feedforward action. In essence, changes in disturbing variables are converted directly into corrective action to the manipulated variable *before* they cause a deviation in the controlled variable.

Feedforward proceeds by direct calculation rather than the trial-and-error method of feedback controllers. This frees it from stability considerations and windup problems. However, to be effective the feedforward calculations must be executed accurately, based on the steady-state and dynamic characteristics of the process being controlled.

In practice, the calculations are limited in accuracy, the process characteristics are never completely known, and not all disturbing variables are measurable. As a result, most feedforward systems require feedback to provide the necessary steady-state accuracy to hold the controlled variable at set point. Figure 6.9 shows how the two control functions are combined. In a well-designed system, typically 90 percent of the corrective action is applied by the feedforward system, with only the remaining 10 percent required of the feedback controller. By reducing the feedback effort tenfold, the feedforward system also will reduce the integrated deviation of the controlled variable by the same tenfold.

Material-Balance Systems   The principal disturbances affecting a column where controls have already been structured optimally according to the directions provided to this point, are feed rate and composition. Of the two, feed-rate disturbances are far greater in both magnitude and frequency. Most columns receive feed from other columns, principally under liquid-level control. Consequently, feed rate can be expected to change continuously and even sinusoidally, reflecting the transient behavior of upstream operations. Even feeds that are controlled from storage are not always constant — changes in production rate can be frequent and are normally introduced stepwise.

By contrast, feed-composition changes tend to vary between relatively narrow limits and at a rate limited by the time constants of upstream capacities. Step changes are only encountered when rates are changed among multiple sources. Therefore, feed composition is used less often as a feedforward input than is feed rate. Additionally, the measuring device, an analyzer, is much more costly and demanding than is the typical flowmeter.

The most accurate feedforward system is that used for external material-balance control because it is independent of variables such as relative volatility and tray efficiency. It uses the familiar material-balance relationships. For a multicomponent column,

$$\frac{D^*}{F} = \frac{z_L - x_L^*}{y_L - x_L^*} = \frac{z_{LL}}{y_{LL}} \tag{6.20}$$

Here the * denotes set points — the controlled variables themselves cannot be used in feedforward control without introducing (positive) feedback. Next, y_L is eliminated by substitution:

$$y_L = 1 - y_H^* - y_{LL} = 1 - y_H^* - \frac{z_{LL}}{D^*/F}$$

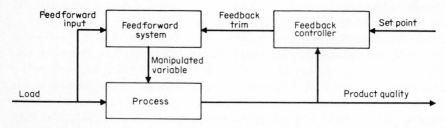

figure 6.9   *A feedforward system can provide most of the correction for measurable disturbances.*

Then the resulting expression is solved for manipulated variable D:

$$D^* = F \frac{z_L + z_{LL} - x_L^*}{1 - y_H^* - x_L^*} \tag{6.21}$$

If the set points are relatively small, e.g., 1 percent, (6.21) may be simplified to

$$D^* = F (z_L + z_{LL}) m \tag{6.22}$$

where m would be the output of the composition feedback controller.

In most cases, there will be no feed analyzer, reducing the feedforward expression to

$$D^* = Fm \tag{6.23}$$

or, where orifice meters are used,

$$D^{2*} = F^2 m \tag{6.24}$$

Then the composition feedback controller must adjust the D/F ratio in response to feed-composition upsets.

A similar derivation is used when B is the best flow to manipulate for composition control:

$$B^* = F \frac{1 - z_L - z_{LL} - y_H^*}{1 - y_H^* - x_L^*} \tag{6.25}$$

Feed composition is left in the form of the light rather than heavy components because they are more readily reported by most analyzers.

Observe that all the above feedforward models feature a linear relationship between the flow rates of feed and product. This assumes that *both* x and y are controlled as indicated in Eqs. (6.21) and (6.25). In order to accomplish this, a second composition control loop is required, manipulating one of the other variables. If D is selected to control y because of favorable relative gain and stream size, boilup would generally be preferred to control x. Because boilup must change proportional to feed rate to control x, a second feedforward loop is required, setting heat input

$$Q^* = Fm_x \tag{6.26}$$

or, where orifice meters are used:

$$Q^{2*} = F^2 m_x \tag{6.27}$$

Here m_x represents the output of the bottom-composition controller, or simply the Q/F ratio, if the loop is not closed. Figure 6.10 shows how a simple feedforward system would be implemented for the DV configuration.

In the event that Λ_{DV} approaches zero and $\Lambda_{SV/B}$ is not particularly favorable, the SD configuration may be the best choice. Then bottom composition should be controlled by B (or D) and top composition by reflux ratio. In this case, only one feedforward path is required because the top-composition controller already manipulates a ratio. This arrangement is shown in Fig. 6.11. Dynamic compensation

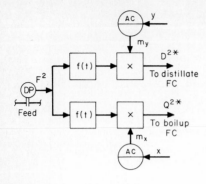

figure 6.10 *The material-balance feedforward system sets one product and one energy flow proportional to feed rate.*

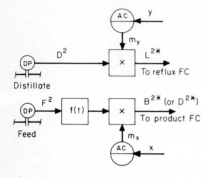

figure 6.11 *Only one feedforward loop is needed for the SD configuration since the top-composition controller manipulates reflux ratio.*

provided by the boxes labeled $f(t)$ is described after the other feedforward systems are introduced.

Controlling a Single Composition   When only one product composition needs to be controlled, and energy is free as from a waste-heat source, the feedforward calculation becomes nonlinear. If y_H is controlled as in Eq. (6.21), then x_L becomes a variable which changes with separation and therefore with feed rate.

A feedforward model may be prepared using the type of information given in Fig. 1.5. Recovery was defined in Eq. (1.15) as

$$R = \frac{D}{Fz}$$

Therefore D may be found in terms of F and z for various recovery factors as a function of V/F. However, V is to be fixed while F varies.

The feedforward calculation solves for D:

$$D = RFz \tag{6.28}$$

where recovery R is a function of V/F and y_H^*. Because of this dependence, the column to which this system is to be applied needs to be evaluated carefully. Recovery data from the column described in Fig. 1.5 for $1 - y = 5$ percent were used to develop the curves of D versus F appearing in Fig. 6.12. Observe that the degree of nonlinearity is a strong function of boilup rate. The broken line in the figure represents 100 percent recovery. If the heat supply is truly free, then boilup can be

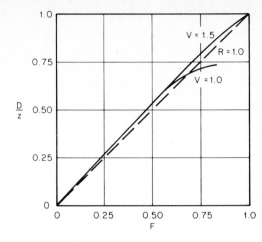

figure 6.12 *Product flow is not linear with feed rate when boilup is fixed.*

maximized at the limit of flooding or condensing so that the type of curve represented by $V = 1.5$ is attained. This curve can be modeled reasonably well by a parabola of the form

$$D = z(mF - kF^2) \qquad (6.29)$$

where m is the output of the composition controller and k is a coefficient adjusted to match the curve toward the upper end of the operating range.

The advantage of the parabolic model lies in its ease of implementation. Figure 6.13 shows the use of the orifice differential-pressure signal as the flow-squared term. Here no feed analyzer is used, the burden of correcting for feed-composition changes being passed on to the product-composition controller, with output appearing as feedback m.

If energy is not free and only one product composition needs to be controlled or can be measured, the linear system of Fig. 6.14 applies. Feedback should be closed around the material-balance variable to correct for changes in feed composition. By contrast, the separation variable may be left without feedback, owing to its low loop gain. This is illustrated in Table 6.1 by the very large gain of the D loop compared to the S loop. If one composition is controlled by manipulating the material balance, the other will be regulated simply by virtue of constant separation.

This system is recommended where no product-composition measurement is available for either product. Feedback may be applied on a calculated basis using

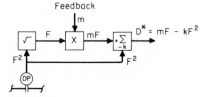

figure 6.13 *The parabolic feed-forward model is easily implemented with analog instruments.*

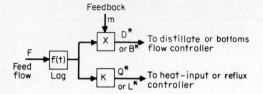

figure 6.14 *Product flow will be directly proportional to feed rate if separation is held constant.*

laboratory analyses. However, to be effective, the feedforward system must be very accurate. Turbine meters then are recommended for feed and product flows.

The SV Configuration   A feedforward system for the SV configuration depends on séparation-factor control and must be derived based on its relationship to L/D as expressed in Eq. (3.58):

$$\frac{D}{Lz} = f(S) \tag{6.30}$$

If D/V is to be manipulated to control y, then Eq. (6.30) must be converted to that form:

$$D^* = \frac{Vz}{f(S) + z} \tag{6.31}$$

Since it is the purpose of the top-composition controller to maintain S constant, its output can be substituted where $f(S)$ appears:

$$D^* = \frac{Vz}{m_y + z} \tag{6.32}$$

To generate the feedforward model required for boilup manipulation, D in (6.31) is replaced by recovery factor from Eq. (6.28). The results is then solved for V^*:

$$V^* = RF[f(S) + z] \tag{6.33}$$

The heat-input set point Q^* is proportional to the desired V^*:

$$Q^* = \Delta H_D V^* \tag{6.34}$$

where ΔH_D is the latent heat of the distillate. The combination of (6.33) and (6.34) contains three constants, but $f(S)$ has already been used as the output of one controller. Therefore, the product of the others is replaced by the output of the bottom-composition controller:

$$Q^* = m_x F(k + z) \tag{6.35}$$

where k is a constant representing D/Lz at base conditions.

In the absence of a feed-composition analyzer, both feedforward equations are simplified:

$$D^* = m_y V \tag{6.36}$$

$$Q^* = m_x F \tag{6.37}$$

Figure 6.15 shows how the system is implemented. Observe that feed flow is only used once as in Fig. 6.11. The feedforward input to the top-composition loop comes from the effect of boilup on column pressure or accumulator level. The top of the column is configured just as shown in Fig. 5.16.

There is one precaution that must be considered with a feedforward system of this type. Should heat input to the column be constrained for any reason, bottom-composition control will be lost but so will feedforward control of top composition. Further increases in feed rate will *not* increase V and therefore no changes in D will be forthcoming without the intervention of the top-composition controller. If boilup constraints are anticipated and feedforward control is still required, then additional features must be included, as described in Chap. 7.

Examination of the *SV/B* configuration in Fig. 5.16 reveals that it does not require a feedforward input from feed rate because compositions are not likely to be disturbed by feed-rate changes. If the feed is a liquid, an increase will propagate downward, raising base level and increasing B. However heat input will rise proportionately to B, driving the light fraction up the column in the correct ratio and at the proper time. Neither does the top composition loop need feedforward for the reason described above for the *SV* system. The component count for the *SV/B* system is even less than for the *SV* system with feedforward because it does not require a dynamic compensator.

If the feed is a vapor, an increase will raise both reflux and distillate in the existing ratio, with the higher reflux eventually reaching the column base where bottom flow and heat input will rise. Again, dynamic compensation does not appear to be required.

Dynamic Compensation   Feedforward action must not only be accurate, it must also be timely. To illustrate, Fig. 6.16 was prepared from Fig. 5.1, truncated to show only the influence of m_1 on loop 2, with the necessary compensation added. Although the steady-state feedforward models have already shown the need for multiplication, the linearized representation is shown for simplicity in illustrating dynamic compensation.

Disturbing variable m_1 must be measured and its signal passed through elements representing the steady-state and dynamic gains that appear in the process. The numerator in the feedforward compensator contains the elements in the disturbing

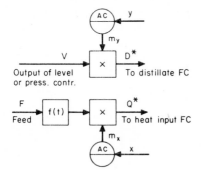

figure 6.15   *The SV configuration requires only a single input from feed flow to achieve feedforward control of both compositions.*

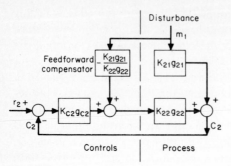

figure 6.16 *The feedforward loop should contain a dynamic model of the process as well as a steady-state model.*

path, while the denominator contains the elements in the manipulated path. If the compensating elements match the process elements exactly, the disturbance will be canceled.

In practice, it is convenient to break the feedforward compensator into its steady-state and dynamic components. The steady-state model should be made as accurate as possible according to the procedures described up to this point. The dynamic compensator can be less exact because its role is transitory and its contribution is limited to rapidly developing disturbances. Accordingly, the dynamic compensation appearing in Fig. 6.16 takes the form of the ratio of two gain vectors:

$$\mathbf{f}(t) = \frac{\mathbf{g}_{21}}{\mathbf{g}_{22}} \tag{6.38}$$

The dynamic model which is representative of most process responses consists of dead time plus a first-order lag. A dynamic compensator based on this model would then include a dead time and a lead-lag function. If the dead time in $\mathbf{g}_{21}$ is longer than that in $\mathbf{g}_{22}$, the compensator should have a dead time equal to their difference. If the reverse is true, no compensation is possible because a time advance cannot be created.

The inverse of the lag in $\mathbf{g}_{22}$ is a lead. When lead and lag are combined in a single device, it has a maximum dynamic gain of the ratio of lead to lag time constants, and an exponential response determined by the lag time ($\mathbf{g}_{21}$).

The need for dynamic compensation for the feedforward system described in Fig. 6.11 can be illustrated by including the column base-level loop in Fig. 6.17. Assume the feed is a liquid. If no dynamic compensation is provided, an increase in feed rate will cause bottom-product flow to increase immediately, although the liquid flow increase to the column base is delayed by the trays below the feed point.

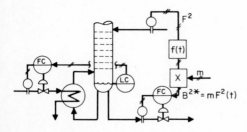

figure 6.17 *The dynamic compensator should match the time response of base level to feed changes.*

The feedforward input will then cause base level to fall before the arrival of more feed causes it to rise. The combination of these two events is shown in Fig. 6.18 (left) to produce a change in boilup that exhibits inverse response. The dynamic imbalance thus created will cause a transient upset in product compositions, possibly more severe than would have been observed without feedforward control.

The right side of Fig. 6.18 shows the effects of compensating with a simple first-order lag. The boilup transient is reduced but not eliminated because the multiple trays produce dead time along with lag. Dead-time compensators are now available in analog as well as digital systems and are very useful in distillation column control. By adding the correct amount of dead time to the feedforward system, the response reversal can be eliminated.

If the feed to the column is a vapor, it will travel upward, leaving liquid flow unaffected momentarily. The increase in feed rate will arrive at the condenser, causing a rise in accumulator level or pressure, and a controller will react by increasing distillate flow. Because of the manipulation of reflux ratio in Fig. 6.11, an increase in reflux flow will take place at the same time. The bottom-product flow change must then wait for the higher reflux to reach the column base before taking place. As a consequence, the dynamic compensator for a vapor feed will require a longer dead time (from column top to base) than was required for a liquid feed (from feed point to base).

Dynamic compensation required for the SV system in Fig. 6.15 is identical. In essence, boilup should not increase until the new liquid flow reaches the column base, whether it comes from a liquid feed or from a vapor feed condensed and refluxed. In all these systems, the dynamic elements g_{22} in the manipulated path are faster than those designated g_{21} in the disturbing path. Hence the compensators require dead time and lag rather than lead action.

The system in Fig. 6.10 shows two dynamic compensators, one for each loop. If the feed is a vapor, it will reach the top of the column quickly, causing overhead vapor composition to begin changing before the effect of a corresponding change in distillate flow can be felt. In this event, the distillate dynamic compensator should have a dominant lead.

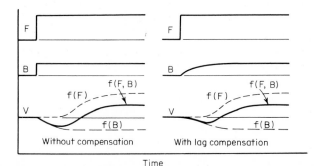

figure 6.18   *Lag compensation can moderate the transient, but further improvement is possible by adding some dead time.*

In practice, dominant lead action is underdesirable because it amplified noise which is always present in flow signals. A preferred method of dynamic compensation uses the reflux accumulator to develop the necessary lead action. This was described in Figs. 4.27 and 4.28, obtained by setting coefficient k to the desired lead-to-lag ratio. (If this system is used, the feed-flow measurement in Fig. 6.10 needs to be linearized.)

If the feed to the column is a liquid, an upset moves preferentially down to the base so that a dead time and lag are required for the heat-input compensator. Distillate flow does not need to change until the resulting change in boilup reaches the condenser. But this response is almost immediate, so the two dynamic compensators can be set the same or combined as in Fig. 6.14.

Dynamic compensation can be adjusted on-line by observing the response of base and accumulator levels to feed-rate changes. If these levels can be maintained reasonably constant following an upset, then internal balances should hold and compositions are not likely to change much. Final adjustment should be based on minimizing transients in product compositions. Reference 9, pages 178 to 186, gives more detail on dynamic compensators and their adjustment.

DECOUPLING

Decoupling is a technique which attempts to combine a system's control signals in such as way that the interaction which naturally exists in the process is canceled. In a decoupled system, individual control loops appear to act independently, notwithstanding the interacting nature of the process under control.

It is possible to apply decoupling in a variety of ways. Because the process is accessible at both its inputs and its outputs, a decoupler could be interposed between the process and controller at either end or in some combination. Decoupling may also be applied in a forward direction, in a backward direction, or in some combination of the two. It may be applied only partially with results which are almost as effective as complete decoupling.

Decoupler Structures   Figure 5.1 illustrated the basic 2×2 interacting process described by four blocks. While a decoupler could be designed to match these four blocks, only two are necessary because only the interacting blocks need to be canceled. The decoupler structure most commonly studied in academic circles has all signals moving in one direction as in the top of Fig. 6.19.

This arrangement suffers from two disadvantages. Consider attempts to place the sytem in operation (initialization) when the two flow controllers are in manual. To balance the set point and measurement for the reflux FC, the operator moves m_y. Having done so, both controllers in the top row may be transferred to automatic. However, balancing the set point against measurement for the heat FC requires the operator to move m_x, which through the decoupler, upsets L^*.

A problem also develops when a constraint is reached. Suppose heat flow is no longer capable of following Q^* because of an overloaded reboiler. Then the two composition controllers will compete through the decoupler for the single remaining manipulated variable without success—one controller will tend to wind up, and the other down.

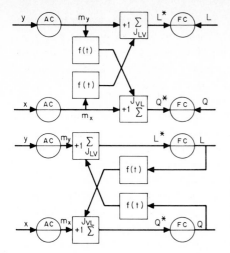

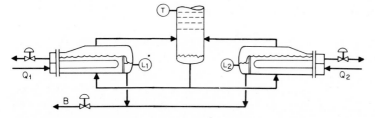

figure **6.19** *The forward decoupling structure (above) is difficult to initialize and fails when constrained; the backward decoupler (below) has neither of these limitations.*

The lower system in Fig. 6.19 avoids these problems. Live measurements are sent to the decouplers, allowing initialization to proceed normally for both loops. Second, failure of either flow-control loop also opens the associated composition loop, thereby avoiding competition.

The functions required for the decouplers are the same as required for feed-forward control as described in the linear system of Fig. 6.16. Elimination of either decoupler will reveal the similarity of the remaining one to the feedforward compensator. Furthermore, elimination of either decoupler makes the two structures of Fig. 6.19 basically identical (except for the interposition of the flow controller). As a result, the steady-state and dynamic decoupler gains for the two structures are identical. Derivation of the decoupler gains appears later in the chapter.

At this point, an example showing the difference between decoupling at controller inputs and outputs (as in Fig. 6.19) is worthwhile. This is illustrated by a column that the author encountered, having parallel kettle reboilers arranged as shown in Fig. 6.20. The two chambers where levels were measured were connected to a common bottom-product valve, which therefore affected both levels. Column temperature was affected equally by both heat-input valves, which drove the levels in opposite directions. Unequal pressure drop across the two vapor-

figure **6.20** *The interaction between liquid levels in parallel reboilers is caused by the common drain valve.*

return lines caused unequal liquid levels. Furthermore, the two levels were connected in a U tube so that they resonated hydraulically.

The three single loops formed by three controllers and valves could not be connected in a stable configuration. Any two loops could be closed, but when the third controller was transferred to automatic, all three loops began to oscillate. The obvious solution was to install independent bottom-product valves for the two reboilers, thereby decoupling the liquid levels. However, this would have required shutting down the column for an extended period, which was unacceptable.

The solution to the problem was the decoupling system shown in Fig. 6.21. Because the single product valve affected both levels equally, their signals were averaged to form the input of a total-level controller ΣLC. (Decoupling was here applied at the controller input.) The temperature controller was sent to both heat-input valves, but balancing was required to keep the levels equal. Therefore, the level signals were sent to a balancing controller ΔLC, one to the measurement input and the other to the set point. This controller's output served to bias the heat-input valves in opposite directions, without changing total heat flow. In this way, the balancing controller did not affect temperature. In operation, the system behaved as three independent, single loops because decoupling had canceled the severe interaction naturally existing in the process.

Decoupling Coefficients All the decouplers shown in Figs. 6.19 and 6.21 are linear. Signals are combined by addition and subtraction, and coefficients need to be applied to the individual inputs. In general, the direct signal path from a controller to its valve should be assigned a coefficient of +1.0. This is the case for all controllers in Fig. 6.19 and for the temperature controller in Fig. 6.21.

The coefficients of the decoupling signals, designated J in Fig. 6.19, are ordinarily low numbers—much less than 1.0—and may be positive or negative depending on the signs of the process gains. If they are not low but approach unity, then interaction is strong as indicated by the need for controllers to affect each other's outputs almost equally. To keep all signals on scale when two inputs of comparable weight are added or subtracted, averaging may be required:

$$c = kb_1 + (1 - k)b_2 \tag{6.39}$$

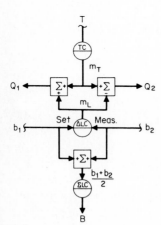

figure 6.21 *The level measurements are combined into a single controller to manipulate B, while the outputs of two controllers are coordinated to manipulate heat input.*

This was done in combining the two level signals in Fig. 6.21; since the reboilers were of identical capacity, k was 0.5. If the coefficients need to be unequal and of opposite sign, negative averaging may be employed by setting $k > 1.0$.

The decoupling coefficients for balancing the levels in Fig. 6.21 could be arbitrarily selected but had to be equal and of opposite sign. To ensure that ΔLC operated about 50 percent scale, each summer included a 50 percent bias:

$$Q_1 = m_T + J(m_L - 0.5)$$

$$Q_2 = m_T - J(m_L - 0.5)$$

Unilateral Decoupling   Decoupling coefficients for the composition loops can be calculated from the relative gains. The relationship can be derived using two subsets which have a common manipulated variable, for example DV and LV. Relative gain Λ_{DV} is really $\bar{\lambda}_{xV}(\Lambda_{DV})$ and Λ_{LV} is $\bar{\lambda}_{xV}(\Lambda_{LV})$. Then $\bar{\lambda}_{xV}$ can have different values, depending on the changes in overhead flows with boilup:

$$\bar{\lambda}_{xV} = \Lambda_{DV}\frac{dL}{dV} + \Lambda_{LV}\frac{dD}{dV} \qquad (6.40)$$

In a material-balance system when the top-composition loop is open, a change in V is totally reflected in L so that dL/dV is 1.0 and dD/dV is zero. On the other hand, if L is fixed and D placed under level or pressure control, the coefficients are reversed. A third possibility is the holding of reflux ratio constant, in which case $dL/dV = L/V$ and $dD/dV = D/V$. This results in constant separation so that $\bar{\lambda}_{xV} = \Lambda_{SV}$. Then

$$\Lambda_{SV} = \Lambda_{DV}\frac{L}{V} + \Lambda_{LV}\frac{D}{V} \qquad (6.41)$$

(This relationship can be verified by substituting the established formulas for the relative gains and obtaining an identity.)

It is possible to use Eq. (6.40) to develop a series of coefficients that will decouple any of the configurations containing V. If we set $\bar{\lambda}_{xV}$ to 1.0, then y will remain constant when V is changed because $\bar{\lambda}_{ym}$ is also 1.0; the decoupling coefficient required to do this is $(\partial m/\partial V)y$ where m is the top manipulated variable. Recalling that dD/dV is $1 - dL/dV$, we can make this substitution in Eq. (6.40) and set $\bar{\lambda}_{xV}$ to 1.0. Then

$$1.0 = \Lambda_{DV}\frac{\partial L}{\partial V}\bigg|_y + \Lambda_{LV}\left(1 - \frac{\partial L}{\partial V}\bigg|_y\right)$$

$$J_{LV} = \frac{\partial L}{\partial V}\bigg|_y = \frac{\Lambda_{LV} - 1}{\Lambda_{LV} - \Lambda_{DV}} \qquad (6.42)$$

This is the coefficient required to manipulate reflux as a function of boilup for the LV configuration. Because Λ_{LV} is usually a relatively high number and Λ_{DV} is ordinarily less than 1.0, J_{LV} tends to approach 1.0. If it exactly meets the process needs, then changes in boilup will not affect the quality of the top product. However, the bottom-composition loop is in effect manipulating both boilup and reflux with nearly the same weight. As Λ_{LV} increases, decoupling becomes more desirable yet also more difficult; because J_{LV} approaches unity, the process is less tolerant of decoupler error.

For the DV subset, (6.40) is solved by substituting for dL/dV:

$$1.0 = \Lambda_{DV}\left(1 - \left.\frac{\partial D}{\partial V}\right|_{y}\right) + \Lambda_{LV}\left.\frac{\partial D}{\partial V}\right|_{y}$$

$$J_{DV} = \left.\frac{\partial D}{\partial V}\right|_{y} = \frac{1 - \Lambda_{DV}}{\Lambda_{LV} - \Lambda_{DV}} \tag{6.43}$$

As (6.43) indicates, J_{DV} is a small number so a decoupler from boilup to distillate can be implemented with minimum difficulty and risk.

To develop the decoupling coefficient for the SV configuration, consider how the D/V ratio will change with V when y is controlled:

$$\left.\frac{\partial(D/V)}{\partial V}\right|_{y} = \frac{(D + dD)/(V + dV) - D/V}{dV}$$

$$= \frac{(D + J_{DV}\,dV)/(V + dV) - D/V}{dV}$$

Then by algebraic manipulation,

$$\left.\frac{\partial(D/V)}{\partial V}\right|_{y} = \frac{J_{DV} - D/V}{V + dV}$$

The decoupling coefficient for D/V as dV approaches zero is

$$\left.\frac{\partial(D/V)}{\partial(V/F)}\right|_{y} = \frac{J_{DV} - D/V}{V/F} \tag{6.44}$$

This coefficient must be negative: raising V to improve separation must lower D/V.

Unilateral decoupling may be desirable even when the relative gain is 1.0 because the RGA does not reveal the presence of partial coupling. Consider, for example, column 1 in Table 5.2. Although the relative gain for the SV/B configuration is 1.13, the principal contribution is from the V/B slope at 8.6 rather than the S slope at 1.0. A change in S will affect y 8.6 times as much as x, yet a change in V/B will affect both equally. It would not be necessary to use a decoupler from D/V to adjust V/B; however, a decoupler from V/B to D/V might be desirable.

The decoupling coefficient for V/B to D/V may be derived in a manner similar to that above:

$$\left.\frac{\partial(D/V)}{\partial(V/B)}\right|_{y} = \frac{(D + dD)/(V + dV) - D/V}{(V + dV)/(B + dB) - V/B} = \frac{(D + J_{DV}\,dV)/(V + dV) - D/V}{(V + dV)/(B - J_{DV}\,dV) - V/B}$$

Again, by algebraic manipulation,

$$\left.\frac{\partial(D/V)}{\partial(V/B]}\right|_{y} = \frac{J_{DV} - D/V}{(V/B)(1 - J_{DV}V/B)} \tag{6.45}$$

This coefficient is also negative.

example 6.1

Evaluate the above four decoupling coefficients for the column described by the following relative gains: $\Lambda_{DV} = 0.62$, $\Lambda_{SV} = 1.91$, and $\Lambda_{LV} = 7.49$; $D/F = 0.495$ and $L/D = 4.35$.

$$J_{LV} = \frac{7.49 - 1}{7.49 - 0.62} = 0.945$$

$$J_{DV} = \frac{1 - 0.62}{7.49 - 0.62} = 0.055$$

$$\frac{D}{V} = \frac{1}{1 + L/D} = \frac{1}{5.35} = 0.187$$

$$\frac{V}{F} = \frac{D/F}{D/V} = \frac{0.495}{0.187} = 2.644$$

$$\left.\frac{\partial(D/V)}{\partial(V/F)}\right|_y = \frac{0.055 - 0.187}{2.644} = -0.050$$

$$\frac{V}{B} = \frac{D/B}{D/V} = \frac{0.495/0.505}{0.187} = 5.24$$

$$\left.\frac{\partial(D/V)}{\partial(V/B)}\right|_y = \frac{0.055 - 0.187}{5.24[1 - 0.055(5.24)]} = -0.0354$$

Figure 6.22 shows how a unilateral decoupler would be applied to an SV/B system. Note that the LC output, taken as V, and the actual flow of B are used to calculate the V/B ratio. This information is more accurate and timely than is the output of the bottom-composition controller.

When a linear decoupler is applied to a process that is fundamentally nonlinear, an ideal match can be expected only at a single operating point. Coefficients should change with operating conditions. Figure 6.23 describes a nonlinear unilateral decoupler designed to prevent changes in heat input from upsetting top composition; feedforward from feed rate is also included. The steady-state calculation is

$$D^* = F\left(m_y - \frac{JF}{Q}\right) \tag{6.46}$$

When multiplied through, the parabolic feedforward model of Eq. (6.29) appears:

$$D^* = m_y F - \frac{J}{Q}F^2$$

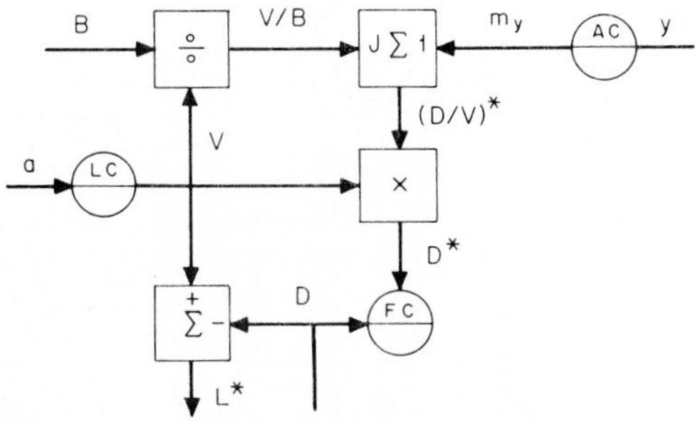

figure 6.22 *To decouple the SV/B system, the output of the accumulator LC (or column PC) is divided by measured bottoms flow; the SV system would be decoupled in the same way but without division by B.*

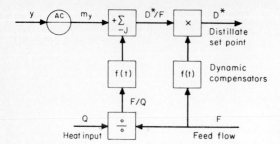

figure 6.23 *This nonlinear decoupler uses the parabolic feedforward model.*

The coefficient of F^2 varies inversely with heat input, which is the kind of relationship required by Fig. 6.12.

To evaluate decoupling coefficient J, consider that

$$J_{DV} = \frac{\partial D}{\partial V}\Big|_y = \frac{\partial (D/F)}{\partial (V/F)}\Big|_y$$

If V is substituted for Q in (6.46), D/F can be differentiated with respect to V/F:

$$\frac{\partial (D/F)}{\partial (V/F)}\Big|_y = \frac{J}{(V/F)^2}$$

Then a solution can be obtained for J:

$$J = J_{DV}(V/F)^2 \tag{6.47}$$

(Note that if a ratio appears in the numerator of a decoupling coefficient, one must also appear in the denominator, as in (6.44).)

The fifth decoupler likely to be used applies to the SD configuration, relevant when Λ_{SD} is the most favorable of the relative gains. Its coefficient can be developed from partial derivatives:

$$\frac{\partial (D/F)}{\partial S}\Big|_x = \frac{\partial (D/F)}{\partial y}\Big|_x \Big/ \frac{\partial S}{\partial y}\Big|_x$$

From the material-balance equation

$$\frac{\partial (D/F)}{\partial y)}\Big|_x = -\frac{z-x}{(y-x)^2}$$

And from the separation equation

$$\frac{\partial S}{\partial y}\Big|_x = \frac{S}{y(1-y)}$$

The derivation is not complete, however, in that S is neither measured nor manipulated but is accessible only through a ratio such as L/D. Using the same method followed in deriving Eq. (6.11),

$$\frac{\partial (D/F)}{\partial (L/D)}\Big|_x = -\frac{(z-x)}{(y-x)^2}\frac{y(1-y)}{S}\frac{nES}{2(1+zL/D)(L/D)}$$

Most of these terms combine into ϵ, defined in Eq. (5.64), leaving

$$\frac{\partial(D/F)}{\partial(L/D)}\bigg|_x = -\left(\frac{z-x}{y-x}\right)\frac{\epsilon}{L/D} \tag{6.48}$$

example 6.2

For the column in example 6.1, $x = 0.04$, $y = 0.95$, and $z = 0.50$; $\epsilon = 0.266$ and $L/D = 4.35$; $\Lambda_{SD} = 0.558$. Then

$$\frac{\partial(D/F)}{\partial(L/D)}\bigg|_x = -\frac{0.46\ (0.266)}{0.91\ (4.35)} = -0.0309$$

Decoupling coefficients should be adjusted in the field for best results. This is done with both controllers in manual. The disturbing variable is stepped and the decoupling coefficient adjusted to return the protected controlled variable to its original position. If a transient deviation develops, dynamic compensation is needed. It is applied in exactly the same manner as in a feedforward system.

Bilateral Decoupling Decouplers come in pairs but both need not be used. The unilateral decouplers described above were intended to protect one controlled variable from upset: it was y in Eqs. (6.42) to (6.46) and x in (6.47). Unilateral decoupling is sufficient to change the relative gains to 1 and 0 because cancelation of any interacting block will do that. Consequently, bilateral decoupling as shown in Fig. 6.19 is not needed unless *both* controlled variables must have the same degree of protection.

Equation (5.33) stated the relative gain in terms of a pair of decoupling coefficients. Therefore, having evaluated one coefficient, the opposite one may be determined from the relative gain. For example,

$$\Lambda_{LV} = \left(1 - \frac{\partial L}{\partial V}\bigg|_y \frac{\partial V}{\partial L}\bigg|_x\right)^{-1} \tag{6.49}$$

Then

$$J_{VL} = \frac{\partial V}{\partial L}\bigg|_x = \frac{1 - 1/\Lambda_{LV}}{J_{LV}} \tag{6.50}$$

$$J_{VD} = \frac{\partial V}{\partial D}\bigg|_x = \frac{1 - 1/\Lambda_{DV}}{J_{DV}} \tag{6.51}$$

$$\frac{\partial(V/F)}{\partial(D/V)}\bigg|_x = (1 - 1/\Lambda_{SV})\bigg/\frac{\partial(D/V)}{\partial(V/F)}\bigg|_y \tag{6.52}$$

$$\frac{\partial(L/D)}{\partial(D/F)}\bigg|_y = (1 - 1/\Lambda_{SD})\bigg/\frac{\partial(D/F)}{\partial(L/D)}\bigg|_x \tag{6.53}$$

Because the V/B curve in Fig. 5.9 has the highest slope of all, a bilateral decoupler for the SV/B system is superfluous.

example 6.3

Calculate the second set of decoupler coefficients for the column in Examples 6.1 and 6.2 and compare them to the first set:

$$J_{VL} = \frac{1 - 1/7.49}{0.945} = 0.917$$

$$J_{VD} = \frac{1 - 1/0.62}{0.055} = -11.1$$

$$\left.\frac{\partial(V/F)}{\partial(D/V)}\right|_x = \frac{1 - 1/1.91}{-0.050} = -9.52$$

$$\left.\frac{\partial(L/D)}{\partial(D/F)}\right|_y = \frac{1 - 1/0.558}{-0.0309} = 25.6$$

Note that the presence of a small coefficient for the first decoupler brings about a much higher number for the second. Decoupling with gains exceeding unity is not practical because the loop gain through the decoupler becomes higher than through the designated loop. Jafarey and McAvoy [10] referred to this property as "degeneracy" and questioned the feasibility of bilateral decoupling in its presence. To the author's knowledge, it has not been done.

Bilateral decoupling of LV systems has been reported, principally on simulated columns. Luyben [11] demonstrated it for two cases: the first required decoupling coefficients of 0.9547 and 0.9488, having $\Lambda_{LV} = 10.62$; the second used coefficients of 0.8518 and 0.8180, having $\Lambda_{LV} = 3.30$. He used the structure at the top of Fig. 6.19. Luyben and Vivante [12] tried it on a laboratory column, using coefficients of 0.568 and 0.635 where Λ_{LV} was only 1.56; results were not outstanding.

Toijala (Waller) and Fagervik [13] used the same structure to decouple a simulated column where ideal coefficients were 0.947 and 0.854; Λ_{LV} was 5.0. They tested the influence of decoupling by gradually increasing the coefficients from 50 percent of their ideal values to 110 percent. The best results without dynamic compensation were achieved between 70 and 90 percent. At 110 percent, the period of oscillation of the two temperature loops increased to about 11 times their natural period. This effect can be explained by inverse response. When the larger coefficient was set at 110 percent of its ideal value, it became 1.042. Thus, more weight was placed on the configuration with the relative gain of -4 than that of $+5$, producing inverse response as shown in Fig. 5.5 and thereby extending the period.

In a subsequent study [14], the author derived the following relative gain expression for a *decoupled* 2×2 process:

$$\overline{\lambda}_{11d} = \left[1 - \frac{(K_{21} + J_{21}K_{22})(K_{12} + J_{12}K_{11})}{(K_{22} + J_{12}K_{21})(K_{11} + J_{21}K_{12})}\right]^{-1} \tag{6.54}$$

where the K's are as they appear in Fig. 5.1 and the J's are the decoupling coefficients. Their ideal values are

$$J_{21} = -K_{21}/K_{22} \tag{6.55}$$

$$J_{12} = -K_{12}/K_{11} \tag{6.56}$$

If *either* is set at its ideal value, $\overline{\lambda}_{11d} = 1.0$. However, certain incorrect values can destroy the response of the system. For example, if $J_{21}J_{12} = 1.0$, the ratio in (6.54) becomes unity, causing the decoupled relative gain to approach infinity. When the coefficients in Ref. 13 were set to 110 percent of their ideal values, their product became 0.979.

If the relative gain of the configuration to be decoupled exceeds unity, the decoupling coefficients will have the same sign. However, if it falls between 0 and 1, they will have opposite signs and their product therefore cannot be 1.0; inferior relative gains will also be inferior when decoupled.

REFERENCES

1. Ryskamp, C. J.: "New Stratgey Improves Dual Composition Column Control," *Hydrocarbon Proc.*, June 1980.
2. Rademaker, O., J. E. Rijnsdorp, and A. Maarleveld: *Dynamics and Control of Continuous Distillation Units,* Elsevier, Amsterdam, 1975, p. 371.
3. Luyben, W. L.: "Feedback Control of Distillation Columns by Double Differential Temperature Control," *Ind. Eng. Chem. Fundam.,* November 1969.
4. Boyd, D. M.: "Fractionation Column Control," *Chem. Eng. Prog.,* June 1975.
5. *Differential Vapor Pressure Cell Transmitter, Model 13VA,* Technical Information Sheet 37-91a, The Foxboro Company, Foxboro, Mass., April 1965.
6. Shinskey, F. G.: "Controlling Distillation Processes for Fuel-Grade Alcohol," *In Tech,* December 1981.
7. Ryskamp, C. J.: "Using the Probability Axis for Plotting Composition Profiles," *Chem. Eng. Prog.,* September 1981.
8. Tolliver, T. L., and L. C. McCune: "Finding the Optimum Temperature Control Trays for Distillation Columns," *In Tech,* September 1980.
9. Shinskey, F. G.: *Process Control Systems,* 2d ed., McGraw-Hill, New York, 1979.
10. Jafarey, A., and T. J. McAvoy: "Degeneracy of Decoupling in Distillation Columns," *Ind. Eng. Chem. Proc. Des. and Dev.,* Vol. 17, 1978, pp. 485-490.
11. Luyben, W. L.: "Distillation Decoupling," *AICHE J.,* March 1970.
12. Luyben, W. L., and C. D. Vivante: "Experimental Studies of Distillation Decoupling,"*Kem. Teollisuus,* August 1972.
13. Toijala, K., and K. Fagervik: "A Digital Simulation Study of Two-Point Feedback Control of Distillation Columns," *Kem. Teollisuus,* January 1972.
14. Shinskey, F. G.: "The Stability of Interacting Control Loops with and without Decoupling," *Proc., IFAC Multivariable Tech. Syst. Conf.,* Univ. of New Brunswick, Canada, July 4-8, 1977.

CHAPTER SEVEN

Controlling within Constraints

Constrained operation is common to most separation units, as it should be. For if at least one variable in a column is not being held at its limit, then separation efficiency or productivity could still be improved. Maximum recovery, for example, can only be achieved at maximum heat input—whether the limit is imposed by reboiler, column, or condenser. Many plant managers are attempting to maximize production by whatever means is available. In this case, constraints are being encountered on all sides. The purpose of this chapter is to locate these constraints, provide protection against their violation, and permit operation to proceed in their presence.

The control systems described thus far for heat input, heat output, and product quality have assumed a freedom to manipulate flow rates as necessary to achieve their objectives. Yet limits exist on all flow rates. At the very extreme, control valves can only range between closed and fully open. In many cases, narrower limits must be imposed to avoid inefficiencies such as flooding and weeping and to protect equipment from overpressure, overtemperature, overflow, etc.

Whenever a limit of any kind is placed on a manipulated variable, the normal function of the control loop is inhibited and control is lost. If loss of quality control is intolerable, then the production rate must be decreased until control is regained. In other cases, it may be possible to restructure the control system by using a different variable to maintain control. In any case, the alternative modes of operation need to be investigated in order to determine the penalties for violating constraints and the remedial action required to keep the plant running.

Finally, the exact location of those constraints should be known at all times to warn the operator of impending crises. With sufficient information about the limits of his plant and its expected disturbances, he should be able to schedule production to make the best use of the resources available to him.

CONSTRAINTS ENCOUNTERED

A constraint is a bottleneck which limits the production capacity of a unit. Because distillation units comprise many elements, there are just as many possibilities for constraints to be encountered.

A constrained situation usually appears as a loss of control. Column pressure rises out of control when the heat load exceeds condenser capacity; steam flow falls below set point when the condensate system is overloaded. In these instances, loss of control is accompanied by a control valve that is fully open — a sure indication of operation against a constraint. If the valve or pump is in fact the limiting element, then it should be changed. But in a well-designed plant, the fully open valve indicates constrained major equipment which is not likely to be changed.

Nonetheless, it is often possible to remove the constraint by remedial action such as venting noncondensibles from a condenser or starting another fan or pump. However, it can be desirable to operate at the constraint, as is done in floating-pressure control. At other times, it becomes necessary to reduce production. But in order to determine the best course of action, a familiarity with the most common constraints is necessary.

Tray Characteristics   The trays in the column have limitations which are more restrictive than those of the reboiler and condenser. Column performance may be measured as overall tray efficiency, which is the ratio of the number of theoretical equilibrium stages observed to the number of trays in the column. For a given tray configuration, efficiency will be maximized only over a specified range of liquid and vapor flow rates. The maximum efficiency and the range over which it extends vary with the type of tray. Perforated trays tend to exhibit efficiency that increases with superficial vapor velocity [1] up to a point and then decreases sharply. Bubble-cap and valve trays have a broader range of high efficiency, but they also suffer a loss at very low or high rates. Causes of these losses in efficiency are examined below in an effort to define the limits of efficient operation.

Liquid on a perforated plate is supported by the flow of vapor through the holes. As vapor flow is reduced, turbulence at each orifice decreases, which hinders vapor-liquid contact and lowers efficiency. Eventually some holes begin to weep liquid, which then substantially escapes vapor contact.

Valve trays are less sensitive in this regard since the area exposed to vapor flow is decreased by the valves as flow falls off. Bubble caps do not tend to weep since liquid is retained by the riser inside each cap. However, combinations of high liquid and low vapor rates can submerge some caps, causing them to dump liquid, while others carry all the vapor load [2]. Since this dumping essentially releases the liquid head, normal operation may again resume until the head returns to its previous level. Consequently tray operation becomes cyclic and is classified as unstable.

To avoid both weeping and instability, low limits must be placed on the liquid and vapor rates. However, the limits are a function of vapor velocity. Consequently, lower-pressure operation, in which vapor density is reduced, can tolerate a lower molar or mass rate of boilup. Reference 2 indicates that minimum vapor rate varies with the square root of vapor density, as might be expected in this turbulent-flow regime.

Figure 7.1 shows a contour of molar or mass boilup of isobutane as a function of column pressure, illustrating the form of a low-boilup constraint. It follows the relationship

$$V = V_M \sqrt{\frac{\rho_{VM}}{\rho_V}} \qquad (7.1)$$

where V and ρ_V are the mass or molar boilup limit and vapor density at any operating pressure, with subscript M referring to maximum or design pressure.

The term *flooding* has been used to describe various conditions which cause a loss of tray efficiency at high vapor rates. In any given column, the actual mechanism could be entrainment, foaming, or downcomer flooding. All are promoted by poor disengagement of liquid and vapor. In downcomer flooding, vapor velocity and hence pressure drop are so high or liquid density so low that liquid is prevented from flowing down the column.

Entrainment and foaming are both sensitive to the difference between the densities of vapor and liquid and to the surface tension of the liquid. Consequently both are related to the critical point of the components since disengagement deteriorates as this point is approached.

Entrainment obviously increases with the velocity of the vapor, which tends to carry liquid droplets from one tray to the next. Tray efficiency is thereby decreased since higher-boiling materials are physically lifted up the column. Liquid loading is also increased because all the entrained liquid must be returned through the

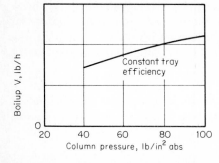

figure 7.1 *The low limit of boilup varies with vapor density and therefore with pressure.*

downcomers. In the extreme, entrainment can be increased to the point where all the liquid cannot be returned, resulting in downcomer flooding.

Whereas the presence of entrainment is not specifically detectable as an inordinate rise in column differential pressure, it can be prevented or limited by controlling differential pressure below the point where entrainment is known to be a problem.

Failure of vapor to become disengaged from the liquid phase causes *foam* to accumulate on the trays and in the downcomers. Its effect is similar to entrainment when the level of foam reaches the tray above. By decreasing liquid density within the downcomers, foam raises liquid levels for a given pressure drop and may lead to downcomer flooding.

When a downcomer is completely full of liquid, a further increase in liquid level produces no increase in downflow because levels in downcomer and on the tray *both* rise. Rising tray level increases the differential pressure encountered by the vapor flow, which forces liquid to back further up the downcomer. From this point onward, the trays continue to accumulate liquid, indicated by both a rising column differential pressure and a falling base level.

The differential-pressure record takes on a different appearance in passing from the normal to the flooded regime as shown in Fig. 7.2. In normal operation, the trays act much like an orifice flowmeter, and the differential-pressure record has a noise level of 1 to 2 percent amplitude. When *downcomer flooding* begins, the noise abates and the differential pressure begins to ramp upward, as it would in a tank being filled. It becomes, in effect, a liquid-level measurement.

The fastest way to stop flooding is to reduce boilup. This removes the force holding up the liquid and allows it to resume downflow. At first, differential pressure will show scarcely any change but then will ramp downward as shown in Fig. 7.2, indicating that the trays are being emptied. If flooding is to be prevented in the future, the differential pressure must not be allowed to rise to that value where the upward ramp began.

Predicting Flooding Limits   Reference 3 presents a correlation by J.R. Fair for vapor velocity at flooding vs. liquid-vapor flow and density ratios for columns with various tray spacings. A correction factor for surface tension below 20 dyn/cm is included. The same correlation is also applied to foaming limits and to predict fractional entrainment.

To demonstrate how flooding limits change with pressure, the cited correlation was evaluated for selected hydrocarbons at temperatures from 80 to 120°F. The data should be representative for columns yielding relatively pure distillate products and rejecting heat into air-cooled condensers. For each tray spacing, Fair's correlation yields

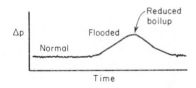

figure 7.2 *The onset of flooding is characterized by a rising differential pressure and a disappearance of noise from the record.*

$$u_{VN}\sqrt{\frac{\rho_V}{\rho_L - \rho_V}} = f\left(\frac{L}{V}\sqrt{\frac{\rho_V}{\rho_L}}\right) \tag{7.2}$$

where u_{VN} = velocity of vapor based on net area, ft/s
        L, V = mass-flow rates of liquid and vapor, lb/h
        ρ_V, ρ_L = vapor and liquid densities, lb/ft^3

The velocity thus found may be converted to mass boilup V by multiplying by vapor density and net area A:

$$V = Au_{VN}\rho_V \tag{7.3}$$

Then the correction factor $(\sigma/20)^{0.2}$ is required for surface tension σ below 20 dyn/cm. Combining all these relationships yields the boilup limit as a function of vapor and liquid densities and surface tension:

$$V = A\left(\frac{\sigma}{20}\right)^{0.2}\sqrt{\rho_V(\rho_L - \rho_V)}f\left(\frac{L}{V}\sqrt{\frac{\rho_V}{\rho_L}}\right) \tag{7.4}$$

Table 7.1 presents values of V/A calculated using ρ_V and ρ_L for the cited hydrocarbons, with 18-in tray spacing and assuming that $L/V = 1.0$. Densities and vapor pressures were taken from Ref. 4; surface tension was calculated using the parachor in Ref. 5.

Note that the allowable boilup for both propylene and propane overhead products passes through a maximum as a function of column temperature. Isobutane and higher-boiling products exhibit a boilup limit increasing with temperature whereas ethane and lower-boiling products have a decreasing limit, as confirmed by Ref. 6.

Three factors contribute to this variation. Increasing temperature (pressure) increases vapor density, thereby reducing velocity for a given mass-flow rate. However, increasing temperature also decreases the difference between liquid and vapor densities and lowers surface tension. The latter effects are more pronounced for the lighter hydrocarbons, since at ambient temperatures they are closer to their critical pressure. These effects more than offset that of decreasing vapor density for propane and propylene above 100°F. Data from the last row of Table 7.1 are plotted against temperature in Fig. 7.3.

Boilup is often measured and controlled as heat input to the reboiler; the two are related by the latent heat of vaporization of the product. Because latent heat increases with falling pressure and temperature, it adds another dimension

TABLE 7.1 Flooding Calculations for Selected Light Hydrocarbons from 80 to 120°F

	Propylene			Propane			Isobutane		
	80°F	100°F	120°F	80°F	100°F	120°F	80°F	100°F	120°F
$p°$, lb/in^2 abs*	174.7	227.6	291.2	143.6	188.7	243.4	53	73	96
ρ_V, lb/ft^3	1.603	2.119	2.778	1.342	1.792	2.347	0.59	0.81	1.063
ρ_L, lb/ft^3	31.52	30.30	28.90	30.59	29.50	28.31	34.2	33.2	32.3
σ, dyn/cm	6.95	5.47	4.04	6.68	5.38	5.15	9.11	7.84	6.79
V/A, lb/h-ft^2	1.160	1.175	1.163	1.067	1.095	1.094	0.905	0.964	1.017

*$p°$ is the vapor pressure of the component at the indicated temperature.

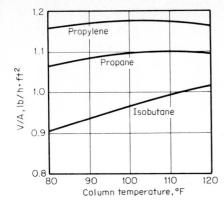

figure 7.3 *Calculated mass vapor flow limits at flooding for selected light hydrocarbons with 18-in tray spacing and L/V = 1.0.*

to the curves representing flooding limits. Values taken for the boilup limits of propylene and isobutane in Fig. 7.3 were multiplied by their latent heats to give heat-input limits and plotted in Fig. 7.4. The horizontal scale was changed to pressure to illustrate the need for protection against flooding during floating-pressure operation.

If the pressure in a column separating isobutane overhead at 140°F were reduced as much as 60 percent under conditions of constant *heat input*, flooding would not be encountered. Normal butane and heavier components, however, will show a positive slope to the curve, as does isopentane in Fig. 7.4. Still, this does not mean that production must be curtailed at lower pressure for n-butane and heavier hydrocarbons. Curves of energy-to-feed ratio against column pressure as described in Fig. 4.23 typically have a slope exceeding that of the limit of heat input at flooding. As a result, it remains possible to increase feed rate with falling pressure while remaining within flooding limits, as described for a deisopentanizer in Ref. 7. This relationship appears again in the discussion of the "Operating Window," below.

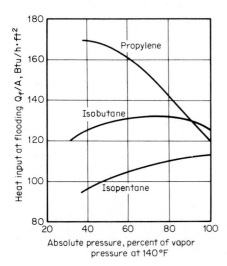

figure 7.4 *Inclusion of the latent heat of vaporization causes the heat input limits to show more variation with pressure than the boilup limits.*

Reboiler Constraints   Heat transfer in the reboiler is almost always governed by free convection, although in special cases forced convection may be applied. Heat is transferred either from condensing steam or a hot liquid to a boiling bottom product. In this chapter, we are concerned only with the limit to heat flow, i.e., its maximum value, which is assumed to be transferred at the maximum flow or condensing rate of the hot fluid. At this upper limit, the hot side of the reboiler can be considered isothermal; i.e., the hot fluid enters and leaves at essentially the same temperature. This assumption greatly facilitates defining the operating limits, since the rate of heat transfer then is primarily determined by the difference in temperature between the heat source and the boiling liquid. (In avoiding integration from inlet to outlet of the hot side of the reboiler, limits are easier to establish without sacrificing much in the way of accuracy.)

The thermal driving force, at the limit, is therefore the temperature difference between the hot fluid and boiling product. If the heating medium is steam, its maximum temperature is the saturation temperature at the available supply pressure. The temperature of the boiling product is a function of both composition and applied pressure. Knowing that no heat will flow when both fluids are at the same temperature allows one point on the boilup-vs.-temperature or boilup-vs.-pressure plot to be marked. The zero boilup point is indicated on Fig. 7.5 as the vapor pressure of the bottom product (in this case n-butane) at the saturation temperature of the steam or inlet temperature of the heating liquid (in this case 212°F).

Figure 7.5 is actually a vapor-pressure-vs.-temperature plot for n-butane. As column pressure is reduced, the boiling point of the bottom product decreases, developing a temperature difference ΔT across the heat-transfer surface. If heat-transfer rate increases linearly with ΔT, the scale of ΔT in Fig. 7.5 could be replaced with an equivalent scale of heat flow.

Changing the heat-source temperature would move the boilup-vs.-pressure curve left or right because heat-transfer rate is primarily influenced by ΔT. The slope of a boilup-vs.-pressure curve would be a function of the area and condition of the heat-transfer surface. With a constant area, the slope should change only as the surface fouls, reducing the boilup attainable at a given pressure.

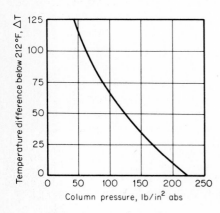

figure 7.5   *The maximum boilup of n-butane using steam at atmospheric pressure is proportional to the above temperature difference.*

For steam-heated reboilers, the control valve may be placed on the steam inlet or the condensate outlet. When manipulating steam into the reboiler, a trap must be provided for the condensate. Maximum heating will be indicated when the control valve is fully open. However, if the condensate trap cannot carry this full load, the steam chest will start to flood, causing the heat-transfer rate to fall because of the reduction in surface available for condensing. In this case, the transient heat-transfer rate is limited by the capacity of the control valve and steam piping, but the steady-state rate is limited by the trap and condensate piping. The limit imposed by condensate-removal capacity is fixed; it depends not on ΔT but on the difference between steam and condensate pressures. It is then a maximum boilup limit independent of column pressure.

With the control valve acting on the condensate, heat transfer is varied by flooding the surface area with condensate. Therefore maximum heat flow will be realized when the steam chest is drained, at which point steam will begin blowing through the valve. This condition will appear when the valve can carry more condensate than the reboiler can condense, and it is more likely to occur in fouled reboilers. If the capacity of the valve and piping is less than the reboiler can condense, some of the tube surface will always be covered. Then the condensate-flow limit described above applies. The operator must be alert to the possibilities of both steam blowing and condensate flooding in these systems.

Condenser Constraints   Condensers have another dimension added to their heat-transfer limit: coolant-supply temperature is usually variable. Air cooling is most variable; refrigerant, evaporative (cooling tower), and river-water cooling are increasingly more constant and reliable. The cooling fluids themselves are usually not manipulated to control heat-transfer rate, except in the case of refrigerants. Thus the temperature rise of the coolant in passing across the heat-transfer surface is usually minimum.

As the heat load approaches zero, the condensing temperature will be the coolant-supply temperature. At this point, the pressure in the condenser will be the vapor pressure of the distillate product at that temperature. To increase the rate of condensation, the condensing temperature must increase. The limit on condensing rate is then determined by the temperature difference between the coolant and the dew point of the distillate product. For a given product composition, the limit on boilup rate that can be accommodated by the condenser is a function of column pressure. Figure 7.6 gives the ΔT across the condenser surface for isobutane condensed against a source of 63°F. Similar to Fig. 7.5, it is essentially a vapor-pressure plot of isobutane.

The greatest influence on the condenser limit is coolant-supply temperature. Ambient temperatures that vary from below freezing to as high as 100°F in temperate climates place severe demands on condensing systems. Wintertime operation requires protection from freezing and shutting down of some of the condensing capacity. But even day-to-night and rain-to-sun variations can impose significant disturbances on a column. Air-cooled condensers which normally reject heat to ambient dry-bulb temperature may operate closer to wet-bulb temperature in the rain. Thus the onset of a sudden rain may cause a pronounced

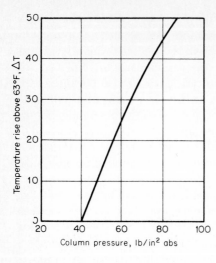

figure 7.6 *The maximum condensing rate for isobutane using a
63°F cooling source is proportional to the above temperature
difference.*

increase in condensing rate, forcing column pressure downward. The condenser-
constraint curve is also shifted by changes in distillate composition. Very small
amounts of noncondensible gases such as methane or nitrogen can severely limit
column operation.

In a floating-pressure system, column pressure will ride the condenser constraint. Increasing boilup rate or coolant temperature will cause pressure to rise to
a new equilibrium level, but it will always be minimum for the existing conditions.
In addition to energy savings, other advantages include reduced reboiler fouling
and increased reboiler capacity. The capacity improvement is obvious, but the reduction in fouling may require some evaluation. A rule of thumb is often quoted to
the effect that reaction rates generally double with every 10°C increase in temperature. If reboiler fouling rate obeys this rule, then an 18°F (10°C) reduction in
reboiler temperature should cut it in half. Reboilers normally operated at summer-
time temperatures year-round ought to require cleaning half as often with floating-
pressure operation.

The Operating Window   Another constraint to be considered is the maximum
allowable operating pressure. Vessels must be protected from overpressure by relief
devices set safely below test limits. Operating pressures must not be allowed to
approach these relief settings very closely or some leakage could result. Consequently each column will have some upper limit of pressure without regard to
boilup rate. This limit will appear as a vertical line on the right side of the boilup-
vs.-pressure diagram of Fig. 7.7.

There may be additional limitations on pressure to ensure transfer of feed to the
tower or products from it. These limits could vary somewhat with flow but not
appreciably. Flow from the reflux pump is also limited, but by suction and discharge heads and vapor pressure. If column pressure floats on the condenser, all
three of these variables should change together. Therefore there should be no net
variation of reflux pumping limit with pressure, except in the case where a significant pressure drop exists between the column and the reflux accumulator.

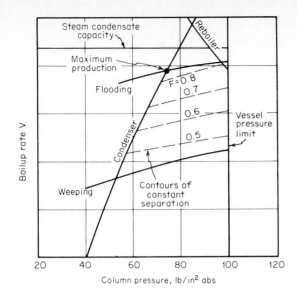

Boilup rate V

Column pressure, lb/in² abs

Steam condensate capacity

Reboiler

Maximum production

F=0.8

Flooding

0.7

0.6

Vessel pressure limit

0.5

Condenser

Weeping

Contours of constant separation

20   40   60   80   100   120

figure 7.7 *This is a typical opera-ting window for a butane splitter, illustrating that production is maximized at the condenser constraint.*

Plotting all the significant constraints on the same figure forms an "operating window." Boilup must always be controlled within this window or along any con-straint if normal operation is to be expected. Figure 7.7 illustrates a typical window for a column separating normal and isobutane.

No vertical scale is given on the figure because actual boilup limits depend on tray and column design, etc. But the relative position of the curves and their shapes are representative. Note that the positions of the condenser and reboiler constraints are variable, the former with coolant-temperature changes, the latter due primarily to fouling. Under changing conditions, maximum boilup could be limited by the condenser, by the reboiler, or by flooding. Whenever the reboiler is not limiting, maximum boilup for the butane splitter will be coincident with maximum pressure. But this is not true for propane or propylene products whose flooding constraints pass through a maximum. For these and lighter products, maximum boilup will be achieved where the flooding curve meets either the condenser or reboiler con-straint or at the maximum point of the flooding curve.

Maximum boilup is not necessarily consistent with maximum *production*, how-ever. Separation is favored at lower pressure because of increased relative volatility, yet Ref. 8 indicates a slight drop in tray efficiency at lower pressure. Nonetheless, as pressure is reduced, less boilup is required to separate a given feedstock.

Within the operating window are contours of constant separation (i.e., constant product qualities), one for each feed rate. If these contours lie parallel to the flood-ing curve, maximum production and most favorable economy will be achieved where the flooding curve crosses the condenser constraint. Only if the slope of the flooding curve is more positive than the contours of constant separation is maxi-mum production to be achieved at maximum pressure. This is likely to be the case only for distillate products heavier than isopentane.

By operating the column represented by Fig. 7.7 at the condenser constraint, a significant reduction in boilup may be realized. Reduction in boilup should have no direct effect on tray efficiency because superficial vapor velocity actually increases somewhat as pressure is reduced along a separation contour. This is borne out by the divergence of separation contours and the weeping line as pressure is reduced. Recall that the weeping line is essentially a contour of constant superficial velocity and therefore virtually constant tray efficiency.

SELECTIVE CONTROL SYSTEMS

Selective control systems are used to impose limits on valve positions or manipulated flows. They can protect equipment or products from unsafe or abnormal situations and maintain control rather than shut down a unit. The principal element used is the signal selector — its function is to select the highest, the lowest, or in certain cases the median of a plurality of signals.

In some systems a controller output will be compared against a fixed signal in a selector — in which case that signal acts as a limit on the controller output. Sometimes multiple signals of the same sort are compared, as when controlling the highest level or lowest pressure in a unit. But most often, controller outputs are compared in an effort to keep all controlled variables on the safe side of certain constraints. Each of these systems is discussed below.

Self-imposed Limits   A few limits to column operation are self-imposed and therefore do not require any overt implementation. The most common of these is the heat-input valve limit. Steam flow to the reboiler could be set to control product quality, set in ratio to the feed, or simply set at a fixed set point. Should the valve be unable to maintain the set point, whatever the reason, the objective of holding that particular flow will be sacrificed.

If quality control were the objective, it would be lost and there is cause for alarm. The operator will have to take remedial action of some kind — either reduce the feed rate, supplement the steam supply, or prepare to rerun the product. In any case, some time may elapse before control is restored. During the time when control is lost, the quality controller — if left in automatic — will continue to index the set point of the steam-flow controller, although the latter is unable to respond. When the overload is corrected, the quality controller, in having been unable to achieve response, may be requesting much more steam than is required to meet specifications, and a large overshoot is likely. This windup of the primary controller in a cascade system can be avoided with the scheme shown in Fig. 7.8.

The primary controller is fitted with what is known as "external feedback"; i.e., its feedback signal is supplied from outside the controller. (Normally, this loop is closed by internal feedback of the output of the controller.) When the deviation between measurement and set point in the flow controller is zero, then integral action takes place in the primary controller as usual. But if a flow deviation persists for any reason, primary integral action stops; the primary controller then has proportional action only, whose bias is the flow measurement Q and whose output is its set point Q^*:

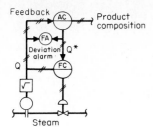

figure 7.8 *The secondary mea-surement can be used as feedback to the primary controller, thereby avoiding windup when control is lost.*

$$Q^* = \frac{100}{P} e + Q \tag{7.5}$$

Then the composition deviation becomes proportional to the flow deviation:

$$e = \frac{P}{100} (Q^* - Q) \tag{7.6}$$

If steam flow is set to control separation, failure to meet the set point could adversely affect the other composition loop as well. Consideration of this effect is taken up later in the chapter.

Another possibility would be the manipulation of steam flow to control column pressure. If pressure could not be maintained, any temperature control loop on the column would be upset. Otherwise, a low pressure would not ordinarily cause a hazard.

If steam flow were used to control base level, failure to reach set point would result in a rising level. Lacking self-regulation, the level would continue to rise, causing flooding of the column. Some additional protection may be required in the way of transferring level control to another variable in this eventuality. A similar consideration applies to the control of accumulator level.

Ordinarily, high and low limits should be placed on heat input as partial protection against flooding and weeping. The limits may be placed on valve position within the controller which manipulates heat input. They will narrow the allowable range of the valve. Loss of control then takes place when demand exceeds these limits, and any transfer of control to auxiliary devices must take place when the limits are reached.

When steam is the heating medium, limits on boilup may be more accurately placed on the steam-flow set point than on valve position. The windup problem of a composition controller setting steam flow will not exist then since the limits would be imposed on its output. But if the composition controller is used for feedback trim of a feedforward system, it is not directly connected to the steam-flow controller. Then to protect against windup, the deviation between the steam-flow measurement and its calculated set point should be used for external feedback. If this deviation is zero, normal integral action takes place; if it is not zero, the composition controller is satisfied by an offset proportional to the steam-flow deviation. The system is arranged as shown in Fig. 7.9.

Again, the primary controller becomes a proportional controller whose bias is replaced by the external feedback signal:

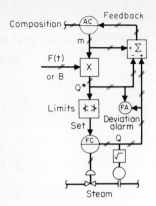

figure 7.9 *Windup of the primary controller in a feedforward system can be avoided by feeding back secondary deviation.*

$$m = \frac{100}{P} e + m - Q^* + Q \tag{7.7}$$

Equation (7.7) then reduces to (7.6).

An alternative configuration calculates the external feedback signal by inverting the feedforward calculation. Where

$$Q^* = mF(t) \tag{7.8}$$

and feedback signal f is

$$f = \frac{Q}{F(t)} \tag{7.9}$$

When $Q = Q^*$, $f = m$, allowing the primary controller to integrate normally. However in the constrained case

$$m - f = \frac{Q^* - Q}{F(t)} \tag{7.10}$$

so that

$$e = \frac{P}{100} \frac{Q^* - Q}{F(t)} \tag{7.11}$$

The difference between the two back-calculations is probably not significant because in the normal condition they produce the same result, and in the constrained condition the loop is open anyway. Figure 7.9 gives the more general solution to the problem, applicable regardless of the complexity of the function between m and Q^*. Observe that the same arrangement applies when heat input is manipulated in ratio to bottom-flow rate as in the SV/B configuration.

Override Controls   Limits on steam flow or other variables do not have to be fixed — they may be automatically adjusted to satisfy measurable constraints. Automatic selector systems are then used to override normal control action to keep constraints from being violated. An example of column differential pressure being used to override composition control of steam flow appears in Fig. 7.10.

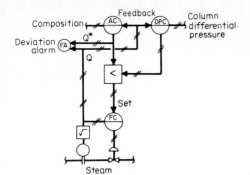

figure 7.10 *Override controls pro-tect against violating measurable constraints (here differential pressure) by automatic selection of the lower steam-flow set point.*

A low selector compares the steam flow required to control composition with that required to control column differential pressure. Normally, composition would be controlled because differential pressure would be less than its set point. The DP controller, in attempting to raise steam flow to satisfy its set point, would increase its output beyond that of the composition controller and would therefore be rejected. Should the differential pressure exceed its set point, however, the output of the DP controller would fall below that of the composition controller and would then be selected to set the steam flow.

Provision must be made to prevent the controller that is not selected from wind-up. This is most readily achieved by feedback of the selected output or steam flow to both controllers. The selected controller sees its own output and therefore has normal integral action. The other controller sees a foreign output and hence behaves as a proportional controller with steam flow as a bias per Eq. (7.5).

Detection of an override is necessary if loss of composition control will result. A deviation alarm on the composition controller could be used to alert the operator, but by the time it is activated the damage has already been done. Instead, a deviation alarm between the steam flow and the output of the composition controller will signal that an override has taken place — before it has a chance to affect composition.

A column-pressure controller may also be used to override heat input. However, it must be a different controller from the one used to manipulate the condenser. The dynamic responses of column pressure to reboiler and to condenser differ, and the controllers would also have different set points. The override controller must be set higher than the condenser controller so that it would function only when the other failed to hold pressure by manipulating the condenser.

When the control valve for a steam-heated reboiler is located in the condensate line, the possibility exists of blowing steam through the valve. Fouling of the heat-transfer surface could cause its maximum rate of condensation to fall below the capacity of the valve. In an effort to increase the steam flow, the valve would open further, but the additional steam might not be condensed. A small condensate tank fitted with a simple level controller as shown in Fig. 7.11 can protect against this condition. If the level controller has proportional action only, it needs no external feedback.

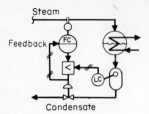

figure 7.11 *A level controller can keep the condensate valve from passing steam; if the controllers are pneumatic, the selector and feedback connection may be omitted, and the LC output used as air supply to the FC.*

A similar override on heat input may be exercised by a column-base level controller. In the event the level falls to the point of exposing reboiler tubes or preventing circulation, the level controller could limit the heat input just as shown in Fig. 7.11. This action would prevent the boilup rate from exceeding the rate at which liquid is entering the column base. Under normal conditions, base level would be controlled at a higher set point by manipulating the bottom-product valve. Two level controllers would then be necessary having different set points and also different mode settings, as was necessary for pressure override.

The feed preheater is also a source of boilup, although its location limits its vapor contribution to the top section of the column. Any override of preheat to prevent flooding should thus be restricted to the trays over which it has direct influence. Therefore, if flooding above the feed tray is a problem, it can be avoided by overriding preheat with differential pressure between the feed and the top of the tower. On the other hand, if a subcooled feed is causing excessive liquid flow below the feed tray and a low vapor velocity above it, preheat could be adjusted to balance the two differential pressures.

Overrides may also be exercised on reflux flow. If reflux is set to control composition, for example, an override from low accumulator level may be necessary to protect against inadequate boilup. This system is the mirror image of the base-level system just described.

Figure 4.13 describes a parallel fuel-air metering system with ratio adjustment from a flue-gas analyzer. Fuel flow is set to maintain the desired outlet temperature, and airflow is expected to follow. In the event that airflow cannot follow because of a limitation in its supply capability, a failure of the damper or fan, or an obstruction of some kind, excess fuel could accumulate. Aside from being inefficient and causing smoke, excess fuel also represents an explosion hazard. To protect against such a possibility, Manter and Tressler [9] developed a two-way selector system for coal-fired boilers. Shown in Fig. 7.12, it is equally applicable to all fuels.

The master controller sets both fuel and air through a pair of selectors. If airflow fails to reach its set point, it will be preferentially selected by the low selector to set fuel flow. Then any of the failures described above will automatically reduce fuel flow to avoid an unsafe condition. It will then be impossible to maintain product temperature, so an alarm must be sounded; furthermore, the master controller must be protected against windup by external feedback from airflow.

The high selector protects against an excess of fuel above that demanded by the master controller. An obstruction preventing the fuel valve from closing, or an operator error, might create such a situation. Airflow will be forced to follow fuel

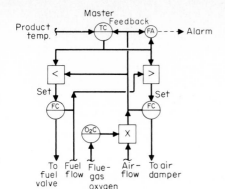

figure 7.12 *The selectors prevent airflow from falling below fuel flow.*

flow in this case. Again, an alarm is actuated and windup of the master controller is prevented by the system in Fig. 7.12.

The fuel-air ratio adjustment is made on the airflow measurement rather than on its set point as in Fig. 4.13. This is necessary because both controllers must have the same set point for the selectors to function. The results are essentially the same in either case.

Meeting Multiple Specifications   Products often have more than one specification, although only one is controllable at the point of discharge. For instance, a certain depropanizer distillate must contain at least 95 percent propane and no more than 2 percent isobutane. The unspecified component is the uncontrollable off-key ethane. If the ethane in the product exceeds 3 percent, the 2 percent isobutane specification cannot be violated and only the purity specification need be satisfied. But with less than 3 percent ethane, either of the two specifications could be limiting.

They may both be satisfied by appropriate reduction in isobutane content. Therefore, a single isobutane controller is needed to manipulate distillate and/or reflux. But its set point will vary with the uncontrolled ethane content. To satisfy both specifications, the isobutane set point y_i^* must be

$$y_i^* \leq 2\%$$

$$y_i^* \leq 5\% - y_e$$

Figure 7.13 shows how y_i^* is calculated from ethane content y_e.

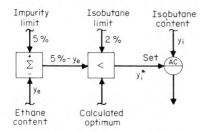

figure 7.13 *The isobutane set point is calculated to satisfy both specifications.*

Another application of this concept appears in Chap. 11. An *optimum* isobutane composition may exist which results in minimum-cost operation. If the optimum lies within the specifications, it should be used as the set point. But under certain conditions of feed composition, heating costs, and product values, the calculated optimum could exceed specifications. Then the specifications must be imposed through selectors as in Fig. 7.13.

VARIABLE STRUCTURES

Considerable importance has been placed on finding the optimum system structure. Whenever a constraint is encountered, however, a degree of freedom available to the control system is removed, and a suboptimum structure must be settled upon. Since control must be maintained over column pressure and liquid levels, one of the composition variables must be sacrificed. In the simplest case, there is no fundamental change in structure, but only the loss of one composition loop, as illustrated in Figs. 7.8 to 7.10.

If the variable threatened with loss of control is pressure or a level, then an alternate configuration must be found wherein its control is maintained and a composition loop sacrificed instead. The system capable of operating in both these conditions has two different structures, hence is termed *variably structured*. Still another possibility is the use of an alternate manipulated variable, in which case no loops need to be sacrificed. However, transferring a controller from one manipulated variable to another also amounts to changing the system structure. Several of these possibilities are examined below.

Backup Control Schemes   Perhaps the simplest method of changing system structure is by means of a second controller on the variable which is subject to loss of control. Pressure control is usually accomplished by manipulating condenser cooling. When cooling is inadequate to meet the heat load on the condenser, pressure will rise above set point. If no auxiliary manipulated variable is available, pressure must either go uncontrolled or take over manipulation of heat input.

In the latter case, a second pressure controller, having a higher set point than the first, would be connected to the heat input controls through a low-signal selector in the same way as the DPC in Fig. 7.10. The advantage of using two pressure controllers is the simplicity of the system and the ability to adjust them separately for their different manipulated variables. The principal disadvantage is loss of pressure control between the two set points. If temperature is used to infer composition, the changing pressure will require some sort of compensation as described in the previous chapter.

It is possible to accomplish the same function using a single pressure controller (PC) as shown in Fig. 7.14. In the normal operating mode, pressure is controlled by manipulating reflux (or distillate) flow from a flooded condenser. Should the heat load exceed the condenser capacity, the level of liquid within it will fall below the lowest row of tubes, and pressure control will be lost. This situation cannot be allowed to continue because in attempting to control pressure, the PC will raise the liquid outflow above the rate of condensing, ultimately cavitating the reflux pump.

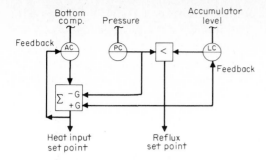

figure 7.14 *A single pressure controller can override heat input when manipulation of condenser cooling is no longer possible.*

Figure 7.14 shows a level controller (LC) manipulating the same flow through a low-signal selector. When the level falls to its set point, whether in the bottom of the condenser or in the accumulator below, the LC will reduce outflow, reestablishing a material balance. To prevent loss of pressure control, the difference between the outputs of the PC and the selector are used to bias the heat input downward. In effect, the pressure controller is transferred to heat-input manipulation whenever a level override is imposed.

The result of this transfer is loss of bottom-composition control, which is the variable sacrificed in Figs. 7.7 to 7.10. The AC will try to retain control, however, unless it is disabled. This is accomplished by applying the heat-input set point (or measurement) to its external feedback connection. Any difference between the output and feedback of the AC will stop integration just as in the case of overrides using selectors. The LC also requires external feedback, but the PC does not since it is always controlling.

When a single controller is used to manipulate two different variables, some provision may be needed to change the loop gain. In Fig. 7.14, it is provided by a common gain adjustment G to both biasing inputs of the summing device. The pressure controller is first adjusted for optimum response when manipulating condenser cooling; then gain G is adjusted when it is overriding heat input.

Auxiliary Manipulated Variables   The undesirable aspect of all the constraint controls presented thus far is the need to sacrifice one of the composition loops. Since this is unlikely to be accepted for very long, further action is required on the part of the operator or the control system. First, the problem must be properly diagnosed, so that the correct action can be taken. Loss of pressure control, for example, could be caused by several conditions: insufficient cooling, excessive production, or accumulation of noncondensible gas.

Either of the first two conditions could be corrected by reduction in feed rate to match the availability of cooling. However, the third should not be accommodated in the same way. If it were, production would continue to be curtailed as more gas accumulates, without correcting the cause of the problem. Here, another variable must be manipulated: a valve to relieve the gas from the condenser.

If the incident is infrequent, manual venting may be adequate. The function is not difficult to automate, however, and the approach is worth pursuing when noncondensibles accumulate frequently. If pressure is controlled by a coolant valve, a condenser bypass valve as in Fig. 4.17, or a drain valve as in Fig. 4.19 (left), loss of control is indicated by an extreme valve position. Then it is possible to sequence a condenser-vent valve to open when that extreme position is reached. For example, a coolant valve could open as the PC output moved from 100 to 50 percent, with the vent valve opening from 50 to 0 percent. In this way venting would only be used when the normal means of control was inadequate.

If pressure is controlled by flooding the condenser with reflux, loss of control is coincident with loss of level in the condenser and is independent of valve position. Some engineers then use a level controller to open the vent valve as in Fig. 7.15. This practice is not recommended because neither loop can be closed without the other. Pressure does not respond to reflux flow nor does level respond to the vent valve. There is, in effect, a single loop enclosing both controllers. Opening the vent valve lowers pressure which causes the PC to reduce reflux, thereby raising level. A steady-state relative-gain array for this process would consist of ones and zeros, with the zeros chosen for control loops.

Figure 7.16 solves the same problem using a variable structure. As long as reflux is under pressure control, the two signals sent to the subtractor are identical and its output will be zero. When the level controller takes over reflux manipulation, a deviation develops between them, opening the vent valve. Again, a common gain G is applied to both signals so that the pressure loop may be properly tuned for manipulation of the vent valve. The hand-control (HIC) station is provided for operator access to the vent valve.

Multiple-Output Systems   Whenever multiple reboilers provide heat to a single column, a decision must be made regarding the distribution of load. Multiple reboilers are generally used in severe fouling service, so that they may be individually cleaned without interrupting operation of the column. Although they may be identical in construction, their staggered cleaning schedule creates a difference in their degrees of fouling at any point in time. Consequently, they cannot all carry the same heat load. In fact, the heat-input valve on the most fouled reboiler could be fully open or limited to maintain condensate level by the system shown in Fig. 7.11.

In any case, there will be variations in duty among the reboilers and variations in the number of reboilers under control. Ideally, the operator should be free to balance the load as he sees fit — without upsetting product quality. Similarly, quality control should not be affected by the number of reboilers in service, in

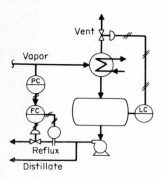

figure 7.15 *This configuration uses nested loops; the vent valve only affects level through the action of the pressure controller on reflux.*

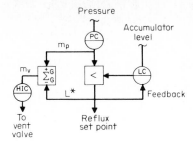

figure 7.16 *The subtractor trans-fers control to the vent valve during a level override.*

automatic, or free of limits. Without some form of compensation, the control-loop gain will change each time a reboiler reaches or leaves a limit, is transferred between automatic and manual, or is removed from or returned to service.

The system shown in Fig. 7.17 is designed to prevent product-quality control from being affected by operator adjustments or imposed limits. Only two reboilers are indicated, but the concept is applicable to any number. The instruments labeled FFC are flow-ratio controllers which allow the operator to vary the load on any reboiler in proportion to the rest. Total steam flow is fed back to a blind (back of panel) controller which manipulates all ratio controllers to satisfy the demand for heat. Should an operator adjust a ratio setting, the steam flow to that reboiler will be affected directly. The resulting change in total flow will be corrected by the blind FC readjusting all individual set points. If the period of oscillation of the individual flow controllers is about 1 s, that of the total-flow controller could be about 3 s. It is quite capable of returning total flow to the demanded level before the disturbance can pass through the reboilers into the column.

When one of the valves reaches a limit or is placed in manual, that flow can no longer respond to the total-flow controller. The total-flow controller will simply continue to move all set points as necessary to meet the demand. Although the gain of the total-flow loop is changed by the unresponsive reboiler, the quality-control loop is unaffected. And because of the fast response of the total-flow loop, its gain change does not pose a problem to the quality loop.

There are other possible versions of the multiple-output control system. For example, the individual ratio-flow controllers could be replaced with ratio stations or bias stations which modify the signals manipulating the valves directly. Then

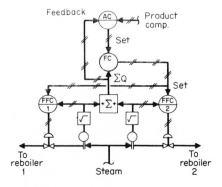

figure 7.17 *This system allows manipulation of individual reboilers without upsetting product-quality control.*

operator adjustments would be applied to valve positions rather than more accurate steam flows. However, removing the individual flow controllers can reduce the period of the total-flow loop to perhaps 1 s.

The feedback from total flow to the product-quality controller prevents windup in the event that all reboilers are in manual control, as during start-up.

Transferring a Quality-Control Loop   Whenever the number of manipulatable variables has to be reduced, a composition loop has had to be sacrificed. In all the examples described to this point, heat input has been overridden because it has the most pronounced effect on column and condenser loading. And since the most common configurations (DV, SV/B, and SV) involve manipulating heat input for bottom-composition control, that loop has been sacrificed.

Yet if there is a choice, the operator may prefer to retain bottom-composition control while sacrificing that of top composition. This requires a different control system structure. In an effort to avoid degrading performance when operating in the normal mode, the optimum structure for that condition should be retained, with transfer provided only during override.

The SV system (with feedforward) of Fig. 6.15 has been modified in Fig. 7.18 to suggest how this might be accomplished. The ratio of heat-input set point to measurement is used to bias the D/V ratio through a dynamic compensator. Thus any event causing heat input to fall below what is needed to control bottom composition will automatically cause D/V to increase proportionately. A gain adjustment k should be set to keep bottom composition constant when this mode is entered. The dynamic compensator needs to provide lead action to offset the sluggish response of bottom composition to reflux. Integral action in the top AC is inhibited by taking its feedback signal downstream from the bias block.

Observe that this system retains feedforward control which would otherwise be lost when heat input failed to respond to its set point. Yet the performance of the bottom-composition loop will be substantially degraded in this mode, even with dynamic compensation. In effect, there is no satisfactory substitute for an optimum basic structure, so constrained operation should be avoided whenever possible or minimized by attacking the source of the problem.

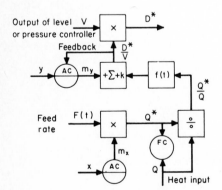

figure 7.18 *Bottom-composition control can be transferred to the top of the column when heat-input control is lost.*

CONTROLLING CONTROLLER OUTPUTS

Selective control systems and those with variable structure impose hard constraints on the operation of the process. When a constraint is reached, action is taken immediately to avoid exceeding it. Loop assignments are changed so that the constraint is honored at the cost of composition control — not a desirable condition, but safe and temporarily expeditious.

This section introduces an alternative concept — the application of "soft" constraints. In effect, an attempt is made to keep *all* loops operating by using additional controllers to reduce production or apply other corrective action when a constraint is approached. These controllers act on the outputs of the regulatory controllers, to keep them in their operating range.

Valve-Position Control   The valve-position controller was introduced in Fig. 2.20 as a part of a floating-pressure control system. Its function was to keep the pressure-control valve near its limit in the steady state but still allow responsive pressure control. It did this by manipulating the pressure set point, so that the two controllers formed a loop within themselves.

In the following applications, the valve-position controller will attempt to keep the assigned valve near its limit by manipulating or overriding *another* variable. If the valve is within the set point of the VPC, no action is taken. Therefore in this role, the VPC acts as a constraint controller. But since its overridden or manipulated variable may have only indirect response on the valve position, there may be an interval of time when the set point is exceeded following an upset.

As an example, consider Fig. 7.19 wherein a VPC is applied to both the condenser valve and the reboiler valve. As long as the reboiler steam valve is below the 90 percent set point, its VPC will be driving upward and therefore rejected by the low-signal selector. But in the event that increasing fouling or an excessive heat load causes the steam valve to pass 90 percent opening, the VPC will reduce its output below the set feed rate and therefore assume production-rate control. In this way, the steam valve will be maintained within its controllable range, and product composition can continue to be controlled by steam flow.

The other VPC operates to keep the condenser valve at or above 10 percent position. (In the case of a condenser bypass valve, this would correspond to 10

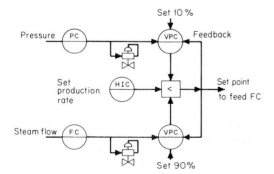

figure 7.19 *Valve-position controllers can be used to reduce production when either the condenser or reboiler approaches full loading.*

percent opening, but in the case of a coolant valve or vapor-throttling valve, this would correspond to 90 percent open since these valves are reverse acting for fail-safe reasons.) When the valve position falls to 10 percent, the VPC will begin reducing feed rate.

The response of the two valve positions to feed-rate manipulation is indirect. Reducing feed rate will tend to affect product composition first, causing the composition controller to reduce heat input. This affects the steam-valve position directly and the condenser valve through a reduced vapor loading. Because a composition loop exists within the valve-position loop, response is quite slow. The presence of a feedforward path from feed rate to heat input, as in Figs. 6.10, 6.14, and 6.15, will improve the response measurably.

Note that Fig. 7.19 shows the two control valves to be equipped with positioners. This is a necessary addition to any system featuring valve-position control. Without a positioner, the presence of a 5 to 10 percent deadband in the valve causes an indeterminate relationship between the output of the controller (which is the input to the VPC) and the true position of the valve stem. By closing the loop around the valve, a positioner typically reduces the deadband by a factor of 20.

It is quite common for columns to be arranged in series, with production rate set at the first column. Then a constraint encountered in a downstream unit may not be effectively accommodated by reducing feed far upstream. This method is therefore limited in its application. In a multiple-column unit, it may be necessary to calculate available capacity on a regular basis so that production may be scheduled with little danger of overloading downstream units. Methods for doing so are presented at the end of this chapter.

Balancing Output and Override   In the override system of Fig. 7.10, control over bottom composition was sacrificed at the encountering of a constraint. The event was not identified by a particular valve position; instead, its occurrence was indicated by the crossing of the AC output and override signals. The difference between these signals can then be used to reduce production in the manner of the valve-position controller described above. Figure 7.20 shows an output controller (OC) overriding column feed rate; its measurement is the output of the composition controller and its set point is the override signal, which represents the maximum steam flow allowed under the constrained condition.

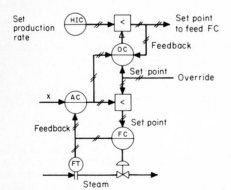

figure 7.20 *The output controller reduces feed rate enough to keep steam flow at the constraint without sacrificing composition control.*

In effect, the OC closes the composition loop by manipulating feed rate in the constrained mode. It operates in series with the AC rather than in parallel, and when the AC is incapable of controlling. By contrast, both VPCs in Fig. 7.19 were set within the operating range of their valves so that the pressure and flow loops remained closed while feed was being curtailed. In this role, the VPC is not likely to encounter stability problems, and integral action can be the dominant mode. But the OC curtails feed when the AC can no longer change steam flow; hence for stability, it needs proportional action.

In the system of Fig. 7.14, control was taken from the AC when the LC assumed manipulation of reflux. Although composition control was sacrificed, the override occurred as the outputs of the pressure and level controllers crossed. If composition control needs to be restored automatically by curtailing feed rate, an output controller must use the PC and LC outputs as measurement and set point. Implementation is then identical to that shown in Fig. 7.20, except that the AC is not connected to the OC.

Again, the loop formed through the output controller is indirect: pressure will only change by reducing heat input through the action of the AC in response to the reduced feed rate. Therefore the OC will have a very long integral time as did the VPCs in Fig. 7.19. Again, the feedforward signal from feed rate to steam demand will improve the response substantially.

ESTIMATING PRODUCTION CAPACITY

While the productivity of a single column may be maximized by the systems described earlier in this chapter, multicolumn management is more complex. For example, a single column may limit the productivity of the entire unit, depending on the configuration of that unit. But other opportunities may exist to allocate feedstocks effectively between parallel trains and even to balance the loading between columns in series by adjusting off-key components. Some of these schemes become possible when enough information is available on the locations of the constraints of all the columns in the unit.

Constraint Projection   To this point, most of the constraints common to distillation have been described and systems for imposing them on the manipulated variables have been presented. For the most part, no control action was taken until the constraint was actually reached. But an equally valuable function would seem to be the prediction of which feed rate will bring about a constrained condition and which constraint will be encountered first. This concept is called "constraint projection," i.e., locating constraints by projecting current operating conditions using mathematical models of the process.

Projection of the reboiler capacity Q_r may be made, based on its current valve position m and the present heat input Q:

$$Q_r = Q + k_r(100 - m) \tag{7.12}$$

Here k_r is a coefficient selected to match the slope of Q versus m over the upper range of valve travel, and m is in percent of full scale. The valve, in combination with its piping, may in fact give a nonlinear relationship between Q and m. Nonetheless, the simplicity of Eq. (7.12) is justified from the standpoint that its accuracy improves as 100 percent is approached. Conversely, as m departs from 100 percent, the constraint becomes less meaningful.

Note that the capacity of the heat-input valve is projected as its heat-input limit Q_r expressed in Btu/h or equivalent steam flow. In most cases, this is the most recognized measure of column capacity regardless of whether the constraint exists in the reboiler, condenser, or column.

For multiple reboilers supplied by a single oil heater, a minimum-energy system like the one in Fig. 2.18 might be applied. Then the position of an individual heat-input valve would not represent an absolute limit on capacity but would be relative to that of the most-open valve. The capacity of the most-open valve would in turn be relative to the temperature of the oil compared to the maximum allowable temperature. For the oil heater, the limit on heat output Q_{oM} would be related to the current heat flow Q_o by the difference between the current oil temperature T_o and its limit T_{oM}:

$$Q_{oM} = Q_o + k_o(T_{oM} - T_o) \tag{7.13}$$

The contribution of the additional capacity of the heater to that of the individual valve can be included by multiplying by the ratio Q_{oM}/Q_o:

$$Q_r = \frac{Q_{oM}}{Q_o}[Q + k_r(100 - m)] \tag{7.14}$$

The condenser limit Q_c can be estimated as a valve-position limit, exactly as done with the heat-input valve, if the pressure is controlled. If the pressure is floating, the valve is normally not limiting, so condenser capacity is then estimated as a function of the difference between current pressure p and its limit p_M. Figure 7.21 shows a typical curve of condenser heat transfer vs. pressure in a column. It is based on the abridged vapor-pressure curve of Fig. 7.6, wherein the temperature rise above the coolant temperature was plotted against vapor pressure for isobutane. A change in coolant temperature causes that curve to shift right or left so that its actual position at any time is quite variable although its slope is constant. Figure 7.21 was obtained by multiplying ΔT from Fig. 7.6 by a constant heat-

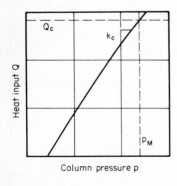

figure 7.21 *The condenser limit under floating-pressure control can be estimated from present operating pressure.*

transfer coefficient and area. If they are constant, then the slope of the curve in Fig. 7.21 is constant. As with the control valve, a linear approximation is sufficient:

$$Q_c = Q + k_c(p_M - p) \qquad (7.15)$$

Again, errors in the slope estimate become smaller as the limit is approached.

The slope k_c could change because of fouling, but this is more likely to occur in a reboiler than a condenser. For a fan condenser, the slope will change with the number of fans in service. If the operator has not started all the fans, however, there can be no danger of a condenser limit being approached, so the question becomes academic. The same is true when rain is falling.

The differential-pressure limit may be also estimated, considering the column as a flowmeter wherein differential pressure varies with the square of boilup. Then the maximum heat input Q_d can be estimated from the ratio of the measured differential Δp to the maximum allowed, Δp_M:

$$Q_d = Q \sqrt{\frac{\Delta p_M}{\Delta p}} \qquad (7.16)$$

Again, the accuracy of the estimate improves as the constraint is approached.

Differential pressure is not the only index of column flooding. When down-comers flood so that liquid is prevented from flowing down the column, its buildup on the trays does raise the differential pressure sharply. But in the case of excessive entrainment or "jet flooding," product quality may begin to deteriorate without a corresponding increase in differential pressure.

To protect against jet flooding, the estimated flood line for a column may be located as was done in Fig. 7.4. Note that each mixture will have its own curve, heavier materials having a more positive slope and lighter materials exhibiting a negative slope. The limits of boilup were determined as described by Eqs. (7.2) to (7.4), with correction then applied for the variation of latent heat with pressure.

For simplicity, a linear flooding model can be constructed from one of these curves by estimating the flooding limit Q_f related to some established reference point Q_M and p_M:

$$Q_f = Q_M - k_f(p_M - p) \qquad (7.17)$$

The slope of the flood line is designated as k_f.

For all these components cited, maximum throughput will be achieved at the minimum attainable pressure, i.e., where the flood line and condenser constraint cross. This was illustrated by the operating window in Fig. 7.7. The window is duplicated in Fig. 7.22 using linear models for both curves.

The flood limit, i.e., where the lines cross, can be found by solving Eqs. (7.15) and (7.17) simultaneously. In Eq. (7.15), however, Q_f and p_f must be substituted for Q_c and p_M; in Eq. (7.17), p becomes p_f. For the condenser, then,

$$Q_f = Q + k_c(p_f - p) \qquad (7.18)$$

and for the column,

$$Q_f = Q_M - k_f(p_M - p_f) \qquad (7.19)$$

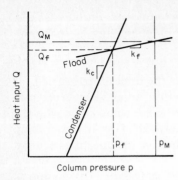

Column pressure p

figure 7.22 *The flood limit is located at the intersection of the flood line and the condenser constraint.*

Combining the two by eliminating p_f yields

$$Q_f = \frac{Q_M - k_f(p_M - p + Q/k_c)}{-k_f/k_c} \tag{7.20}$$

Using Eq. (7.20), the flood limit may be estimated for any operating condition of p and Q, given the other terms as constants.

Estimating Column Capacity Once all the constraints have been located, the calculated limits must be compared to determine which will be the first to be imposed:

$$Q_l = \text{lowest of } (Q_r, Q_c, Q_d, Q_f) \tag{7.21}$$

Then the fractional load on the column is

$$\text{Load} = \frac{Q}{Q_l} \tag{7.22}$$

where Q is the current heat flow into the reboiler.

If Q/Q_l is less than 100 percent, the column has additional capacity to process feed–the margin between Q/Q_l and 100 percent. But when a constraint is reached, the override controls or self-imposed limits will prevent Q from rising above Q_l. Consequently, Eq. (7.22) is only valid up to 100 percent capacity.

What would provide a useful measure of overcapacity depends on what strategy is applied when capacity is exceeded. If control of one of the product compositions is simply forsaken, a measure of overcapacity can be obtained by comparing the current (limited) heat input to that needed to maintain control. In Figs. 7.9 and 7.10, the deviation between desired heat input Q^* and actual Q can be used:

$$\text{Load} = \frac{Q^*}{Q} \tag{7.23}$$

Note that (7.23) is not valid for operation *below* 100 percent capacity. To provide a relationship applicable to both operating regimes, Eq. (7.22) must be combined with (7.23):

$$\text{Load} = \frac{Q}{Q_l} \frac{Q^*}{Q} = \frac{Q^*}{Q_l} \tag{7.24}$$

During undercapacity operation, Q follows Q^*, whereas above 100 percent capacity, Q follows Q_l; therefore, (7.24) satisfies both conditions.

When the composition controller is in fact not manipulating heat input because of imposed limits, Q^* may not exactly represent the demand, however. Equation (7.6) describes how the deviation between measurement and set point for the primary controller relates to that of the secondary controller. Substituting the deviation in product quality $x - x^*$ for e in (7.6) gives

$$x - x^* = \frac{P}{100}(Q^* - Q) \tag{7.25}$$

In actual practice, a deviation $x - x^*$ can be corrected by an appropriate change $Q^* - Q$, although the proportionality is not necessarily the proportional band of the feedback controller.

The heat input required to meet specifications can be estimated from those specifications, current composition, and current heat input. First, consider that the ratio of the required to actual heat input is the ratio of required to actual boilup:

$$\frac{Q^*}{Q} = \frac{V^*}{V}$$

However,

$$\frac{V}{F} = \frac{D}{F}\left(1 + \frac{L}{D}\right)$$

Therefore the boilup ratio can be calculated from the material balance and the separation model Eq. (3.59), evaluated at desired and actual conditions.

To illustrate the method, consider a column of 50 theoretical trays, separating an equimolar binary mixture having a relative volatility of 1.40. Let y be controlled at 0.95, and x vary from 0.01 to 0.05 but desired at $x^* = 0.02$. Using the above procedure, the relationship between actual and desired heat inputs as a function of actual bottom composition was calculated and plotted in Fig. 7.23. Since the re-

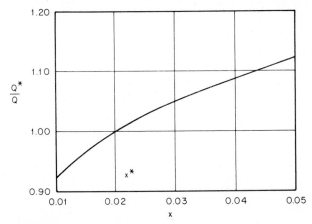

figure 7.23 *Overload may be estimated from the deviation sustained in product quality.*

lationship is reasonably linear above x^*, overload estimation can be reduced to the expression:

$$\frac{Q^*}{Q} = k_x(x - x^*) \tag{7.26}$$

or

$$\frac{Q^*}{Q} = k_y(y^* - y) \tag{7.27}$$

depending on which composition is sacrificed.

If the estimate of capacity is reasonably accurate, the operator should be able to make a corresponding adjustment in feed rate to approach 100 percent capacity. If the capacity estimate were to be 87 percent, for example, the operator should be able to increase feed 13 percent (of value) without losing control. Or if 105 percent capacity is indicated, control should be restored by reducing feed rate by 5 percent (of value). All the estimates become more accurate as 100 percent capacity is approached.

REFERENCES

1. Anderson, R. H., G. Garrett, and M. Van Winkle: "Efficiency Comparison of Valve and Sieve Trays in Distillation Columns," *Ind. Eng. Chem. Proc. Des. Dev.*, vol. 15, no. 1, 1976.
2. Van Winkle, M.: *Distillation,* McGraw-Hill, New York, 1967, pp. 527–532.
3. Reference 2, pp. 525, 526.
4. *Matheson Gas Data Book,* 4th ed., The Matheson Company, East Rutherford, N. J., 1966.
5. *Technical Data Book–Petroleum Refining,* 2d ed., pp. 10–11, American Petroleum Institute, Washington, D. C., 1970.
6. Smuck, W. W.: "Operating Characteristics of a Propylene Fractionating Unit," *Chem. Eng. Prog.,* June 1963.
7. Shinskey, F. G.: "Energy-Conserving Control Systems for Distillation Units," *Chem. Eng. Prog.,* May 1976.
8. Doig, I. D.: "Variation in Operating Pressure to Manipulate Distillation Processes," *Aust. Chem. Eng.,* July 1971.
9. Manter, D., and R. Tressler: 1961, *Instrument Soc. Of America.* "A Coal-Air Ratio Control System for a Cyclone Fired Steam Generator," ISA Paper, 1-CI-61, 1900.

PART THREE

Complex Processes

CHAPTER EIGHT

Multiple-Product Processes

Only rarely are distillation columns operated singly in production processes. Chemical plants and petroleum refineries consist of many operating units, each of which typically includes several columns. Petroleum refining begins with the separation of crude oil, a complex mixture of hydrocarbons, into several fractions; in turn, each is refined further. Included are various types of chemical reactors needed to upgrade less-valuable products. And each reaction tends to produce a spectrum of components which require further separation.

It is not suprising then that distillation columns appear in groups. To improve the thermal efficiency of these operations, successive separations are often combined into a single column from which several products are withdrawn. Where purity specifications do not allow this practice, energy integration between columns is becoming more common.

All these operations are more complex than the simple two-product column studied to this point. However, the tools developed to understand and control the two-product column still apply and are extended to multiple-product units in this chapter.

MULTIPLE-COLUMN UNITS

Whenever a multicomponent feedstock must be split into several relatively pure fractions or a cut must be taken from the heart of the feed, more than one column is ordinarily required. Control strategies devised for a single column often have to be adjusted to satisfy the demands of other columns. When the product from one column is the feed to another, for example, wide fluctuations in its flow must be avoided. And as has already been demonstrated, the control of the quality of certain products may have to be implemented one or more columns upstream.

A classic design problem exists in ordering the sequence of separations for a multicomponent feed. A feedstock consisting of k major components requires $k - 1$ columns for complete separation. But they may be arranged in a variety of ways. Reference 1 describes a computer-aided design procedure for selecting the optimum (least cost) arrangement, both with and without energy integration. This means of arriving at an optimum arrangement is beyond the scope of this text. However, common arrangements are presented so that their individual and collective control problems may be pointed out.

Light-Ends Fractionation   Perhaps the most common application of multiple-column fractionation is the separation of light hydrocarbons. Natural-gas liquids, virgin naphthas, and the light products from cracking reactions all contain a spectrum of hydrocarbons from ethane through gasoline fractions. The components are often removed sequentially, beginning with the lightest. This practice probably gave rise to the nomenclature common in refineries, where towers are named according to the light component removed in them, e.g., deethanizer, depropanizer, etc.

Since the lightest components are the most difficult to condense, they are usually removed first. Then downstream columns may be operated at lower pressure and without refrigeration. Isomers which are relatively difficult to separate are often split out together, however, to be separated alone. These relationships are only qualitative, since the best arrangement depends heavily on relative concentrations as well as on opportunities for energy integration.

The case of a multicomponent feed given in Example 3.4 will be developed further at this point. Already considered was the column splitting between n-butane and isopentane—the debutanizer. The two products from this column may be further fractionated in a sequence such as that shown in Fig. 8.1. The author makes no claims as to whether Fig. 8.1 represents an optimum arrangement: it is simply used to illustrate control problems associated with distillation trains.

Each product from the train may have two specifications—a higher- and a lower-boiling component. Thus the propane product may have specifications on ethane and butane content, the isobutane product may have specifications on propane and n-butane content, etc. For the lightest and heaviest products—in this

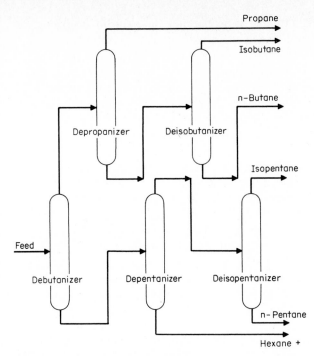

figure 8.1 *This is one possible configuration for separating a
mixture of propane through gasoline.*

case propane and hexane (and heavier)—only one specification is controllable,
however. If the ethane limit on propane must be met, it must be met in the feed.

Products from the last two columns are the most difficult to control. The amount
of n-butane in the deisobutanizer distillate and isobutane in the bottom product are
both controllable at the deisobutanizer. But the percentage of isopentane in the
n-butane can only be adjusted at the debutanizer—two columns upstream.
Closed-loop feedback control of this variable is simply not feasible—the delays
distributed through the three towers are excessive. Instead, the isopentane content
of the debutanizer distillate must be controlled in ratio to its n-butane content in the
manner described in Eq. (1.11). In this way, the two components will be in
approximately the correct ratio when they emerge from the deisobutanizer.

It may be desirable to operate below the isopentane limit. If the remaining specification on n-butane is simply a purity limit without regard to the isobutane content, reducing the isopentane level will allow the isobutane to increase. This has the
benefit of reducing the separation required of the deisobutanizer, but at the same
time it increases that of the debutanizer. If the isopentane and isobutane command
different selling prices, their value difference also enters into the operating cost.
Thus a two-column optimization problem exists with the debutanizer overhead
composition as the cost-determining element.

This is not the only possibility. The amount of n-butane in the isopentane product must also be controlled at the debutanizer, and its level affects the separation

of the deisopentanizer as well. In the same manner, propane in the isobutane product affects separation in both the depropanizer and the deisobutanizer.

Another problem common to multiple-column systems is that a bottleneck in any column can limit unit production. It is possible to shift the loading somewhat between columns through the same mechanism as suggested for optimization. If, for example, the deisobutanizer alone has reached an operating constraint, some of its separation load may be shifted to the debutanizer by reducing isopentane in its distillate. At the same time, some loading could be shifted to the depropanizer by reducing propane in the deisobutanizer feed. These relationships are mentioned at this point only to suggest the complexities of operating a distillation train. Actual optimization procedures are developed in Chap. 11.

Parallel and Serial Configurations   The arrangement of columns in Fig. 8.1 is somewhat unusual in that the first column is a splitter whose products are further separated serially. More often, the entire separation is conducted serially, and parallel trains of similar serially connected columns are also common.

The capacity estimates developed in Chap. 7 should be very useful in helping to balance the feed streams to parallel trains of columns. While it might seem that this is the sort of task which only has to be done once, this is rarely the case. Few parallel trains are identical either in capacity or details of construction. Quite often, the second train was built after the first and is larger, even having a different source of heating and cooling. Sometimes the columns of the second train are added one at a time, in which case constraints are changed each time a new column is added. In the final analysis, the trains tend to have different limits, changing with ambient conditions, feed composition, etc., often requiring cross flow at selected points between them. Then rebalancing may be required often and the capacity estimates will be found quite useful.

Columns aligned serially within a train may also be balanced. This opportunity appears in the multiple-column unit in Fig. 8.1. The off-key components in the debutanizer's products, i.e., propane in the distillate and isopentane in the bottom, have a profound effect on the separation requirement if a total purity specification has to be met on each product. Thus an increase in propane content requires that less n-butane be allowed in the isobutane product if the specification of, say, 95 percent purity is to be met. Then a reduction in separation at the depropanizer would allow more propane to reach the deisobutanizer, where the separation must subsequently be increased. There may be an economic incentive to reach an optimum propane content based on its value relative to isobutane and the cost of separation in each tower. But if there is not, the amount of propane leaving the depropanizer can be adjusted to balance the loading of the two columns. By adjusting the isopentane content of the debutanizer overhead, it is also possible to balance its load against the deisobutanizer. Consequently, without adding or removing any intermediate streams all three of these columns can probably be operated at the same capacity. Following this line of reasoning, the heavy-ends columns of the same unit could also be balanced.

The load distribution for this or any multicolumn unit is heavily dependent on feed composition. So it is important for the operator to know the percent loading on each column if he is to maximize the capacity of the unit. Then he can use relationships such as those given in Eq. (7.26) and Fig. 7.23 to solve for the *composition* change which will bring about a required change in capacity. Serial balancing can then be done on a calculated basis.

Production Scheduling   There is no need to wait until a constraint is reached before making an adjustment to the charge rate or to intermediate compositions. An analysis of the charge stock at any point in time can be used to predict the unadjusted loading on all the columns. Then if the estimate indicates a potential overload on any column, an intermediate composition may be found to distribute the load more equitably. As in any feedforward system, the adjustment must be programmed on a timed basis to coincide with the arrival of the new feedstock to the affected column.

No new technology is needed to reach this goal. The material balances that have been used throughout this book apply, as do the relationships between boilup and separation. Where floating pressure is applied, the effect of pressure on separation for anticipated ambient conditions should be taken into account. Then the maximum advantage will be gained from diurnal variations in condenser duty. A plant can then anticipate a token production increase every night. Columns equipped with water-cooled condensers can be adjusted to accept some of the load of columns with air-cooled condensers during the day, relinquishing it at night.

Managing a multicolumn unit in this manner may seem too complex at first thought. Although the required calculations can be made by hand for one or two columns, beyond this a computer is required. A computer can be programmed with a multicomponent model of each column in the unit, using the multicomponent material and separation equations described in Chap. 3. Having calculated all flow rates and boilup rates for each tower, the computer may adjust feed rate, cross flows, and intermediate compositions to maximize the production of the unit. Nonlinear programming is the method needed to adjust the multiple parameters against the boilup constraints successively until maximum production is achieved. A computer can converge on a solution in minutes, whereas it may take the separation unit itself hours to respond completely to a single change in feed rate. The time lags in the various columns may, in fact, prevent equilibrium from ever being reached within the diurnal cycles that affect the plant. This is a natural limit to the prospects of multicolumn optimization.

Units with Reactors   Chemical reactors pose special problems for distillation columns. In most chemical plants, the reactor is at the head of a train of columns which are designed to separate its effluent into various product and waste streams. In this situation, the columns must accept whatever the reactor discharges, which could vary considerably in rate, composition, and temperature, as a function of time and product targets. Catalyzed reactors typically change the component distribution in their products with catalyst age. Where frequent regeneration is required, column-feed compositions will change as often as reactors are brought on- and off-line.

Some reactors, particularly in petroleum refineries, require prefractionation of a feedstock. Then upstream columns encounter the problem of having to feed a reactor at a constant rate, which is usually deemed necessary to secure stable and efficient conversion. If the bottom product from a column, for example, must feed the reactor under flow control, it cannot be used to close the column material balance or to control composition. In this case it may be necessary to control base level with column-feed rate. When a reactor must be fed from a column, an extra large storage capacity in its base or accumulator is advisable. The alternate practice of diverting material from the column to a storage tank and then to the reactor is discouraged by the need to cool and depressurize into the tank, followed by pumping and heating out of the tank.

Most reactors operate well below 100 percent conversion of feed to product. Then their effluent contains unconverted feed which may require recycling after being separated from the product. Then the specification placed on the recycled stream is not fixed but is a function of the penalty imposed by other components returning to the reactor. This is an optimization problem discussed more in depth in Chap. 11.

Recycle around a reactor can pose problems in inventory control, too. Figure 8.2 shows a simplified flowsheet of a sulfuric-acid alkylation unit as an illustration. The acid catalyzes the reaction of isobutane with an olefin such as isobutylene to form an isooctane:

$$
\underset{\substack{|\\C}}{C} - \underset{\substack{|\\C}}{C} - C + C = \underset{\substack{|\\C}}{C} - C \longrightarrow C - \underset{\substack{|\\C}}{C} - C - \underset{\substack{|\\C}}{C} - C \tag{8.1}
$$

Equation (8.1) is but one of many possibilities. Butenes are so difficult to separate from one another that the olefin stream may contain 1-butene and 2-butene as well. Other olefins, if present, will also participate. And as is typical of many organic reactions, side reactions also take place, yielding a mixture of highly branched paraffins, principally in the 6 to 9 carbon range. If the reaction is properly controlled, the alkylate is a high-octane motor fuel.

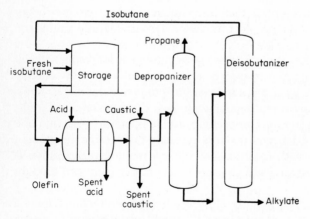

figure 8.2 *A simplified flowsheet of a sulfuric-acid alkylation unit.*

One of the control problems is to keep the olefin components from reacting with each other. They have a strong tendency to polymerize, and the polymer is a poor fuel since it is unsaturated. Polymerization is minimized by diluting the olefins with isobutane, in a ratio of 1 part in 5 to 10, and by operating at low temperature with refrigeration. The reactor effluent then contains only 1 part of alkylate in 4 to 9 parts of isobutane, which must then be separated and recycled. Figure 8.2 shows a deisobutanizer making that separation.

In a recycle system, care must be taken to avoid accumulation of components not participating in the reaction. Thus a depropanizer is necessary to remove propane and lighter hydrocarbons which will build up even if present only in minute amounts in the feed. Normal butane is also unreactive and must be removed with the alkylate if no other outlet is provided. In the next section, a sidestream will be added to remove it.

Principal specifications on the isobutane recycle stream are propane and n-butane. Both take up reactor space and therefore reduce production. Isobutane losses with the alkylate are undesirable since it raises the vapor pressure excessively. More is said on this subject under multiproduct towers.

Proper sizing and control of the storage tank are essential to the successful operation of any recycle system. For example, consider the alkylation unit in a steady state with the storage tank half full. To increase production by 20 percent would require a temporary 200 percent increase in fresh isobutane if the level in the storage tank were to remain constant, assuming a 10:1 ratio of isobutane to olefin. Similarly, a 20 percent decrease in production would require a 200 percent decrease in fresh isobutane, which would clearly not be possible even if the temporary 200 percent increase were. Eventually most of the increase or decrease in isobutane will be returned as recycle from the deisobutanizer. But this takes time because levels in the reactor and column trays must change to accommodate the new flow rate. It is therefore impossible to control the level in the storage tank in the unsteady state. Furthermore, any attempt to control it by executing the gross manipulations in fresh feed estimated above would place unrealistic demands on the feed-processing unit.

If the flow variations imposed on the fresh feed are to be no greater than the changes in reactor feed, the storage tank must be large enough to absorb the entire change in inventory of the balance of the system. Therefore the storage tank level must be low when the production rate is high and high when production is low. The level should not be tightly controlled at midscale. On the contrary, if fresh feed is set proportional to tank level, a reasonably high feed rate will be maintained when the level is low, but only enough to keep the level from falling further. It is not necessary to add enough to raise the level to midscale—this overstresses the upstream feed-processing unit.

These basic rules apply whenever a storage vessel is located between columns or processing units. The intended function is usually to absorb upstream fluctuations while providing a nearly constant feed rate to the next unit. These vessels are often called *surge tanks*, descriptive of their role in absorbing surges. In the steady state, outflow must match inflow, but rapid inflow fluctuations should not be passed on.

Consequently, level in the surge tank cannot be held within narrow limits since this will shrink the effective volume of the vessel to those limits. If the level controller manipulates the outflow, the level should be high when the flow is high and low when the flow is low. In this way, the vessel will be best able to absorb upsets in the direction of more normal operation or of the opposite extreme. For further dicussion on this subject, the reader is directed to Ref. 2.

PARTIAL CONDENSERS

At this point it is appropriate to reexamine the role of partial condensers in distillation. A demethanizer and a deethanizer are presented in Figs. 4.20 and 4.21. In the first, the distillate is removed in the vapor phase, while in the second, it is subsequently condensed by refrigeration. Both of these systems had the property of strong interaction among column pressure, accumulator level, and distillate composition, not common to columns with total condensers.

After the nature of these interactions has been resolved, the problem of removing distillates simultaneously in two phases can be addressed. The application of floating-pressure control to this column is then presented.

Interaction between Level and Pressure   If overhead product composition is a significantly slower responding variable than column pressure or accumulator level, then its controller will not upset either level or pressure appreciably in manipulating its valve. In this light, the prospect of three loops interacting in the overheads of the columns in Figs. 4.20 and 4.21 can be reduced to a 2×2 problem.

First, the interaction between top and bottom compositions needs to be evaluated to determine the basic system configuration. The procedure developed in Chap. 5 only considers the composition subsets, assuming levels and pressure were controlled but making no assignments for their loops. Second, once the composition assignments have been made, then the level and pressure interaction can be examined for the remaining unassigned manipulated variables.

As an example, consider the demethanizer in Fig. 4.20; the structure shown is for a DV configuration in that distillate is already assigned to control top composition. That leaves condenser cooling and reflux to control pressure and accumulator level. As the cooling is increased by evaporating more refrigerant, pressure will fall and level rise by condensing more vapor — both variables respond in approximately the same degree. Increasing reflux will cause level to begin falling, but its effect on pressure is limited to its degree of subcooling.

If the overhead product is a relatively pure binary mixture, the difference in temperature between bubble point and dew point will be quite small, so that the temperature of the reflux will be close to that of the top tray. Then an increase of reflux flow will not condense a significant increment of vapor, and pressure will not be noticeably affected. If the mixture is a relatively impure binary, which could be the case in the ethane-propane product taken from the deethanizer in Fig. 4.21, then the reflux temperature could be substantially below the vapor temperature. Then increasing its flow will condense additional vapor and cause pressure to fall. Similarly, if the overhead vapor contains a noncondensible gas (hydrogen might be present in the demethanizer), reflux will tend to be subcooled below the vapor temperature.

In either case, the incremental flow of vapor condensed by the increase in reflux cannot be more than a small fraction of the increase in reflux flow itself, in accordance with Eq. (4.3). The ratio of specific heat to latent heat is in the range of 0.0025 to $0.0055°F^{-1}$ for hydrocarbons at $100°F$. Consequently, subcooling by as much as $20°F$ will only cause an increase in vapor condensation of 5 to 11 percent of the increase in reflux flow. By contrast, an increase in condenser cooling removes an equal increment of vapor for every increment of liquid produced.

On this basis, reflux should not be used to control pressure in a system with a partial condenser. Then when D and V are used to control compositions as in Fig. 4.20, level should manipulate reflux, and pressure be controlled by cooling as shown. If an SV system is used, separation would be manipulated through the reflux ratio. Reflux would still be used to control accumulator level, with distillate set in ratio to it, with the top-composition controller adjusting the ratio. Level and pressure assignments would remain as in Fig. 4.20.

Other possibilities would be SB and LB configurations. With bottom flow used to control bottom composition, distillate vapor would have to be placed under pressure control to close the external material balance. Reflux or reflux-to-distillate ratio would control top composition, and condenser cooling would be left to control accumulator level; base level would have to be controlled by boilup.

This same line of reasoning applies to the deethanizer in Fig. 4.22; it differs from the demethanizer only in having a distillate condenser. Both these systems as shown suffer from the indirect response of reflux to distillate flow. A change in distillate flow first must affect column pressure, then the PC must adjust cooling to produce more or less condensate, then the LC must respond to change the reflux accordingly. Both will be improved by biasing reflux set point from distillate flow measurement as was done in Fig. 4.27; the fact that distillate is not drawn as a liquid from the reflux accumulator does not matter. In fact, the SV/B system in Fig. 5.16 applies as shown, regardless of the phase of the distillate.

Overhead Products in Both Phases   Having examined the relationships among the variables when either a liquid or a vapor product is withdrawn, it is now possible to predict the results of withdrawing both at the same time. This practice is common when total condensation is impossible because of volatiles in the feed and the principal distillate product is a liquid. In essence, the condenser becomes a one-stage stripper.

The equilibrium between liquid- and vapor-product composition is dictated by condenser pressure and temperature as before. Second, their relationship to top-tray vapor composition is affected by the flow rate of the vapor product relative to top-tray vapor rate, again as seen before. But because the two products have different compositions, they tend to affect the column external balance differently. When the entire overhead product was withdrawn either as a vapor or a liquid, the mole fraction of the lightest component varied with feed composition as in Eq. (3.6). But by splitting a vapor stream away from the liquid product, the mole fraction of the lightest component in the liquid can thereby be controlled.

The flow of vapor product is an additional manipulated variable and is properly assigned to control the vapor pressure of the liquid product. In this sense, the pressure on this column is more significant than it is on columns having total condensers. But it requires condensate temperature to be coordinated with it. All too often, partial condensers are left without cooling controls, allowing condensate temperature to float with coolant temperature and heat load. Then vapor pressure will also vary although static pressure is controlled.

One solution to the problem is to add a condensate temperature controller to manipulate heat removal. Controlling both this temperature and pressure then fixes the vapor pressure of the liquid product. Alternatively, the temperature may be left uncontrolled, with the pressure set point adjusted in relation to it, as shown in Fig. 8.3.

The temperature measurement is characterized to an equivalent vapor pressure representing the desired composition. Then as coolant and ambient temperatures change, column pressure is adjusted accordingly. A differential-vapor-pressure cell would provide the same function even though it is not normally applicable to ternary systems such as this. A problem may arise in finding the correct filling solution.

The relationship between pressure and temperature is nearly linear over the range typically encountered with condensers. Furthermore, the slopes of the vapor-pressure curves for various compositions tend to be quite similar, although they are displaced from one another. Consequently, a bias adjustment should be available to recalibrate the functional relationship for desired changes in composition.

The instrument configuration shown in Fig. 8.3 provides true floating-pressure control. There is no restriction on the condenser, so the condensate is always as cool as possible. Hence the pressure in the column will always be as low as possible, maximizing relative volatilities and column capacity and minimizing the energy required for separation. Pressure drop and therefore power loss across the vapor valve are also minimized. This can be an important consideration when pressure is developed by a feed-gas compressor — minimum column pressure allows reduced compressor power, increased flow, or a combination of both.

However, positive feedback in the pressure loop needs addressing. As temperature falls, a proportionate reduction in pressure set point will cause liquid to flash, further lowering temperature. The floating-pressure system for the total condenser in Fig. 2.20 has a valve-position controller whose integral time could be adjusted as needed. A long time constant was recommended to maintain pressure stability and avoid upsetting compositions during cooling disturbances. In Fig. 8.3,

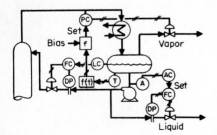

figure 8.3 *Overhead products in both phases require vapor-pressure control.*

this stability is provided by a dynamic compensator $f(t)$ in the form of a first-order lag. It should be set in the 10- to 20-min range.

The heavy key in the liquid distillate is shown controlled by its own flow, as was also true in the two previous examples. Whether this is the optimum configuration depends on the relative gains of the xy subsets, as with those columns. Although vapor pressure is a composition variable, it is so much faster than the other composition loops that it does not have to be included in their relative-gain array. Pressure and level do not interact greatly because vapor flow has little effect on accumulator level, and reflux flow has only token effect on pressure, as described for the other partial condensers.

SIDESTREAM COLUMNS

There is a wide variety of columns with sidestream products, and they are becoming more common as the need for energy efficiency grows. In some refineries, two or three columns have been combined, with consequent elimination of some reboilers and condensers.

Yet the efficiency is not gained without some cost. Sidestream columns are more difficult to characterize, and their additional variables complicate the issue of finding the optimum control-system structure. (Remember that the number of possible single-loop configurations varies factorially with the number of loops.) Furthermore, internal liquid and vapor rates are altered by the sidestreams and are generally not measurable. Side coolers, sometimes necessary, add still another dimension.

In most of these columns, however, there will be a single reboiler and condenser which establish liquid and vapor traffic through the column. It is this liquid-to-vapor ratio which determines the separation factors between the various components and affects all at once. The added manipulated variables—the sidestream flows—are all product streams and therefore enter principally into the external material balances. Finding the optimum system structure depends on relating product compositions to those streams having the greatest influence over them.

Minor Sidestreams   One of the most troublesome problems encountered in distillation is posed by the presence of a component boiling between the light and heavy keys. The intermediate component tends to accumulate within the column, reducing the concentrations of the key components on several trays and thereby rendering those trays less effective in separating them. A steady state is reached when the intermediate component leaves with either or both products, but this is usually less than satisfactory.

In certain columns, a component may concentrate to the point of reaching a solubility limit. This can happen in an ethanol-water column, where higher alcohols known as *fusel oil* form intermediate-boiling azeotropes with water. Unless withdrawn, either continuously or periodically, these alcohols can accumulate to the solubility limit, whereupon they form a second liquid phase. This creates an unstable mode because both liquid phases exert their full vapor pressure. As a result, vapor evolution momentarily doubles, driving flow upward and dropping column temperature. Flooding usually then carries the fusel oil overhead. If a temperature controller is manipulating boilup, it may react by further increasing vapor flow. In any case, the fusel oil tends to leave overhead, after which the column may return to normal for a few hours or a few days.

An intermediate-boiling component should be removed continuously to limit its accumulation. If there is a specification on it in either major product, then side-stream flow should be manipulated to control that concentration. If there is no such specification, sidestream withdrawal is still important to unload that component from the trays. In this role it can be used to limit or optimize the reflux ratio needed to separate the keys. Much will depend on the destination and treatment of the sidestream.

If the presence of the intermediate component has a measurable effect on column temperature, that loop can be closed. However, a liquid sidestream will tend to have more effect on a temperature below the point of withdrawal than above. This relationship is examined further under the subject of "Fractionating Crude Oil" later in the chapter.

In the case of fusel oil, a sidestream is withdrawn and cooled, which reduces its solubility in water. A decanter is provided to allow the oil to float to the surface. It is then withdrawn to control the position of the interface while the remaining solution is returned to the column, as shown in Fig. 8.4. If no suitable means of measuring the intermediate component exists, then its presence may be regulated by simply setting sidestream flow in ratio to column-feed rate.

Pasteurizing Columns   This term is given to distillation columns where the principal distillate product is taken a few trays down from the top. The purpose is to

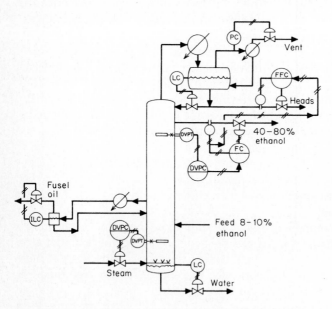

figure 8.4   *A small flow of "heads" is removed in proportion to the ethanol to regulate its concentration in the product; fusel oil is removed at the solubility limit set by the sidestream cooler.*

remove a small quantity of light impurity overhead, thereby limiting its concentration in the principal distillate. It is common in ethanol-water distillation, as shown in Fig. 8.4, to remove aldehydes or "heads" from the ethanol. In beverage production, they affect product quality, and in the manufacture of absolute or fuel-grade alcohol, they interfere with dehydration downstream.

The column uses open stream to strip the ethanol from the water, owing to the low concentration of the feed and the high relative volatility at low ethanol concentration. A differential-vapor-pressure cell filled with water senses the presence of ethanol four or five trays up from the bottom and is used to control the flow of steam.

Because the ethanol in the water leaving is in the parts-per-million range, compared to the relatively impure ethanol product, Λ_{DV} for this column is very nearly 1.0. Correspondingly, experience has indicated very little interaction between the two composition control loops.

A second DVP cell filled with product of the desired quality is located just below the draw-off tray. Vapor pressure there is controlled by manipulating ethanol product flow. Heads will accumulate in the top of the column and must be withdrawn to limit their concentration in the ethanol. Lacking a measurement of their concentration the withdrawal of heads is simply set in ratio to the flow of ethanol. If the ratio is too low, the product will be contaminated; if too high, an excessive quantity of ethanol will leave with the heads stream. This stream is usually sent to a heads column where the ethanol is recovered and the heads concentrated as a by-product.

Another common tower of the refining or pasteurizing type is the ethylene fractionator, shown in context with other columns of its unit in Fig. 8.5. Its feed contains a small amount of methane along with the principal impurity, ethane. The ethylene fractionator actually has two final products: the distillate containing virtually all the methane in the feed, and the refined product, high-grade ethylene, leaving as a sidestream.

Actually, more of the methane could have been removed in the demethanizer but at the cost of higher boilup or increased ethylene losses there. Withdrawing the

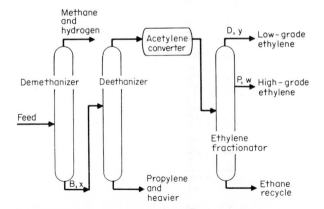

figure 8.5 *Virtually all the methane from the bottom of the demethanizer must leave with the low-grade ethylene.*

high-grade product as a sidestream allows its methane content to be controlled precisely at the point of withdrawal. This is an important feature in a multicolumn unit. But having a market for low-grade ethylene adds another dimension to the picture — the yield of that product is tied to the operation of the demethanizer.

To illustrate this relationship, assume that all the methane and ethylene leaving the demethanizer are recoverable. Then a methane balance yields

$$Bx_1 = Dy_1 + Pw_1 \qquad\qquad (8.2)$$

where
$$\begin{aligned} B &= \text{demethanizer bottoms flow} \\ D &= \text{low-grade ethylene flow} \\ P &= \text{high-grade ethylene flow} \\ x_1, y_1, w_1 &= \text{methane contents of associated streams} \end{aligned}$$

A similar balance may be drawn on the ethylene, indicated by subscript 2:

$$Bx_2 = Dy_2 + Pw_2 \qquad\qquad (8.3)$$

If the two equations are combined by eliminating B, the relationship of the D/P ratio to demethanizer product composition appears:

$$\frac{x_1}{x_2} = \frac{y_1 D/P + w_1}{y_2 D/P + w_2} \qquad\qquad (8.4)$$

Equation (8.4) has a twofold significance. It indicates how the D/P ratio must be adjusted to control y_1 or w_1 for an imposed x_1/x_2 ratio. Second, it gives the x_1/x_2 ratio needed to satisfy a specification on y_1 or w_1 and to provide the desired D/P ratio to meet market demands. In actual practice, D must be manipulated to control w_1. However, the x_1/x_2 ratio should also be adjusted at the demethanizer so that the demands for both products may be satisfied.

example 8.1

Estimate the ratio of methane to ethane in the demethanizer bottom product to satisfy the following specifications:

$$y_1 = 2\% \qquad w_1 = 0.1\%$$

$$y_2 = 98\% \qquad w_2 = 99.7\%$$

$$\frac{D}{P} = 0.2$$

$$\frac{x_1}{x_2} = \frac{0.02(0.2) + 0.001}{0.98(0.2) + 0.997} = 4.19 \times 10^{-3}$$

The relationship among the compositions on the right side of (8.4) is a function of the separation in the top section of the ethylene fractionator:

$$S = \frac{y_1/w_1}{y_2/w_2} = \frac{y_1/y_2}{w_1/w_2} \qquad\qquad (8.5)$$

This statement is required for a complete description of the top section.

example 8.2

Starting with the conditions stated above, determine how the D/P ratio must be adjusted to change w_1 to 0.15 and 0.05 percent, assuming a constant separation.

$$S = \frac{0.02/0.98}{0.001/0.997} = 20.35$$

Solving for y_1 for new values of w_1 requires substituting $1 - y_1$ for y_2 and $0.998 - w_1$ for w_2, along with rearranging (8.5):

$$y_1 = \frac{Sw_1}{Sw_1 + w_2} = \frac{Sw_1}{Sw_1 + (0.998 - w_1)}$$

For $w_1 = 0.15\%$:

$$y_1 = \frac{20.35(0.0015)}{20.35(0.0015) + 0.998 - 0.0015} = 0.0297$$

For $w_1 = 0.05\%$:

$$y_1 = \frac{20.35(0.0005)}{20.35(0.0005) + 0.998 - 0.0005} = 0.0101$$

Next the D/P ratio may be found by rearranging (8.4):

$$\frac{D}{P} = \frac{w_2 x_1/x_2 - w_1}{y_1 - y_2 x_1/x_2}$$

For $w_1 = 0.15\%$,

$$\frac{D}{P} = \frac{0.9965(4.19 \times 10^{-3}) - 0.0015}{0.0297 - 0.9703(4.19 \times 10^{-3})} = 0.104$$

For $w_1 = 0.05\%$,

$$\frac{D}{P} = \frac{0.9975(4.19 \times 10^{-3}) - 0.0005}{0.0101 - 0.9899(4.19 \times 10^{-3})} = 0.618$$

The results are summarized in Table 8.1

The control-system configuration for the ethylene fractionator depends on the flow rates and compositions of the three product streams. Typically D is much smaller than P, so its principal contribution is to control the methane content of P and is then manipulated for that purpose. Then, as in the ethanol-water column, interaction between the remaining composition loops can be estimated as if the section above the sidestream did not exist.

TABLE 8.1 Effect of Flow Ratio on Methane Content of Ethylene Products

Flow ratio D/P	High grade w_1, %	Low grade y_1, %
0.618	0.05	1.01
0.200	0.10	2.00
0.104	0.15	2.97

In most of these processes, the high-grade ethylene stream is much purer than the ethane recycle. In this light, Λ_{PS} would tend to be very low, such that P should not be used to control its own ethane content. However, the combination of a very high reflux ratio and large number of trays also tends to make Λ_{SV} and possibly $\Lambda_{SV/B}$ very high. The most-favorable relative gain may then be Λ_{SB}. If the bottom-product flow is similar to or less than P, it should be manipulated to control its ethylene content; then the ethane concentration in P should be controlled by manipulating (internal) reflux ratio, i.e., the ratio of internal reflux to P. Because of the difficulties in measuring and controlling internal reflux, this problem is examined next.

Major Sidestreams   The ethylene fractionator seems to fit the description of a column whose largest flow is the sidestream. (This was not true of the ethanol-water column whose bottoms flow was the largest.) Following the guidelines of Chap. 4 for manipulating material flows with maximum accuracy, the largest flow should be used to close the mass balance. This poses a rather difficult problem for sidestream columns because the sidestream is taken from the column itself and therefore has no direct effect on liquid levels.

In general, sidestreams removed above the feed trays are taken from the liquid phase. Figure 8.6 illustrates two possibilities. On the left, a "chimney" tray traps all the liquid, allowing vapor to pass through without contact. The liquid is withdrawn and split between product and reflux, the latter being distributed below the chimney tray. Because all the liquid is removed, the flow rates of both sidestream and reflux are measurable. The "partial trap-out" tray differs from a conventional tray only in providing some additional head for the withdrawal of product; reflux overflows internally and cannot be measured.

Because the flow of a major sidestream may make up a sizeable portion of the liquid reaching the trap-out tray, some provision must be made for maintaining an adequate flow of reflux below that point. For the total trap-out tray, this would be accomplished by placing the returned stream under flow control and using the LC shown to manipulate the sidestream.

Figure 8.7 shows how this might be accomplished with a partial trap-out tray. In this example, the flow rates of distillate and bottoms are too small for level control but instead are used to control compositions. However, both reflux and heat input cannot be used solely for level control because then the internal vapor and liquid traffic would be indeterminate. To solve this dilemma, the accumulator LC sets both reflux and sidestream flow. The two flows are coordinated to hold a constant internal reflux below the sidestream. Internal reflux is controlled by feedforward because it is unmeasurable. The product-flow set point P^* that will provide the

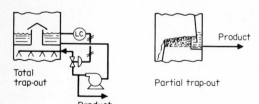

Total
trap-out

Partial trap-out

Product

Product

figure 8.6 *A total trap-out tray requires an external-reflux loop, whereas a partial trap-out tray simply overflows.*

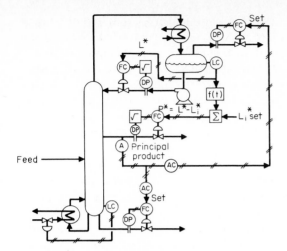

Feed

figure 8.7 *A column whose major product is withdrawn as a liquid sidestream requires internal-reflux control below that tray.*

desired internal reflux L_i^* is calculated from the flow of reflux requested by the level controller:

$$P^* = L^* - L_i^* \tag{8.6}$$

A rising accumulator level will then increase reflux and sidestream flow by the same amount, thereby keeping internal reflux constant while also satisfying the overall material balance. A dynamic compensator is included to simulate the hydraulic delay between the top of the column and the sidestream, to provide a dynamic balance.

In Fig. 8.7 the light impurity in the sidestream is controlled by manipulating distillate flow while the heavy impurity is controlled by manipulating the bottoms flow. The ratio of internal reflux to product flow determines the separation achieved between the components and thereby the amount of product lost with the distillate and bottoms. If one wished to keep this ratio constant, thereby saving energy during reduced production, then a solution must be obtained for P^* in terms of L^* and the desired ratio $(L_i/P)^*$:

$$\left(\frac{L_i}{P}\right)^* = \frac{L^* - P^*}{P^*} = \frac{L^*}{P^*} - 1$$

Solving for P^*,

$$P^* = \frac{L^*}{1 + (L_i/P)^*} \tag{8.7}$$

In effect, Eq. (8.7) is a simple ratio system in which sidestream flow is set in direct ratio to reflux. Again, a dynamic compensator is needed for balancing.

The principal limitation in a ratio system such as this is the possibility of complete loss of reflux should feed flow be interrupted. In this event, P^* could be calculated both by (8.6) and (8.7), using a low-limit setting for L_i^* in Eq. (8.6). Then the lower of the two results would be selected to set sidestream flow, thereby ensuring that internal reflux would not fall below L_i^* upon loss of feed.

A similar problem is encountered when the major product is removed as a vapor sidestream. Again, being the largest stream, it should be used to control a liquid level. Here again, distillate and bottom flows would be manipulated to control the light and heavy impurities, respectively, in the sidestream. If the sidestream is to be used for base-level control, then heat input must be manipulated to maintain the necessary vapor flow above the side draw. If this is not done, a rising base level may open the vapor valve so much that insufficient flow is left to support the liquid on the upper trays; if they begin to weep, base level will continue to rise until the trays are empty.

In Fig. 8.8 differential pressure across the top section of the column is controlled by heat input. Then a rising level will open the vapor valve, but the resulting drop in differential pressure will cause heat input to increase, driving the level downward.

If a differential-pressure measurement is not available, base level might be controlled by setting heat input in ratio to sidestream flow. This arrangement is similar to a reflux-ratio system, tending to keep separation constant over a broad range of flow rates. To protect against loss of feed, there should be a low limit on the heat-input set point. A vapor sidestream may also be manipulated directly to control accumulator level, in which case reflux flow would be fixed or set in ratio to distillate. This leaves base level to be controlled by heat input, which was also necessary in Fig. 8.7. If inverse response prevents this loop from being closed, it may become necessary to control base level with reflux and to set heat input in ratio to either the feed or the sidestream flow.

Determining Configurations   At this writing there is no differentiable process model for sidestream columns similar to that used for calculating relative gains of two-product columns. If plant data, such as those presented by Mosler [3], or rigorous simulation, as described by Doukas and Luyben [4], are available, relative gains can be calculated. The applicable procedure is to perturb selected manipulated variables and measure their effects on the significant controlled variables. The gains for each combination are arranged in a $\mathbf{K}$ matrix, which can be converted into a relative-gain array following the procedure described in Eqs. (5.27) and (5.28).

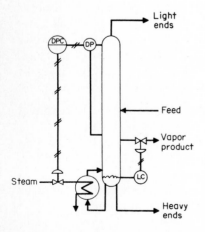

figure 8.8 *When the major product is removed as a vapor sidestream, vapor flow above the point of withdrawal must be controlled.*

example 8.3

The column simulated by Doukas and Luyben [4] separated a mixture of 10 percent benzene, 45 percent toluene, and 45 percent o-xylene in 40 theoretical trays. The benzene, containing 5 percent toluene, was taken overhead; toluene with about 5 percent of each of the other components was removed 13 trays down; and the xylene bottom product contained 5 percent toluene. Reflux ratio was 20.

The gains reported for the column are arranged below:

$$
\mathbf{K} = \begin{array}{c} y_t \\ w_b \\ x_t \end{array}
\begin{array}{|ccc} L/D & P & V \\ -1.986 & 5.24 & 5.984 \\ 0.020 & -\ 0.33 & 2.380 \\ 0.374 & -11.30 & -9.811 \end{array}
$$

where w is the composition of the sidestream whose flow is P, and subscripts t and b refer to toluene and benzene concentrations.

This matrix is then inverted to give:

$$
\mathbf{H} = \begin{array}{c} L/D \\ P \\ V \end{array}
\begin{array}{|ccc} y_t & w_b & x_t \\ -0.550 & 0.296 & -0.264 \\ -0.020 & -0.315 & -0.089 \\ 0.002 & 0.374 & -0.010 \end{array}
$$

When the elements of these two matrices are multiplied (after transposing H), the relative-gain array results:

$$
\Lambda = \begin{array}{c} y_t \\ w_b \\ x_t \end{array}
\begin{array}{|ccc} L/D & P & V \\ 1.093 & -0.104 & 0.012 \\ 0.006 & 0.104 & \mathbf{0.890} \\ -0.099 & \mathbf{1.000} & 0.098 \end{array}
$$

The best selections are indicated in bold type. Substitution of xylene content in the sidestream for benzene content reduces the 1.000 above to 0.858, and the 0.890 to 0.749. This is not enough to affect the decision as to structure.

Lacking such information, the general conclusions developed from the study of two-product columns may be applied. The sidestream column can be considered as a two-product column over its larger section. For example, the ethanol-water column in Fig. 8.4 and the ethylene fractionator in Fig. 8.5 were essentially two-product columns between the sidestream and bottom. Separation must be maintained between those two points, and it determines the quality of the purest product. The separation in the remaining section of the column is then already established, so that control there is entirely by material balance.

A different orientation appears in the alkylation deisobutanizer shown in simplified form in Fig. 8.9. The column was introduced earlier as a two-product separation in Fig. 8.2. However, refiners have discovered that they could use it to separate mixed butanes as well, and thereby eliminate the splitter which had prepared fresh isobutane feed for the alkylation unit.

The principal separation burden exists between n-butane in the distillate, and isobutane in the sidestream, owing to their low relative volatility. Because the alkylate must flow from the upper feed tray past the side draw, its concentration in the n-butane product is not really controllable. The key controlled variables are isobutane in the sidestream and n-butane in the bottom product.

The concentration of isobutane in the sidestream is typically about 1 percent, whereas 4 percent n-butane is allowed in the distillate. This main section of the column, where the bulk of the separation takes place, may be evaluated as a two-product for purposes of estimating relative gains. Given the above product purities and a feed containing 80 percent isobutane, if 80 theoretical trays are used to separate the butanes at a reflux ratio of 1.25, $\delta = -0.202$ and $\mu_{V/F} = -9.20$; then $\Lambda_{DV} = 1.02$. (The negative slope for the V/F curve is the result of a low reflux ratio relative to the large number of trays.)

In recent years, operating practice has been to move the upper feed point higher—many of these columns are now fed at the top, with reflux discontinued. This was a natural result of a progressive increase of isobutane-to-olefin ratio in the reactor and subsequent increase in the isobutane circulation rate through the top of the column. The loss of reflux as manipulated variable has removed the ability to control the n-butane content in the distillate—it now floats on feed composition.

Below the sidestream, compositions are governed by material balances. The concentration of n-butane in the alkylate product is determined by sidestream flow; insufficient flow will force n-butane out the bottom because it has no other exit.

In summary, the isobutane concentration in the sidestream should be controlled by manipulating the ratio of boilup to upper feed, and the n-butane concentration in the alkylate controlled by manipulating the ratio of the sidestream to the lower feed, which is the source of n-butane. Both top and bottom products may be under liquid-level control (in absence of a manipulable reflux).

Although energy efficient, multiple-feed, multiple-product columns such as this are compromises from which pure products are not readily obtainable.

DISTILLATION OF CRUDE FEEDSTOCKS

The most common crude feedstock is, of course, crude oil. It belongs in a class by itself not only because of its wide boiling range but because of the admixture of

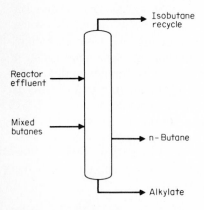

figure 8.9   *Being able to withdraw n-butane as a side product allows the addition of a mixed-butane feed.*

components it contains. A typical crude oil is composed of nearly 50 percent aromatic hydrocarbons (many of multiple rings) along with cycloparaffins as well as straight-chain and branched paraffins. Effluents from fluid catalytic-cracking units and hydrocracking units are by comparison simpler mixtures, the former being rich in olefins while the latter are not.

Crude oil also is rich in low-volatile hydrocarbons — asphalts and waxes which require vacuum distillation. But the subject at hand is multiple-product distillation, characteristic of the pipestill or atmospheric crude column. This unit is similar in many respects to the main fractionator for catalytic-cracker and hydrocracker effluents. Consequently, the problem of crude-oil fractionation is discussed here, considering that the other separations are less difficult.

The Atmospheric Tower   Figure 8.10 describes the essentials of an atmospheric tower for separating crude oil into a typical set of products. The feed is heated by the overhead condenser, two side coolers, and one or more product streams before entering the fired heater. At this point, the temperature is raised as high as possible short of cracking the feed, usually about 700°F but varying with the quality of the crude. The feed then enters the tower between 50 and perhaps 70 percent vaporized, again depending on the crude composition. There is no reboiler as such; the liquid portion falls through the few trays in the bottom section, where it is stripped of additional volatiles with steam. Because there is little vapor flow in this bottom section, the diameter is reduced.

If reflux were provided only at the top of the tower, vapor and liquid rates would vary drastically from top to bottom. The liquid flow is lower at the bottom because the sidestream products are all liquids. The vapor flow is greater at the top because

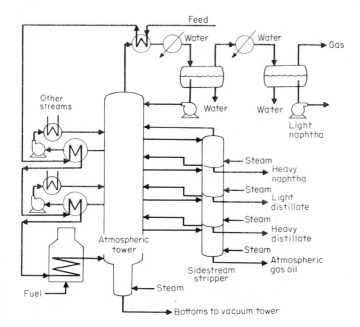

figure 8.10   *All the heat used in crude-oil distillation enters with the feed.*

the large quantity of sensible heat at the bottom is converted into latent heat at the top. In addition, the steam used for sidestream stripping also accumulates in the top.

To distribute liquid and vapor loading more equitably, heat is generally removed at two points in the column as shown. These "pumparound" loops reduce vapor flow and increase liquid flow at the same time. However, the effect is virtually a step change in distribution. Vapor and liquid flow rates are highest just below the tray where cold liquid is returned, and flooding is therefore most likely to occur at that point.

Steam stripping is used to increase the recovery of volatile components by lowering their partial pressure in the vapor at a given liquid temperature. The theory behind steam stripping is analyzed in detail in the next chapter and so will not be developed here. Van Winkle [5] indicates that up to 10 percent additional recovery is possible by steam stripping, using from perhaps 0.3 to 0.5 lb steam per gallon of product, depending on the number of plates in the stripper. In excess of 1.0 lb/gal is necessary to increase the recovery of volatiles from the topped crude by a comparable amount. Steam usage beyond these values achieves relatively little incremental recovery of volatiles.

Product Specifications   A single pure component will, of course, completely vaporize at one temperature. If a binary mixture is distilled batchwise through many trays with a high reflux ratio, the boiling point will change progressively from a lower to a higher limit as the lighter component is removed. This method of distilling results in what is known as the true boiling point (TBP) of each component. As more components are added to a mixture, the plot of temperature vs. percent distilled takes on the appearance of a staircase with one step for each component. But because the reflux ratio and number of trays are less than infinite, some rounding of the steps is always evident. A crude-oil sample, or even one of its fractions, contains so many components that the TBP curve is smooth, as shown in Fig. 8.11.

A second method of evaluating a complex mixture is the American Society for Testing Materials (ASTM) method. A flask without plates is used to distill the sample under very carefully controlled conditions. The separation between compo-

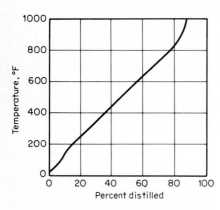

figure 8.11 *A true-boiling-point (TBP) curve for a typical crude oil.*

nents is not so great as with the TBP method, and a different curve of temperature vs. percent distilled will be obtained. Van Winkle [6] offers well-documented comparisons of these methods and others used to determine the composition and behavior of complex mixtures.

The reason for describing these testing methods is that the specifications on the products fractionated from crude oil are often based on their results. Samples of the crude feed and each fraction are distilled in laboratory or automated apparatus. Products are then characterized on the basis of boiling-point range. Although it would be convenient to characterize products based on the full 0 to 100 percent boiling range, it is not practical to do so. The TBP and ASTM curves are probably least reliable at the extremes, so the temperatures at which 5 and 95 percent of the sample are distilled are most often used. Although the 50 percent distilled point gives a measure of the average boiling point of a mixture, it does not indicate purity. Specifying the 5 and 95 percent ASTM or TBP temperatures is equivalent to specifying a light and heavy impurity of a three-component mixture.

Typically a crude feed will be split into about seven products in the atmospheric column. Table 8.2 lists the commonly recognized nomenclature of a set of those products along with their approximate boiling ranges and destinations.

The ultimate uses and therefore the boiling ranges of the products listed will change from one refinery to the next. They depend heavily on refinery facilities, market, and the characteristics of the crude. A refinery with a market for aromatic chemicals, for example, may carefully cut heavy naphthas from several sources for reformer feed. Another refinery without this outlet will concentrate on gasoline blending stocks. Whether a light distillate is to be used for commercial jet fuel, military jet fuel, or heating oil will determine its specifications as to boiling range and flashpoint. By the same token, diesel fuel and heating oil have different specifications.

As in other multiproduct towers, separation is not individually adjustable for each product. In fact, it is scarcely adjustable at all. Consequently only one specification can be met on each product. The chosen specification is usually the 95 percent distilled point or "end point" as it is sometimes called. The only way two specifications on a single product can be controlled is by sacrificing one on an adjacent

TABLE 8.2 Typical Products of an Atmospheric Crude-Oil Column

Product	5–95% ASTM, °F	Carbon atoms per mole	Use
Gas	<100	<5	Liquefied gas and fuel gas
Light naphtha	80–180	5–6	Gasoline blending
Heavy naphtha	180–380	7–9	Reforming to aromatics or gasoline
Light distillate	380–550	10–16	Jet fuel and heating oil
Heavy distillate	550–650	16–20	Diesel fuel and heating oil
Atmospheric gas oil	650–750	20–30	Cracking to olefins and gasoline
Bottoms	>700	>30	Vacuum-distilled to cracker feed, lube oil, residual fuel, asphalt

product. For example, to control both the 5 and the 95 percent distilled temperatures for the heavy naphtha, no control can be exercised over the 95 percent point of the light naphtha. Overlap, if any, between the end point of a light product and the initial boiling point of the next heavier product is essentially fixed by tower design and percent vaporization of the feed.

The heaviest distilled product — atmospheric gas oil — may have no end-point specification. Although production of this stream should be maximized, liquid cannot be totally removed from the column. If it is, there will be no fractionation or scrubbing action below that point and nonvolatiles can be carried upward by the vapor into the gas-oil stream. Since these materials can interfere with cracking of the gas oil, some liquid must be returned to wash them down the tower. This liquid, which is flashed from the feed but not taken as a distillate, is called *overflash*. It typically amounts to perhaps 2 to 4 percent of the feed.

From an economic standpoint, it is desirable to minimize the overflash, which is the same as maximizing the yield of gas oil. The best approach to this problem is to specify the amount of nonvolatiles permitted in the gas oil as measured by color. Then gas-oil flow can be raised to the level at which color bodies begin to appear, indicating insufficient overflash.

Feedforward and Decoupling   For purposes of mathematical analysis, the product breakdown for crude distillation may be likened to a component breakdown for a simpler mixture. A given feedstock may contain — as indicated by TBP or ASTM distillation — a certain percentage of light naphtha, heavy naphtha, etc., in accordance with their specified boiling ranges. The column can do no better than to yield those same percentages as products. Poor separation will cause overlap between adjacent products. Separation can be maximized by heating the feed as high as practicable and using enough stripping steam. For a fixed separation, the flow rates of all the products are determined wholly by material balance. The flow of each product should then be set in ratio to the feed rate.

Nowhere is interaction between compositions more apparent than in an atmospheric crude-oil column because of the multitude of products and their similar flow rates. If the end point for the heavy-naphtha product rises too high, for example, its controller will reduce the flow to force the higher-boiling components further down the column. However, the components rejected from the heavy naphtha will tend to displace heavier components in the light distillate, reducing its end point. The light-distillate end-point controller will then have to increase the flow of light distillate to return the end point to its previous value. Any delay in this control action could allow the disturbance to propagate downward into heavier streams. Tests conducted on crude-oil columns have verified the existence of this partial interaction. For example, an incremental increase of 3 percent in the flow of light naphtha will typically raise the end points of all products between 20 and 30°F. A similar increase in the flow of heavy naphtha raises its own end point and all those below it but has no effect on the light-naphtha product. Disturbances are thus propagated downward but not upward.

Decoupling is therefore applied in the downward direction as shown in Fig. 8.12. Each product-flow controller has as its controlled variable the sum of its own side-

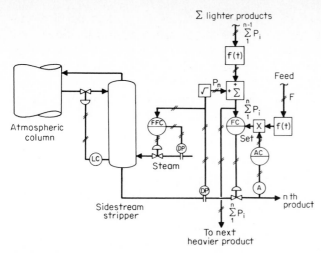

figure 8.12 *Crude-column decoupling requires application in a downward direction only.*

stream plus all lighter products. Then if any change is made to a flow higher in the column, it will be reflected in the inputs to all lower flow controllers and cause opposite adjustments to all lower sidestreams. With appropriate dynamic compensation in the form of dead time plus lag, compositions of lower products should then be immune to disturbances in higher streams.

Note that Fig. 8.12 includes a dynamically compensated feedforward signal from feed rate to each product-flow controller. The product composition controller then sets the ratio of the total flow of light material (from that point upward) to the column feed. Change in feed composition and product specifications require that ratio to be readjusted. In addition, the flow of steam to each sidestream stripping column is set in ratio to the flow leaving it.

Heat Recovery Heat is recovered from the crude-oil column wherever possible. Most flowsheets are far more complex in the way of heat-recovery schemes than Fig. 8.10, but in every case, cooling is directed first to the crude feed, then to other process users, and finally to cooling water. An important objective in operating the atmospheric column is to reject as little heat to cooling water as possible. Any incremental heat-flow increase to cooling water — all other things being constant — must be made up by fuel at the charge heater.

A tradeoff exists between heat recovery and separation, however. As more heat is recovered at the *lower* pumparound loop and less is lost to the reflux condenser, liquid flow between these points will be reduced. This reduces reflux ratio and hence separation between the components of the lighter products. Ryskamp [7] devised the system shown in Fig. 8.13 to maintain reflux at some designated optimum flow.

In this system, the overhead vapor temperature, which determines the end point of the light naphtha, is controlled by manipulating reflux. If the flow required to do this is more than the designated optimum, the reflux FC will increase the set point

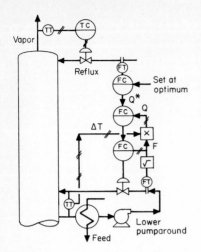

figure 8.13 *Heat removed from the bottom pumparound is adjusted to develop the optimum flow of reflux.*

Q^* to the lower heat-flow FC, to improve heat recovery. Similarly, if reflux is too low, separation suffers, and the reflux FC will reduce bottom heat recovery.

Flooding is most likely to develop where cold liquid is returned at the *upper* pumparound because both vapor and liquid internal flow rates are maximum at that point. To maximize separation from that point downward, internal traffic should be controlled just below the flooding limit. Wade and Ryskamp [8] developed a computer program to calculate these internal flow rates from external flows, temperatures, and specific-gravity information. The set point for the upper-pumparound heat-flow controller was then positioned to keep the column at a designated percent of flooding.

HEAT-RECOVERY STRATEGIES

The increasing cost of fuel and regulations placed on discharge of water and waste heat have spurred innovation in heat recovery. There are many ways to improve the thermodynamic efficiency of distillation, some of which are introduced in Chap. 2. At this point we are concerned with the control of schemes involving heat recovery because they tend to remove a degree of freedom and at the same time present an additional source of disturbances to a distillation unit. Each process arrangement has its own peculiar characteristics.

Series and Parallel Cascading   A common arrangement of columns for separating a ternary mixture appears in Fig. 8.14. This has been used for producing high-purity methanol from a feed containing ethanol and water — ethanol leaves the bottom of the second column. Because steam introduced at the trim reboiler is used only once, its flow should be minimized or discontinued altogether. However, this removes a degree of freedom because the columns then have the same heat flux.

Pressure in column 2 is controlled in a conventional manner, floating on the cooling source if practical. Pressure control for column 1 presents additional problems. There is a temptation to use the vapor valve feeding column 2 to control

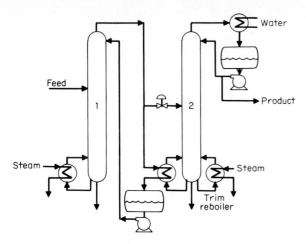

figure 8.14 *The pressure in column 1 is dependent on that in column 2 and the heat transferred between them.*

pressure in column 1, but that loop will interfere with product-quality control. The pressure is a function of the temperature at which condensing takes place in the condenser-reboiler. This, in turn, varies with heat load, compositions on both sides of the heat-transfer surface, and pressure in column 2. If pressure in column 1 is held constant *without* interfering with heat transfer, then the composition of the feed to column 2 must change with heat load.

In essence, the pressure-control problem is the same for both columns — the controller *must* interfere with heat transfer. Alternatively, column 1 could be left without pressure control and allowed to float on column 2. This would result in maximum relative volatility while at the same time eliminating a control loop and therefore would seem to be the most-effective strategy.

In a similar installation featuring a deisobutanizer heated by the vapor from a debutanizer, Ryskamp [9] fixed the heat-transfer rate between the two columns by setting the flow of condensate leaving the common heat exchanger. The pressure in column 1 was then controlled by manipulating heat input. However, this required that the composition of its bottom product had to be controlled by feed rate, a scheme which has limited application. The top products from both columns were controlled by D/V ratio. In the event that heat flow from column 1 is insufficient to provide the separation needed by column 2, the auxiliary reboiler must be used to control bottom-product quality.

Parallel cascading is described for a propane-propylene separation in Fig. 2.24. Here both columns split the same feed into identical products, using a common heat flux. But because they are operated at different pressures, the columns must have different numbers of trays or different reflux ratios, or both.

In any case, control of pressure in column 1 is the same problem as for Fig. 8.14 and should be approached in the same way. Ideally, its pressure should float in column 2 which should float on the condenser, so that they both are as efficient as possible and their relative volatilities change simultaneously.

Composition controls for column 1 would be determined by a relative-gain analysis, probably resulting in an SV or SV/B configuration. The top composition for column 2 can be controlled identically, but its bottom composition cannot. Because heat input to column 2 cannot be manipulated independently, its bottom composition can only be controlled by feed rate. This does not face the same limitation as mentioned for the series-cascaded columns above, because the feed to column 2 can be balanced against that to column 1, leaving total feed constant. Heat input to column 1 should be set proportional to its feed rate or its bottom flow. Then if separation in column 2 needs adjusting, a change in its feed rate will also result in an opposite change in heat input, providing corrective action from two directions.

This system was examined by Frey et al. [10] to determine the interactions between the four composition loops. They discovered the need to decouple column 2 from the influence of the bottom-composition controller of column 1 which manipulated heat input. This was achieved by manipulating the feed rate to column 2 in ratio to the heat input to column 1. When this was done, the 4×4 RGA was reduced to two independent 2×2 arrays whose relative gains were nearly identical. Both columns used reflux ratio to control top composition and V/F to control bottom composition; in column 1, V was manipulated relative to F, while in column 2, F was manipulated in proportion to V.

Another possibility these authors investigated was the introduction of the bottom product from column 1 into the base of column 2. This allows recovery of some sensible heat while eliminating the requirement to control both bottom streams separately. Reflux ratio was used to control both top-product compositions and V to control the single bottom composition. If the feed rates are not properly balanced, the separation achieved in the columns will differ, causing unequal bottom compositions. The optimum feed ratio will produce equal bottom compositions; once that value is found, it should remain constant unless the columns themselves change characteristics relative to each other.

Heat Recovery from a Separate Unit   Heat is often recovered from a reactor effluent stream for columns which are separating its products, or even for a column in an unrelated process. Such a source tends to be unregulated and is typically supplemented by a manipulable source. Control over heat input then requires a total-flow measurement, such as differential pressure across the column. Alternatively, a heat-flow calculation such as shown in Fig. 8.13 could be made for both sources, and their sum could serve as the manipulated flow. A third possibility would have the sum of reflux and distillate set to control vapor loading, and let pressure or accumulator level manipulate heat input.

Another situation develops when the heat removed from one unit and that used by another must *both* be controlled. An example of this situation is shown in Fig. 8.15, where heat rejected from a crude-oil column is used to reboil a debutanizer. Heat flow from the crude column can be calculated from total flow and temperature entering and leaving the column, as shown in Fig. 8.13; a total oil-flow controller would manipulate the valve through the feed preheater.

Heat input to the debutanizer could be determined from column differential pressure or calculated from the flow of oil to its reboiler and its inlet and outlet

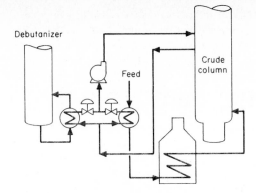

Debutanizer

Feed

Crude column

figure 8.15 *Heat removed from the crude column and that sent to the debutanizer both require control.*

temperatures; the valve to its reboiler would be manipulated. In the absence of a suitable indication of heat flow, the reboiler valve could be manipulated to control debutanizer pressure or accumulator level as noted above.

Coupling between the two loops would be minimal. Changes in heat flow from the crude column can be achieved by moving the feed preheater valve with little upset to the debutanizer. And if the preheater valve is used to control total flow, then changes in oil diversion to the debutanizer will not bring about changes in total heat removal. Preheater outlet temperature will be affected however, requiring an adjustment to the firing rate of the feed heater. If this disturbance is severe or frequent, fuel flow can be adjusted as a function of preheater outlet temperature by feedforward action.

REFERENCES

1. Rathore, R. N. S., K. A. VanWormer, and G. J. Powers: "Synthesis Strategies for Multicomponent Separation Systems with Energy Integration," MIT Industrial Liaison Program, April 1974.
2. Shinskey, F. G.: *Controlling Multivariable Processes,* Instrument Society of America, Research Triangle Park, N. C., 1981, pp. 35–47.
3. Mosler, H. A.: "Control of Sidestream and Energy Conservation Distillation Towers," presented at AIChE Workshop on Industrial Process Control, Tampa, Fla., November 1974.
4. Doukas, N., and W. L. Luyben: "Control of Sidestream Columns Separating Ternary Mixtures," *In Tech,* June 1978.
5. Van Winkle, M.: *Distillation,* p. 359, McGraw-Hill, New York, 1967.
6. Reference 5, pp. 127 ff.
7. Ryskamp, C. J., H. L. Wade, and R. B. Britton: "Improved Crude Unit Operation," *Hydrocarbon Proc.,* May 1976.
8. Wade, H. L., and C. J. Ryskamp: "On-line Calculation of Crude Tower Tray Loading," *Hydrocarbon Proc.,* November 1977.
9. Ryskamp, C. J.: "New Strategy Improves Dual Composition Control," *Hydrocarbon Proc.,* June 1980.
10. Frey, R. M., M. F. Doherty, J. M. Douglas, and M. F. Malone: "Controlling Thermally Linked Distillation Columns," *Ind. Eng. Chem. Process Des. Dev.,* to be published.

CHAPTER NINE

Absorption, Stripping, and Extractive Distillation

Absorption is the process of capturing or condensing certain components in a gaseous stream by a liquid absorbent. Stripping is the reverse: absorbed components are removed or desorbed from the liquid. The two processes may be conducted independently of one another, such as the absorption of sulfur dioxide from flue gas or the stripping of hydrogen sulfide from a "sour-water" stream. More often, however, they are combined to recover certain components selectively from a gas stream. In this way, natural-gas "liquids," i.e., ethane, propane, butane, and pentane, are commonly recovered from the gas by absorption in oil followed by stripping.

Some absorbents are chosen for their selectivity: for example, monoethanol-amine is used to remove carbon dioxide. In some cases it is possible to affect the absorption or desorption by altering the chemical composition of the absorbent. In every case, regardless of whether a physical or chemical equilibrium is involved, the condition of the absorbent must be altered to shift the equilibrium between absorption and desorption.

These processes are very similar to distillation. In fact, the absorber resembles the top section of a distillation column since the feed enters the bottom and there is no reboiler; the absorbent is analogous to the reflux. As will be seen, the vapor-liquid equilibria are essentially the same as in a distillation column.

The stripping column, on the other hand, may take on any one of several configurations, having a reboiler, a condenser, both, or neither. Its simplest configuration appears like the bottom of a distillation column, as the liquid feed enters at the top; open steam, stripping gas, or reboiled vapor enters at the bottom.

When open steam is used as the stripping medium, another possibility arises. If the liquid in contact with the steam is immiscible with water, an unusual vapor-liquid equilibrium appears. This is sufficiently different from those relationships already described that a separate section later in the chapter is devoted to steam stripping.

Also included in this chapter is a presentation of extractive distillation. As will be seen, the extractive-distillation column differs from a conventional unit only in the addition of an extractant stream. Since its function is similar to that of an absorbent, consideration in this chapter is natural. Furthermore, the stripping column used to recover the extractant is identical to that used in absorbent recovery.

PHYSICAL ABSORPTION AND DESORPTION

This section considers the absorption of components of a vapor mixture by a non-selective solvent and their release from solution. Equilibrium here is a physical rather than a chemical phenomenon, related to the vapor pressures of the components under various conditions. The solvent is essentially an inert carrier with the function of reducing the concentrations of the absorbed components in the liquid phase below what they would be in conventional distillation. This enhances recovery of volatile components without resorting to cryogenic operation.

Material Balances The flowsheet for a simple absorber appears in Fig. 9.1. Feed gas enters at mol rate V and concentration z. The gas engages a counter-current flow of absorbent at mole rate L and composition w; product compositions are y and x, either one of which may be controlled, but usually not both.

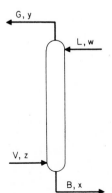

figure 9.1 *This representation may be used either for an absorber or a simple stripping column.*

Overall and component balances are

$$V + L = G + B \tag{9.1}$$

and

$$Vz_i + Lw_i = Gy_i + Bx_i \tag{9.2}$$

where G and B are mol rates of gas and liquid effluents, and subscript i denotes any of the individual components. The only manipulable flow is L, V being given, and G and B being dependent variables. Because of their dependence, it is customary to eliminate G and B from the material balance by basing compositions on inlet flows alone. Then (9.2) can be rewritten

$$V(z_i - Y_i) = L(X_i - w_i) \tag{9.3}$$

where Y_i and X_i are concentrations corrected to inlet flow rates.

The correction factor required varies with the amount of material transferred between vapor and liquid phases. Since

$$VY_i = Gy_i$$

$$y_i = Y_i \frac{V}{G} \tag{9.4}$$

The amount of material transferred is $V - G$, which is the total amount described by Eq. (9.3) for all components:

$$V - G = V \sum (z_i - Y_i) \tag{9.5}$$

Combining the last two expressions gives y_i in terms of Y_i and z_i:

$$y_i = \frac{Y_i}{1 - \sum (z_i - Y_i)} \tag{9.6}$$

By a similar manipulation,

$$x_i = \frac{X_i}{1 + \sum (w_i - X_i)} \tag{9.7}$$

The simple stripping column has the same flowsheet as the absorber, Fig. 9.1. The principal difference is that $G > V$ and $L > B$. In any case, the same equations apply because the algebra is bidirectional. The distinguishing feature between the two operations is the shift in vapor-liquid equilibria.

Vapor-Liquid Equilibria   In a distillation column, components having equilibrium vaporization ratios (K factors) exceeding unity tend to concentrate in the vapor phase, while those having lower values concentrate in the liquid. The same relationship governs absorption and stripping.

Absorption takes place when the vapor pressure of the component is below the total pressure at the temperature of the solvent, or $K_i < 1.0$. To reverse the process, pressure is dropped and/or temperature increased so that $K_i > 1.0$. The selection between components is therefore based on their different K values. For example,

n-butane is preferentially removed from a natural-gas stream at 300 lb/in² abs and 50°F because its K value is 0.12 compared to methane at 8.0.

An isothermal absorber with an infinite number of stages could develop exit concentrations that approach equilibrium with inlet concentrations:

$$y_i \to K_i w_i \tag{9.8}$$

$$x_i \to \frac{z_i}{K_i} \tag{9.9}$$

However, real absorbers operate well away from these conditions and are best modeled by means of the absorption factor A:

$$A_i = \frac{L}{K_i V} \tag{9.10}$$

Each component i has its own absorption factor. Product compositions then are related to absorption factors and the number of theoretical stages n by the Kremser-Brown-Souders equation [1]:

$$\frac{z_i - Y_i}{z_i - K_i w_i} = \frac{1 - A_i^{-n}}{1 - A_i^{-(n+1)}} \tag{9.11}$$

First, individual values of A_i are calculated from K_i at the temperature and pressure of the absorber along with the selected value of L/V. Knowing n and the composition of the two inflows, all significant values of Y_i may be calculated. Then corresponding values of X_i can be determined from material-balance Eq. (9.3). Finally, values of y_i and x_i may be calculated using (9.6) and (9.7).

example 9.1

A stream of natural gas enters an absorber at 300 lb/in² abs and 50°F. Feed compositions and K values are given below. Estimate product compositions given four theoretical stages and an L/V ratio of 0.5.

	z_i	K_i	w_i
Methane (1)	0.908	8.0	0
Ethane (2)	0.0504	1.46	0
Propane (3)	0.0190	0.417	0.0002
n-Butane (4)	0.0100	0.120	0.0005

$$A_1 = \frac{0.5}{8.0} = 0.0625, \qquad \frac{1 - A_1^{-4}}{1 - A_1^{-5}} = 0.0625$$

$$A_2 = \frac{0.5}{1.46} = 0.342 \qquad\qquad = 0.339$$

$$A_3 = \frac{0.5}{0.417} = 1.199 \qquad\qquad = 0.865$$

$$A_4 = \frac{0.5}{0.120} = 4.167 \qquad\qquad = 0.997$$

$$Y_1 = 0.908 - 0.0625(0.908) = 0.8512$$

$$Y_2 = 0.0504 - 0.339(0.0504) = 0.0333$$

$$Y_3 = 0.0190 - 0.865[0.0190 - 0.417(0.0002)] = 0.0026$$

$$Y_4 = 0.0100 - 0.997[0.0100 - 0.120(0.0005)] = 0.00085$$

$$
\begin{aligned}
\Sigma(z_i - Y_i) = \; & 0.908 - 0.8512 \\
& 0.0504 - 0.0333 \\
& 0.0190 - 0.0026 \\
& \underline{0.0100 - 0.00085} \\
& 0.09945 \qquad 1 - 0.09945 = 0.901
\end{aligned}
$$

$$y_1 = \frac{0.8512}{0.901} = 0.945$$

$$y_2 = \frac{0.0333}{0.901} = 0.0370$$

$$y_3 = \frac{0.0026}{0.901} = 0.0029$$

$$y_4 = \frac{0.00085}{0.901} = 0.00094$$

$$X_1 = \frac{0.908 - 0.8512}{0.5} = 0.1136$$

$$X_2 = \frac{0.0504 - 0.0333}{0.5} = 0.0342$$

$$X_3 = \frac{0.0190 - 0.0026}{0.5} + 0.0002 = 0.0330$$

$$X_4 = \frac{0.0100 - 0.00085}{0.5} + 0.0005 = 0.0188$$

$$
\begin{aligned}
\Sigma(X_i - w_i) = \; & 0.1136 \\
& 0.0342 \\
& 0.0328 \\
& \underline{0.0183} \\
& 0.1989
\end{aligned}
$$

$$x_1 = \frac{0.1136}{1.199} = 0.0948$$

$$x_2 = \frac{0.0342}{1.199} = 0.0285$$

$$x_3 = \frac{0.0330}{1.199} = 0.0275$$

$$x_4 = \frac{0.0188}{1.199} = 0.0157$$

The same procedure can be applied to a simple stripping column. The stripping factor is the inverse of the absorption factor, thereby reversing the sign of the exponent and calculating liquid composition:

$$\frac{w_i - X_i}{w_i - z_i/K_i} = \frac{1 - A_i^n}{1 - A_i^{n+1}}$$

(9.12)

Multistage Absorbers   As in any equilibrium process, multiple stages of vapor-liquid contact will improve the separation between the overhead and bottom products. Consequently, absorption and stripping columns are typically fitted with trays or packing to provide several equilibrium stages just like distillation columns. However, in the case of absorbers at least, all stages are not equally effective. Owens and Maddox [2] discovered that 80 percent of the mass transfer in an absorber takes place in the top and bottom stages. Although their survey indicates that most absorbers contain 6 to 10 theoretical stages, only the terminal ones are particularly effective. The ineffectiveness of interior stages is due to an unusual temperature profile. Temperatures of the terminal trays are related closely to the temperatures of the absorbent and feed gas. But the absorption which takes place in these stages and within the column causes internal temperatures to rise, so that the terminal trays are coolest. As a result, the equilibrium in the interior trays is shifted away from absorption. This situation can only be corrected with interstage cooling like that used in crude-oil distillation. If more than 6 to 10 stages are needed, this cooling must be provided.

Absorbers and most stripping columns have an overhead product in the vapor phase. Therefore, the properties of partial condensation as described earlier apply. If the absorbent or stripper reflux temperature is variable as a function of coolant temperature or heat load, column pressure should be adjusted accordingly. The ideal program of pressure vs. temperature would keep the K value of the key component constant. Then the concentration of that component in the off-gas is not likely to change with temperature. If the feed and absorbent concentrations and temperatures are constant, then holding a constant L/V ratio should regulate the compositions of both product streams. Should either inlet temperature rise, L/V should be increased to compensate for the increase in K values. Similarly, a rise in the concentration of the less-volatile components in the feed should be accommodated by an increase in L/V. This change ought to be detectable as a rise in mid-column temperature caused by more absorption. All these disturbances can be countered by a mid-column temperature controller manipulating L or L/V.

Another equilibrium stage can be added if part of the absorbent is mixed with the gas feed upstream of the feed cooler, as shown in Fig. 9.2. Absorbent used in this

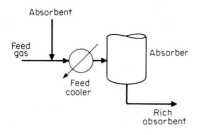

figure 9.2   *Mixing some absorbent with the feed entering the cooler can add another equilibrium stage and increase the capacity of the absorber.*

fashion is often called a "sponge oil." In absorbing some of the components from the vapor, it releases heat which is removed by the cooler. Additionally, the presence of the liquid in the cooler promotes heat transfer. Thus part of the absorption is already complete upstream of the absorber without raising feed temperature. Furthermore, both the liquid and the vapor loading in the absorber are reduced: vapor loading, in that some of the feed is already condensed; liquid loading, in that part of the absorbent bypasses the tower. This practice is especially helpful in increasing the capacity of an already heavily loaded absorber.

Properties of the Absorbent   Absorbers have an additional degree of freedom which distillation columns do not — the choice of the absorbent. This choice is important even in nonselective systems. In these systems, *molar* concentrations of volatile components absorbed in a solvent are essentially independent of the properties of the solvent. However, Treybal [3] points out that their *weight* concentrations then vary with the *molecular weight* of the solvent. Consequently, heavier oils tend to recover fewer pounds or barrels of volatile hydrocarbons per pound or barrel of oil. They consequently require higher circulation rates and consume more energy than lighter solvents. The principal disadvantage of lighter absorbents is that their volatility tends to favor somewhat higher loss with the effluent vapors. Absorbents with a combination of low molecular weight and low volatility are thereby preferred. Water, for example, makes a good absorbent for some vapors.

To illustrate the effect of molecular weight, Table 9.1 gives the equilibrium concentrations of methane and n-butane in the vapor and liquid leaving the top tray of an absorber at 300 lb/in^2 abs and 50°F. Weight percentages of both components in 500°F boiling-point oil (molecular weight 200) are compared to those achieved using heptane (molecular weight 100). Where heptane is used as the absorbent, the vapor leaving the top tray contains 0.2 mol % heptane under the stated conditions. Loss of oil is much lower if entrainment is nil. However, the oil flow required to reach the equilibrium mole percent given is almost twice as great as that required for heptane, which would result in proportionately heavier entrainment losses.

Reducing the molecular weight of the absorbent has the obvious effect of requiring a lower liquid-vapor mass ratio in the absorber. This increases the capacity of a tower for both vapor and molar liquid flow rates. Thus, towers that are capacity-limited can actually have their production rate *increased* while using a *lower* mass flow of absorbent just by reducing its molecular weight.

When absorption and stripping are combined as in Fig. 9.3, the stripping column also benefits by the reduced molecular weight of the absorbent. Both its higher vapor pressure and its lower flow rate reduce the energy required to heat it to the boiling point. By the same token, less sensible heat must be removed as the lean

TABLE 9.1 Equilibrium Concentrations of Vapor and Liquid at 300 lb/in² abs and 50°F

		Mol %		Wt %	
	K	Vapor	Liquid	In oil	In heptane
Methane	8.0	97.9	12.2	1.3	2.4
n-Butane	0.1	2.1	21.0	8.2	15.0

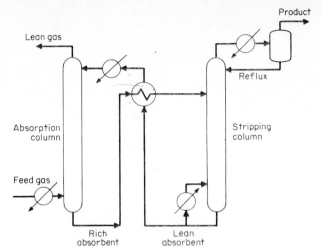

figure 9.3 *This combination of absorption and stripping circulates an absorbent in a closed cycle.*

absorbent is returned to the absorber. If the absorber described in Table 9.1 operates in conjunction with a stripping column held at 100 lb/in² abs, the 500°F oil would have to be heated to 650°F to boil whereas the heptane would boil at 360°F. The sensible heat added by the reboiler and removed by the cooler to return the absorbent to 50°F would be 3.5 times as great for the oil as for the heptane, assuming equal molar flow rates. More than just the sensible heat must be transferred, since the product must be first condensed and then vaporized. But the sensible heat represents a total loss, and its reduction can greatly improve the thermal efficiency of the unit.

It is possible to use the heavier liquids recovered from natural gas as the absorbent. The vapor temperature at the top of the stripping column is the boiling point of the reflux. By adjusting this temperature, reflux composition and hence absorbent composition can be varied.

Ideally, the system should be operated in such a way that absorbent does not have to be added or withdrawn. If liquid product is not removed as fast as it enters, it will accumulate. A means for balancing absorbent inventory recommended by C. J. Ryskamp is illustrated in Fig. 9.4. The flow of the lean absorbent is controlled at the absorber; its inventory is maintained at the stripper base and is controlled by adjusting overhead vapor temperature. Any accumulation will cause the level controller to raise overhead temperature, thereby increasing the withdrawal of product. A decline in absorbent inventory will be countered by lowering the temperature set point and increasing reflux, with less product withdrawn.

Reboiler outlet temperature determines the boiling point of the lean absorbent, which influences the losses of volatiles out the absorber. If this temperature is raised, the volatile content of the absorbent will be reduced and hence losses as well. However, this action consumes more reboiler heat and also increases the cooling that must be applied to the lean absorbent.

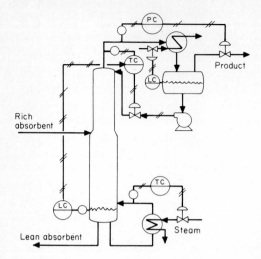

figure 9.4 *The base-level control-ler regulates absorbent inventory by adjusting the dew point of the overhead product.*

If the pressure of the stripper is allowed to float on condenser temperature, both overhead and reboiler temperature measurements should be pressure-compensated.

Cryogenic Stripping   Higher liquid recovery from a natural-gas stream is attainable through cryogenic processing, illustrated in Fig. 9.5. The feed gas at a high pressure is chilled against overhead vapor from the stripping column. At this point, some of the heavier liquid components condense and are separated from the vapor and fed to the column. Flashing will occur as the liquid pressure is dropped through the control valve — a recovery turbine here would recover work and produce less flashing.

Gas leaving the separator is passed through either an expander or a bypass valve to condense liquid. In the process of crossing through the critical region, some

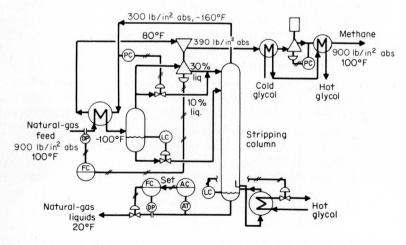

figure 9.5 *Light hydrocarbons may be recovered from natural gas by using expansion of the chilled feed to produce reflux.*

liquid will condense in the valve because of the Joule-Thompson effect. But the expander will condense more liquid and recover work as well. A look at the pressure-enthalpy diagram for methane in Fig. 9.6 will demonstrate this point. The gas expands adiabatically through the valve but nearly isentropically through the expander. Houghton and McLay [4] report expansion efficiencies of 78 to 85 percent across a flow range of 40 to 120 percent of design conditions.

The power recovered by the expander is delivered to a compressor mounted on the same shaft for recompression of the reheated overhead gas. The power recovered falls short of that needed to restore the gas to its original pressure by the combined efficiencies of compressor and expander. Supplementary compression is usually used to restore the gas to its original pressure for transmission. The discharged gas must then be cooled, and the heat of compression thereby recovered is more than enough to reboil the stripping column. Since this energy is returned to the overhead vapor, it does not leave the system; consequently, the work of compression must ultimately be removed by cooling against air or water. A glycol-water mixture is commonly used as the medium to transfer heat from the compressor to the reboiler because the boiling point of the bottom product is typically below the freezing point of water.

The fractionation achievable in the stripping column is a function of the boilup-bottoms ratio. But the boilup is wholly dependent on the liquid condensed in cooling and expanding the feed gas. Therefore, maximum liquefaction should be pursued to maximize recovery of the condensible gases in the feed. This goal is only achievable if no throttling through control valves is allowed. Feed flow should be controlled by adjusting the nozzles in the expander as described in Ref. 4. The bypass valve should remain closed. Column pressure can be controlled by manipu-

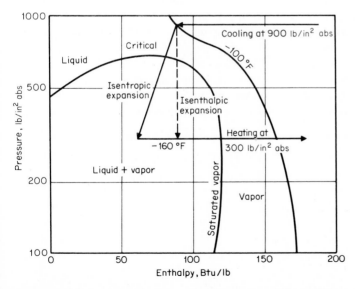

figure 9.6 *Isentropic expansion condenses more liquid than adiabatic expansion and recovers power as well. (Enthalpy of gas is zero at absolute zero temperature.)*

lating the bypass valve since this valve takes power from the expander and therefore adds energy to the column. For a given pressure drop, the expander can condense typically 20 to 30 percent of the gas while the valve will condense less than 10 percent.

If the valve is left closed, column pressure will float on the gas-to-gas heat exchanger, feed-gas temperature, and the supplementary compressor. Increasing the feed rate tends to augment the temperature drop across the heat-transfer surface, raising the expander feed temperature and column pressure as well. However, with the bypass valve closed, the column will always be at its minimum attainable pressure, providing the maximum relative volatility, power recovery, and condensate recovery attainable at any given feed rate. A floating-pressure control system, as described in Fig. 2.20, should be used to actuate the bypass valve.

The objective of the column is to maximize recovery of bottom product with a specification on its methane concentration or methane-ethane ratio. Although only 30 percent of the feed gas may be liquefied, the quantity of recoverable hydrocarbons is much less. As a result, bottom-product flow is smaller than the boilup rate, which leads to the control-loop arrangement shown in Fig. 9.5. This particular arrangement minimizes the effects of the heat balance on product quality—an important consideration in this application, where liquefaction is variable, as is the temperature of the heating medium.

CHEMICAL ABSORPTION AND DESORPTION

This section considers selective absorption based on chemical as well as physical equilibria. For example, certain specific components may be removed from a gas stream by an absorbent having a chemical attraction for them. In this way a polar component may be selectively removed from a mixture of less polar gases by a polar solvent. A notable example is the selective absorption of carbon dioxide from a mixture of carbon monoxide, hydrogen, and nitrogen by a solution of an ethanolamine in water. The ethanolamines are water-soluble, volatile organic bases having an affinity for acid gases. Other acid gases such as sulfur dioxide or hydrogen sulfide would also be absorbed if present. The ethanolamine, being volatile, is recovered in a reboiled stripping column—the circuit is essentially that shown in Fig. 9.3. Reference 5 gives operating guides to setting flow rates, compositions, and temperatures.

Chemical Equilibria   In the preceding example, absorption is selective whereas desorption is physical. The next group of systems to be studied are those in which equilibria are shifted between absorption and desorption by chemical means. These systems all feature acid-base reactions, where the absorbed or stripped components are acid or basic gases such as carbon dioxide or ammonia. The acid gases are absorbed by basic solutions and desorbed by the addition of a nonvolatile acid. Conversely, basic gases are absorbed by acids and stripped in the presence of a nonvolatile base.

Consider the problem of absorbing ammonia from a mixture of other gases, using water as the absorbent. The concentration of free NH_3 in the water solution is determined by the water temperature and the partial pressure of NH_3 in the gas. A chemical equilibrium also exists between the NH_3 in solution and the NH_4^+ ion:

$$NH_3 + H_2O \overset{K_B}{\rightleftharpoons} NH_4^+ + OH^- \tag{9.13}$$

Because of this ionization, the total ammonia in the liquid phase will always exceed the free ammonia in equilibrium with the gas.

The ionization equilibrium is governed by an ionization constant K_B, defined as the ratio of the product of the ions to the free ammonia in solution:

$$K_B = \frac{[NH_4^+][OH^-]}{[NH_3]} \tag{9.14}$$

The brackets indicate an expression of concentration of the component within. The units customarily used are gram ions per liter of solution, known as normality (N).

Let the total ammonia in solution be identified as x_B:

$$x_B = [NH_3] + [NH_4^+] \tag{9.15}$$

Then, from (9.14),

$$x_B = [NH_3]\left(1 + \frac{K_B}{[OH^-]}\right) \tag{9.16}$$

In any aqueous system, the hydroxyl-ion concentration $[OH^-]$ is determinable as a function of the hydrogen-ion concentration $[H^+]$:

$$K_W = [H^+][OH^-] \tag{9.17}$$

Constant K_W is known as the ionization constant for water and is 10^{-14} at 25°C. Finally, the hydrogen-ion concentration is measurable as solution pH:

$$[H^+] = 10^{-pH} \tag{9.18}$$

Inserting the last two expressions into Eq. (9.16) yields x_B as a function of $[NH_3]$ and pH:

$$x_B = [NH_3](1 + 10^{pK_W - pK_B - pH}) \tag{9.19}$$

Here pK_W and pK_B are defined as $-\log K_W$ and $-\log K_B$, respectively, to place them in the same terms as the measurable pH. For ammonia, pK_B is 4.75 and pK_W is 14 at 25°C.

pH Adjustment Note that when the exponent in Eq. (9.19) is zero, $x_B = 2[NH_3]$; that is, half the total ammonia is ionized. For ammonia at 25°C, this condition occurs at pH 9.25. Every unit increase in pH reduces the ammonium-ion concentration by one decade such that at pH 10.25, $x_B = 1.1[NH_3]$. Each unit decrease in pH increases the ion concentration one decade such that at pH 8.25, $x_B = 11[NH_3]$. Thus the equilibrium concentration of total ammonia in solution can be shifted by a factor of 10 by adjusting solution pH between 8.25 and 10.25. Because this relationship is extremely nonlinear, it is presented best in tabular form (see Table 9.2).

TABLE 9.2   Effect of pH on the Absorption of Ammonia at 25°C

pH	$x_B/[NH_3]$
7	178.8
8	18.78
9	2.79
10	1.179
11	1.018
12	1.002

As a gas is absorbed, heat is evolved which raises the temperature of the liquid, thereby tending to impede further absorption. Similarly, as ammonia is absorbed, the pH of the solution rises, also impeding further absorption. To maintain the pH at a favorable level, an acid reagent must be added to the solution at a rate equal to the rate of absorption.

Stripping involves shifting the equilibrium by raising temperature, lowering pressure, and, in the case of ammonia, elevating the pH with a nonvolatile base. The amount of base required to release the ammonia is equal to the amount of acid used to absorb it. Control over pH of the stripping operation is both more important and more difficult than for absorption. Table 9.2 shows that the pH during absorption, say 7 to 8, has a pronounced effect on ammonia concentration; conversely, variations in ammonia concentration and hence acid usage have relatively little effect on pH. During stripping, the pH must be controlled above 9 where the ammonia content is minimal and pH is much more sensitive to the flow of the base. Furthermore, insufficient base will leave residual ammonia and any excess is wasted.

The set of titration curves in Fig. 9.7 demonstrates this point. A 1.0 N solution of ammonia will have a pH of 11.62 at 25°C. Addition of acid to that solution will lower the pH along the $x_B = 1.0$ N curve by generating ammonium ions. To free the ammonia from the ionized state, caustic must be added to return the pH to 11.62. Similarly, weaker solutions of ammonia absorbed in acid must have their pH adjusted to 11 or thereabouts to neutralize the acid and thereby free the ammonia. Observe that the slope of each curve is maximum at the point of zero excess

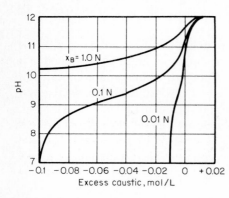

figure 9.7   *Solution pH should be raised above 11 to strip ammonia effectively.*

caustic — the "end point." This confirms that pH is most difficult to control at the end point. The slope is least at the half-neutralization point, where $pH = 14 - pK_B$.

The end point may be approximated as halfway between the half-neutralization point and the end of the pH scale. Ammonia is half-ionized at pH 9.25; the average of 9.25 and 14 is pH 11.625. Exact calculation of the end-point pH requires the trial-and-error solution of Eq. (9.20) for bases:

$$x_B = 10^{pH-pK_W}(1 + 10^{pH-pK_W+pK_B}) \tag{9.20}$$

and Eq. (9.21) for acids:

$$x_A = 10^{-pH}(1 + 10^{pK_A-pH}) \tag{9.21}$$

Note that the end point is also the pH of a solution of the gas in pure water.

The Acid Gases   Absorption and stripping are commonly conducted on hydrogen chloride, chlorine, hydrogen fluoride, hydrogen sulfide, carbon dioxide, sulfur dioxide, and the nitric oxides. Of these, only hydrogen chloride completely ionizes in solution — ionization of the others varies with pH, as was the case with ammonia.

Hydrogen chloride absorbs so readily in water that a mole of water may be vaporized for every mole of HCl absorbed adiabatically. Hydrochloric acid forms a maximum-boiling azeotrope at about 20 wt % HCl, the highest concentration obtainable adiabatically. To produce commercial 31% HCl, external cooling must be applied. Its great solubility is indicated by a partial pressure of HCl gas of only $0.02 \ lb/in^2$ in equilibrium with the 31 percent acid solution at 70°F. Under the same conditions, ammonia would generate a partial pressure of $5.8 \ lb/in^2$.

Absorption of chlorine in water forms the weak hypochlorous acid as well as hydrochloric acid:

$$Cl_2 + H_2O \rightarrow HCl + HClO \tag{9.22}$$

Similarly, absorption of nitrogen dioxide yields strong nitric acid and weak nitrous acid:

$$NO_2 + H_2O \rightarrow HNO_3 + HNO_2 \tag{9.23}$$

The presence of the strong acid tends to depress the pH of these solutions and therefore reduce the solubility. Addition of caustic first neutralizes the strong acid, then the weak. Stripping requires the addition of enough acid to regenerate both the strong and the weak components of the gas. The end point is essentially the pH of the strong component.

Many of the acid gases ionize in two steps; for example,

$$SO_2 + H_2O \overset{K_1}{\rightleftharpoons} HSO_3^- + H^+ \tag{9.24}$$

$$HSO_3^- + H_2O \overset{K_2}{\rightleftharpoons} SO_3^{2-} + H^+ \tag{9.25}$$

Only the first step is important from an absorption-desorption point of view, in that it determines the concentration of gas dissolved but not ionized. A second end point appears midway between the two steps at a pH which is the average of the two pK values. At this point, half the acid is neutralized; i.e., step 1 is complete. Table 9.3

**TABLE 9.3 Ionization Constants and
End Points of Various Acid Gases**

Gas	Weak acid	pK	End-point pH
Cl_2	HClO	7.5	$-\log [HCl]$
CO_2	CO_2	6.35, 10.25	4.0, 8.3
HF	HF	3.17	1.8
H_2S	H_2S	7.0, 12.9	4.0, 10.0
NO_2	HNO_2	3.2	$-\log [HNO_3]$
SO_2	SO_2	1.8, 6.8	1.0, 4.3

lists the pK values for the weak acids formed, along with their estimated end points. Because end-point pH varies with concentration, the values given in Table 9.3 are typical for the solubilities of the particular gas. Where a strong acid is formed, e.g., with chlorine and nitrogen dioxide, the end-point pH is determined entirely by the concentration of that acid.

Figure 9.8 describes the successive stripping of hydrogen sulfide and ammonia from a refinery sour-water stream. The water contains a nearly neutral solution of ammonium sulfide resulting from treating petroleum feedstocks. Since the effluent is aqueous, live steam may be used as a stripping medium, set in ratio to the feed. The pH of the stripped effluent may be high and therefore require adjustment before disposal. It cannot be recycled because it contains sodium sulfate, which will accumulate with reuse. But the sulfide and ammonia pollutants have been removed by the strippers and may be recovered for further processing. Titration curves and details on the measurement and control of pH for this application are given in Ref. 6.

Flue-Gas Desulfurization   Selective absorption between two acid gases is possible if their end points are far enough apart. A common problem in industry is that of absorbing a few hundred ppm of SO_2 from a flue gas while not absorbing the

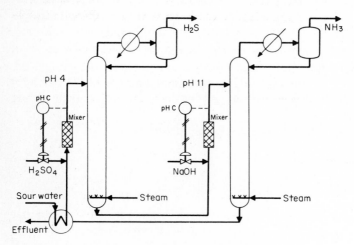

figure 9.8 *Refinery "sour water" may be stripped of hydrogen sulfide and ammonia successively by appropriate adjustment of pH.*

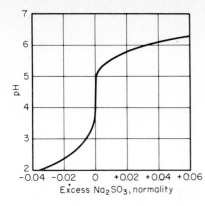

figure 9.9 *Sulfur-dioxide absorption is essentially complete at pH 4.3.*

16 percent CO_2 present. Absorption of CO_2 consumes reagent unnecessarily, increases the sludge volume, and promotes scaling, all of which are undesirable.

Table 9.3 shows that the second end point for SO_2 and the first end point for CO_2 are nearly coincident in the pH range of 4 to 4.5. Therefore, in this range absorption of SO_2 is essentially complete while that of CO_2 is virtually nil. Figure 9.9 shows a titration curve of SO_2 with a basic solution of sodium sulfite in a partial pressure of 0.2 atm CO_2. The end point for SO_2 absorption appears to be pH 4.3. The rich absorbent in this sodium-based process is then reacted with lime to raise the pH and precipitate calcium sulfite. The clarified solution, which is principally sodium sulfite, is then recycled to the absorber.

Table 9.4 demonstrates how little CO_2 is absorbed in comparison to SO_2 at various pH values despite its high partial pressure. The flue-gas composition is typical of scrubber effluents. Only a small fraction of the absorbed SO_2 (3.65×10^{-4} mol/L) exists as the dissolved gas under the above conditions — the balance is ionized. The concentration of absorbed CO_2 (3.14×10^{-3} mol/L) as the dissolved gas is actually greater, but these pH levels do not favor its ionization.

Absorbed gases can also be transferred to the solid state by precipitation as an insoluble product. If a calcium-based scrubbing medium is used (either lime or limestone), the calcium ions will combine with sulfite ions to form a precipitate:

$$Ca^{2+} + SO_3^{2-} \underset{}{\overset{K_S}{\rightleftharpoons}} CaSO_3 \downarrow \qquad (9.26)$$

Here K_S is known as the solubility-product constant and is defined as

$$K_S = [Ca^{2+}][SO_3^{2-}] \qquad (9.27)$$

TABLE 9.4   Solution Concentrations in Equilibrium with Flue Gas Containing 200 ppm SO_2 and 16% CO_2 at 120°F

pH	SO_2, mol/L	CO_2, mol/L
5	0.588	0.00328
6	6.70	0.00454
7	149.5	0.0172

Again, solution pH determines the absorption of SO_2 and its conversion to SO_3^{2-} ions. Consequently, this reaction is also characterized by a titration curve of pH versus lime addition similar in some respects to Fig. 9.9. However, CO_2 absorption interferes with the precipitation because it forms $CaCO_3$, with a solubility 2.5 orders of magnitude lower than $CaSO_3$. In the presence of CO_2, or if limestone ($CaCO_3$) is used as a reagent, the solubility limit of $CaCO_3$ is reached at about pH 6.7—the pH cannot be increased above that point. Consequently it must be controlled at or below 6 in this system to avoid carbonate formation.

STRIPPING WITH STEAM

Open steam is used as a stripping or reboiling medium in two principal situations:

1. When the bottom product is water
2. When the volatiles are immiscible with water and a minimum stripping temperature is desired

The first situation was demonstrated in Fig. 9.8. Other familiar applications are the stripping of relatively small quantities of methanol, ethanol, or acetone from water. The principal advantage in using open steam is to eliminate the reboiler. Its only drawback is a dilution of the bottom product. If a feed is rich in methanol, for example, a proportionally large amount of steam would be necessary to distill it overhead. Then the small quantity of water in the feed would be augmented by the larger quantity of steam condensate. Since equilibrium in the column is on a mole-fraction basis, increasing the water flow in the bottom product would increase the rate of methanol loss.

A Single Liquid Phase   Steam stripping of a water-immiscible liquid may be conducted with or without a liquid aqueous phase present. If no aqueous phase is present, the steam acts just like any stripping medium to reduce the partial pressure of the other components and thereby promote their vaporization. Following condensation, the water may be readily decanted from the other liquid phase. This is why steam is injected into the bottom of the atmospheric crude-oil tower. The crude oil has already been heated to its highest allowable temperature ($\sim 700°F$). By diluting the vapors with steam, further vaporization is possible. No aqueous phase can exist since the oil temperature is well above the boiling point of water. In fact, the steam as introduced is below the oil temperature and this produces some cooling. The additional vaporization it brings also causes a reduction in temperature.

The vaporization achievable by steam injection is limited by two effects. First, cooling by the steam and the additional vaporization reduce the temperature of the liquid. Second, the loss of volatile components from the liquid lowers their partial pressure in the vapor. As a result, the amount of additional hydrocarbon vaporized per pound of steam falls off sharply as the steam rate is increased. Van Winkle [7] has plotted this relationship for several sets of conditions common in crude-oil fractionation.

Steam stripping will be more effective in recovering additional volatiles when its temperature is higher relative to the boiling point of the oil and when the boiling range of the oil is reduced. This is borne out by graphs in Ref. 7, which show more than twice the recovery for kerosene as for atmospheric bottoms, when stripped with 0.5 lb steam per gallon.

The molar ratio of steam w to oil o in the vapor is the ratio of their partial pressures:

$$\frac{y_w}{y_o} = \frac{p_w}{p_o} \tag{9.28}$$

Since the oil vapor is in equilibrium with its liquid, its partial pressure is also its vapor pressure:

$$\frac{y_w}{y_o} = \frac{p_w}{p_o^\circ} \tag{9.29}$$

Two Liquid Phases The partial pressure of the steam cannot be increased indefinitely. When it reaches its vapor pressure at the temperature of the system, water will condense, forming a second liquid phase. This will happen in the overhead condenser for the crude-oil column, and it could happen on some of the top trays as well.

When two liquid phases exist in equilibrium with a vapor, an entirely new relationship appears: *each liquid phase exerts its own vapor pressure independent of the other.* This situation can develop whenever steam is used to distill an immiscible liquid, depending on how much steam is used in comparison to the externally applied heat. It is also a property of heterogeneous azeotropes, which are discussed in detail in the next chapter. Water is not an essential ingredient in these systems — any two volatile immiscible liquids will do. However, water is the most common liquid and the condensate of our most universal heating medium, steam.

Open steam is used to distill certain heat-sensitive products at temperatures well below their normal boiling points without using vacuum. For example, turpentine, with an atmospheric boiling point of 309°F, may be distilled at 204°F by open steam.

The theoretical limits of the molar ratio of steam to oil in the vapor is reached when an aqueous phase exists. The partial pressure of the steam cannot exceed the vapor pressure of water at the existing temperature. Then

$$\frac{y_w}{y_o} = \frac{p_w}{p_o^\circ} \le \frac{p_w^\circ}{p_o^\circ} \tag{9.30}$$

Nonvolatiles in the oil phase dilute the volatile component, thereby reducing its vapor pressure. If the concentration of the volatile component in the oil phase is x_o, then from Raoult's law

$$p_o = x_o p_o^\circ \tag{9.31}$$

This alters (9.30) to read:

$$\frac{y_w}{y_o} = \frac{p_w}{x_o p_o^\circ} \le \frac{p_w^\circ}{x_o p_o^\circ} \tag{9.32}$$

The mole ratios described by Eqs. (9.30) and (9.32) represent the minimum amount of steam capable of vaporizing an amount of oil. Ellerbe [8] notes that the vaporization efficiency of open steam is much less than 100 percent. The efficiency limit is essentially the sort that impedes establishment of vapor-liquid equilibria in general—insufficient vapor-liquid contact. Consequently only 60 to 90 percent of the oil theoretically capable of being distilled per pound of steam may in fact be distilled. Fortunately, the low molecular weight of water compared to most distilled oils still results in a reasonably high yield in pounds of product per pound of steam.

Heat Balances   Steam distillation will always use more energy than conventional distillation by the amount of steam which must be condensed with the distillate. As more external heat is applied, less open steam is needed. The minimum vaporization temperature will be maintained as long as an aqueous liquid phase exists. Above this point no open steam is condensed. Therefore, at this point all the heat required to vaporize the oil must come from the external heat source and whatever superheat may be available in the open steam. The superheat in the open steam may be considerable if the boiling point of the oil is low. But in situations where the oil is composed of high-boiling components such as fatty acids, high-pressure steam is necessary and superheat may be an unaffordable luxury.

Note that the higher the boiling point of the oil, the more steam will be required to distill it. In the case of turpentine, the maximum amount that may be distilled per pound of water is 1.33 lb if two liquid phases exist [8]. Roughly 0.17 lb additional steam will condense in vaporizing the turpentine. So in this example more than *six* times as much steam is used as is actually necessary to distill the turpentine. If the 0.17 lb of steam were used in a reboiler, the total heat load would not change because an aqueous phase would still exist. As the reboil heat increases, so that the boiling point rises above that of water, there will be a single liquid phase. Then the boiling point will vary with the amount of open steam used per pound of turpentine vaporized by the reboiler.

Since under these conditions no open steam condenses, the weight of oil vaporized W_o is determined solely by reboiler steam W_r:

$$W_o = \frac{W_r \Delta H_w}{\Delta H_D} \tag{9.33}$$

Here ΔH_w is the heat given up per pound of reboiler steam. The mole ratio of water to oil in the vapor is then related to the weight ratio of stripping to reboiler steam:

$$\frac{y_w}{y_o} = \frac{W_w/M_w}{W_o/M_o} = \frac{W_w}{W_r} \frac{M_o}{M_w} \frac{\Delta H_w}{\Delta H_D} \tag{9.34}$$

Here M represents the molecular weight of the species indicated by the subscript. There is really no optimum ratio in this problem. As with the crude tower, as much reboil heat should be applied as possible—to whatever temperature limit applies to the product. Then vaporization at that temperature is made possible by injecting the appropriate amount of open steam.

EXTRACTIVE DISTILLATION

Extractive distillation is a true distillation process which uses a solvent to improve the relative volatility of the components of interest. It can be used to separate homogeneous azeotropes or simply close-boiling components which do not form an azeotrope. The solvent is selected to have a particular affinity for one component or class of components. In this respect it acts much like a selective absorbent.

Two steps are required for complete separation of two components. In the first, the extractant attaches itself to one component while the other is distilled. A second column then strips the captured component from the extractant. Consequently the extractive distillation process bears a distinct relationship to the conventional absorber-stripper combination. Extractive distillation is commonly used to separate close-boiling, nearly ideal mixtures of olefins and paraffins as well as less ideal mixtures such as acetone and methanol.

The Extractant Circuit   The first column shown in Fig. 9.10 has neither sufficient trays nor reflux to make the required separation between the key components by conventional distillation. But the extractant introduced near the top preferentially lowers the vapor pressure of component B, sweeping it to the bottom of the column. The few trays above the point of extractant introduction are needed to separate it from component A. From the extractant entry point downward, the column performs essentially like a conventional unit except that the liquid rate tends to be much greater than the vapor rate.

Typically, very little of the extractant boils in the first column — most of the vapor is provided by component B. But since the vapor rate may be substantially greater than the flow of component B leaving the column, there is a sizable change in composition across the bottom tray. This can be demonstrated with a simple material balance, using Fig. 9.11. The flow if extractant L_E is essentially the same

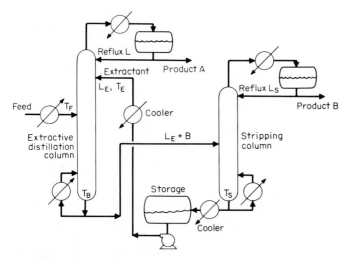

figure 9.10   *The extractant captures component B, which is then released in the stripping column.*

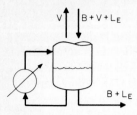

figure 9.11 *Since little of the extractant vaporizes, the liquid composition changes dramatically across the bottom tray.*

everywhere in the column. But the flow of component B in the liquid phase decreases from the first tray to the bottom-product stream by the vapor rate V. Thus the fraction of component B in the bottom product is

$$x_0 = \frac{B}{L_E + B} \tag{9.35}$$

whereas the composition leaving the first tray is

$$x_1 = \frac{B + V}{L_E + B + V} \tag{9.36}$$

Typically $B \ll V \ll L_E$, which makes $x_0 \ll x_1$. This step change in composition is indicated by a step change in temperature.

In its role of enchancing relative volatility, the extractant flow can be set in proportion to the feed or, more precisely, in proportion to the captured component in the feed. This will bring about a uniform concentration of extractant in the feed to the stripping column. Increasing the extractant flow has an effect similar to that of increasing reflux — by diluting component B further, it depresses its vapor pressure and improves separation.

The stripping column is essentially identical to those reboiled stripping columns operated in conjunction with absorbers. Its goal is to produce an extractant-free distillate while returning a minimum of component B to the extractive-distillation column. Relatively few trays are required when the boiling point of the extractant is much higher than the distillate. This is, in every respect, a conventional distillation process. The only unusual feature would be the sizeable temperature gradient observable — bottom temperature will approach the boiling point of the extractant. The amount of component B recycled with the extractant influences the separations in the extractive column in a manner similar to variations in feed composition. Thus an increase in the concentration of component B in the extractant requires an increase in extractant flow to offset it.

Because the boiling point of the extractant tends to be considerably higher than that of the components to be separated, it is usually cooled in the recycle loop. The cooling could be eliminated if the towers were operated at different pressures. This is not usually possible, however, since the two components which are separated must both be condensed and their boiling points must be close together (or extractive distillation would not have been necessary).

The Extraction Mechansim The extractant acts to produce a nonideal liquid mixture. The relative volatility of a given liquid mixture may be expressed as a function of vapor pressure and activity coefficients γ:

$$\alpha_{AB} = \frac{\gamma_A \, p_A^{\circ}}{\gamma_B \, p_B^{\circ}} \qquad (9.37)$$

For an ideal mixture, these activity coefficients are unity. When the components of a nearly ideal mixture such as n-butane and 1-butene have virtually the same vapor pressure, separation is difficult. Then an extractant is useful that can alter one activity coefficient relative to the other.

The activity coefficients for a nonideal liquid mixture can be approximated by van Laar's equation [9]:

$$\log \gamma_i = \frac{Z_{iE}}{[1 + (Z_{iE} x_i / Z_{Ei} x_E)]^2} \qquad (9.38)$$

In Eq. (9.38) subscript E is used to identify the extractant and i may refer to either of the components to be separated; the coefficients Z_{iE} and Z_{Ei} are van Laar's binary constants, some sets of which are given in Ref. 9. For the case of infinite dilution,

$$\log \frac{\gamma_A}{\gamma_B} = Z_{AE} - Z_{BE} \qquad (9.39)$$

Using data from Ref. 9 activity coefficients and relative volatilities for the n-butane/1-butene system were evaluated with furfural as an extractant (see Table 9.5). Calculations were made for conditions at the extractant entry point ($T = 150°F$, $x_B \to 0$) and near the bottom of the column ($T = 200°F$, $x_A \to 0$).

The vapor pressure of 1-butene is actually greater than that of n-butane—in fact their ratio at 150°F is 0.853. But if sufficient furfural is present, it raises the activity coefficient of n-butane enough for it to become the more volatile component. The 2-butenes are slightly less volatile than n-butane and are therefore somewhat easier to separate in the extractive column. In a multicomponent mixture, n-butane and 1-butene are the keys, with isobutane lighter and the 2-butenes heavier than the keys.

Because the extractant concentration has such a strong influence over the relative volatility, a certain concentration must be reached before the separation can be achieved at all. This is particularly true of the system illustrated here in that α crosses 1.0 by the addition of the extractant. Like the L/D ratio in a conventional column, maximum separation is achievable only with an infinite ratio of extractant to product.

TABLE 9.5   Estimated Relative Volatilities of n-Butane/1-Butene Mixtures in Furfural

Furfural-hydrocarbon mole ratio	T, °F	n-Butane		1-Butene		Relative volatility, α_{AB}
		x_A	γ_A	x_B	γ_B	
∞	150	0	10.0	0	5.75	1.48
20	150	0.05	8.15	0	5.75	1.21
	200	0	8.12	0.05	4.43	1.56
10	150	0.10	6.70	0	5.75	0.99
	200	0	8.12	0.10	3.94	1.76

Selecting the Extractant   Van Winkle [10] has compiled lists of extractants effective in improving the relative volatility of several systems, including aromatic-paraffin and olefin-paraffin mixtures. Most of the solvents selective for aromatics are aromatic in nature, such as nitrobenzene, but some are alcohols such as propylene glycol. Many of the solvents selective for olefins contain nitrogen, such as nitriles, amines, and nitro-compounds; oxygenated organics including ketones, ethers, and alcohols are also effective. This reference compares relative volatilities of the hydrocarbons for certain solvent-dilution factors. For example, diluting an unspecified butane-butene mixture 3:1 by volume with furfural is listed as raising the relative volatility to 1.40 at 158°F.

Naturally, the extractant must be miscible with the components to be separated, in the proportions that exist in the column. If it is not, the extracted component is not in fact extracted, since it is able to exert its own vapor pressure. Water, for example, is capable of extracting olefins from paraffins, but its solubility in these hydrocarbons is so low that its effectiveness is severely limited.

It is possible to use an extractant that forms a high-boiling azeotrope with the extracted component. In fact, this approach would be quite effective, although it would present a problem in recovering the extractant. An extractant which forms a low-boiling azeotrope will pass overhead — this becomes a different type of distillation, which is covered in detail in the next chapter.

The volatility of the extractant is an important consideration. A very volatile extractant will require more trays above the entry point and a greater reflux in both the extractive column and the stripping column. However, increasing the difference between its boiling point and that of the product means that more sensible heat must be removed prior to recycling. This sensible heat is a net energy loss — it contributes nothing to the separation but still must be provided by the two reboilers.

Some practitioners have found that the volatility of an otherwise effective extractant can be adjusted by adding another component. A commonly used mixture is a solution of about 4 percent by weight of water in furfural; this is enough to depress the boiling point nearly to that of water. The amount which may be added is low because of a solubility limitation — the 4 percent solution is saturated with water at about 60°F.

Provision must then be made for withdrawing any excess water which may have been added above the solubility limit. Even within this limit the effect is considerable in reducing the sensible heat loss to cooling. However, the presence of the additional component alters the temperature-composition relationship in the bottom of the stripping column. Rather than control temperature there, steam flow should be set in ratio to feed or bottoms rate; temperature control at the top should manipulate reflux.

If a water layer is always maintained in the furfural storage tank, and if the temperature of the tank is controlled by the stripper bottoms cooler, a constant water content in the extractant can be attained. In this case further cooling prior to extraction cannot be applied because this would allow water to separate from the mixture.

The addition of water to the furfural apparently does not detract from its effectiveness as an extractant and in fact enhances it. Reference 10 indicates that dilution of a butane-butene mixture by a factor of 3.7 by volume with furfural containing 4 wt% water improves the relative volatility to 1.78 at 128°F. By contrast, aniline does not seem to show an improvement in selectivity when a similar amount of water is added.

Heat Balances   The relative volatility of the components to be separated in an extractive-distillation column is a function of their dilution by the extractant. If the feed to the column is a liquid, dilution by the extractant is less below the feed tray than above. This discontinuity can be avoided by completely vaporizing the feed in a preheater, which is common practice in these separations.

The energy which must be supplied to the extractive-column reboiler (Q_E) serves to vaporize the reflux L and increase the sensible heat of the extractant L_E and bottom product B. Credit is given for condensing component B from the feed:

$$Q_E = (L - B)\Delta H_D + L_E C_E(T_B - T_E) + BC_B(T_B - T_F) \tag{9.40}$$

Here ΔH_D is the heat of vaporization of the components to be separated; C_E and C_B are the specific heat of extractant and bottom product, respectively, and T_E and T_B are their temperatures; T_F is feed temperature.

In the stripping column, heat Q_s must be provided to vaporize reflux L_s and distillate B and to supply the sensible heat gained in raising the extractant flow to its boiling point T_s. Some heat is recovered in cooling product B to reflux temperature T_L:

$$Q_s = (L_s + B)\Delta H_D + L_E C_E(T_s - T_B) - BC_B(T_B - T_L) \tag{9.41}$$

Combining these two equations gives the total heat Q which must be supplied to the system. If T_L is close to T_F, the last terms in the two equations above will cancel, yielding

$$Q = (L + L_s)\Delta H_D + L_E C_E(T_s - T_E) \tag{9.42}$$

Separation between components A and B in the extractive column can be improved by

1. Increasing reflux flow
2. Increasing extractant flow
3. Reducing extractant temperature

The last variable has a pronounced effect on internal reflux and its distribution. If extractant temperature T_E is below the boiling point T of the components at the point of entry, some vapors will be condensed, increasing L below that tray by ΔL:

$$\Delta L = \frac{L_E C_E(T - T_E)}{\Delta H_D} \tag{9.43}$$

If extractant temperature exceeds the component boiling point, ΔL will be negative, indicating a *lower* reflux below the extractant feed tray. Since reducing T_E increases internal reflux, it tends to increase separation below that tray for a given external reflux.

The relationship between separation and heat input is different for each of the three variables listed above. The effect of the reflux-component B ratio on separation between the component is some function of the extractant concentration:

$$\frac{\partial S}{\partial L/B} = f\left(\frac{L_E}{L}\right) \tag{9.44}$$

The effect of extractant temperature is essentially the same in that it adjusts internal reflux:

$$\frac{\partial S}{\partial T_E} = -\frac{L_E C_E}{B \Delta H_D} \frac{\partial S}{\partial L/B} \tag{9.45}$$

The effect of extractant flow is conversely expected to depend on the L/B ratio:

$$\frac{\partial S}{\partial L_E/B} = f\left(\frac{L}{B}\right) \tag{9.46}$$

Note that these functions are highly nonlinear and interacting. From Table 9.5 it can be seen that no amount of reflux can make the required separation if insufficient extractant is present.

A choice must be made regarding which variable should be adjusted to control separation. It is not possible to base the selection on the partial derivatives given in the preceding equations because there is no common denominator. The common denominator that applies most universally is the energy input to the system. The expense in terms of energy requirements must be evaluated for the three manipulations. From (9.42),

$$\frac{\partial Q}{\partial L} = \Delta H_D \tag{9.47}$$

$$\frac{\partial Q}{\partial T_E} = -L_E C_E \tag{9.48}$$

$$\frac{\partial Q}{\partial L_E} = C_E(T_s - T_E) \tag{9.49}$$

The value of the three manipulations can then be determined by dividing the set of (9.44) to (9.46) by (9.47) to (9.49). As it happens, the improvement in separation attained for a given increment in heat input is the same when manipulating either L or T_E:

$$\left.\frac{\partial S}{\partial Q/B}\right|_{L_E} = \frac{\partial S}{\partial L/B} \frac{1}{\Delta H_D} \tag{9.50}$$

Actually, manipulating L would be slightly more beneficial than T_E in that it increases reflux above the point of extractant entry as well as below. For manipulation of extractant flow,

$$\left.\frac{\partial S}{\partial Q/B}\right|_{L,T_E} = \frac{\partial S}{\partial L_E/B} \frac{1}{C_E(T_s - T_E)} \tag{9.51}$$

The choice of the variable to manipulate for control of separation would fall on the one with the greater effect on separation per unit heat input as evaluated in (9.50) and (9.51).

If reflux manipulation is more efficient, reflux flow should be set at maximum and extractant temperature at minimum so that extractant flow is no greater than required to achieve the desired separation. Should extractant flow have the greater effect on separation per Btu, its flow and temperature should be high, with just enough reflux provided to keep the extractant out of the overhead product.

Steam flow to the extractive distillation column should be manipulated to control the content of component A at or near the bottom. Again, temperature is not an indication of composition, owing to the preponderance of extractant. Temperature control using reflux at the top may be adequate to keep extractant from being taken overhead. Then component B in the distillate would be controlled by manipulating extractant flow; its concentration does depend however, on how much component B remains in the lean extractant leaving the stripping column. In absence of a top-composition analyzer, extractant flow must be set in ratio to column feed rate.

REFERENCES

1. Perry, R. H., and C. H. Chilton: *Chemical Engineer's Handbook,* 5th ed., McGraw-Hill, New York, 1973, p. *14–12.*
2. Owens, W. R., and R. N. Maddox: "Short-Cut Absorber Calculations," *Ind. Eng. Chem.,* December 1968.
3. Treybal, R. E.: *Mass-Transfer Operations,* McGraw-Hill, New York, 1955, p. 207.
4. Houghton, J., and J. D. McLay: "Turboexpanders Aid Condensate Recovery," *Oil Gas J.,* March 5, 1973.
5. Butwell, K. F., D. J. Kubek, and P. W. Sigmund: "Amine Guard III," *Chem. Eng. Prog.,* February 1979.
6. Shinskey, F. G.: *pH and pIon Control in Process and Waste Streams,* Wiley-Interscience, New York, 1973, pp. 67–69.
7. Van Winkle, M.: *Distillation,* McGraw-Hill, New York, 1976, p. 359.
8. Ellerbe, R. W.: "Steam-Distillation Basics," *Chem. Eng. (N. Y.),* March 4, 1974.
9. Reference 1, pp. *13–6–13–7.*
10. Reference 7, pp. 464–469.

CHAPTER TEN

Azeotropic Distillation

An azeotrope is a mixture of two or more volatile components having identical vapor and liquid compositions at equilibrium. Its composition, therefore, is not changed by distillation—it is a constant-boiling-point mixture. This property precludes its being separated by simple distillation. The azeotropic mixture could boil either at a higher or lower temperature than its components, depending on the nature of the system. Many acids such as hydrochloric and nitric form maximum-boiling-point mixtures with water. However, minimum-boiling azeotropes are far more common and are used as examples throughout this chapter.

If one attempts to distill a mixture of two components forming an azeotrope, only one of the components can be removed, the azeotrope becoming the other product. Since the azeotrope cannot be separated by conventional distillation, other methods, such as extraction, must be combined with distillation. In every case, a column is required for *each* pure component to be recovered. Thus to separate a binary azeotrope requires two columns.

In addition to the boiling-point classification, azeotropes may be either homogeneous or heterogeneous. Heterogeneous azeotropes separate into two liquid phases when condensed from the vapor. Homogeneous azeotropes do not, which makes their separation much more difficult. Nonetheless, several methods have been developed to bring about their separation. One method for separating homogeneous azeotropes has already been covered — extractive distillation. These specialized distillation processes are not reserved for azeotropes: they may be applied to any mixture that is close boiling. Whether a mixture is simply extremely difficult to fractionate (as butane-butene) or impossible (as the methanol-acetone azeotrope), extractive distillation can be effective. Another classic is the acetic acid–water system — it does not form an azeotrope but is so difficult to separate that it is treated as such. And, as will be seen, a true azeotrope is used to force the separation.

There seem to be so many ways to separate these nonideal mixtures that not all can be covered in a single chapter. But the more common methods are described, each illustrated by an example from industry.

BINARY HETEROGENEOUS AZEOTROPES

The heterogeneous azeotrope can actually be easier to separate than many more-ideal mixtures. The very immiscibility of the liquid phases condensed from the vapor effectively breaks the azeotrope. This allows complete separation of the components in two columns or partial separation in one column. Quite pure products may be made with comparatively few trays and little energy, owing to the limited solubility of the two liquid phases. This property is even used to advantage to assist the separation of other close-boiling mixtures and homogeneous azeotropes. The approach used is to add a third component forming a low-boiling heterogeneous azeotrope with one of the other components. But before ternary systems like this can be presented, a simple binary heterogeneous system must be examined.

The Furfural-Water System   Furfural and water form a binary heterogeneous azeotrope described by the phase diagram given in Fig. 10.1. Any liquid mixture in the center of the diagram will separate into two phases, the compositions of which are represented by the nearly vertical curves. Reducing the temperature decreases the mutual solubility of the components, enhancing their separation.

The lighter (aqueous) layer may be separated by fractional distillation into water and vapor of the azeotropic composition. Although the difference between the boiling points of these two products is small (212 versus 208°F), their relative volatility is substantial. Vapor-liquid equilibrium data taken from Ref. 1 indicate that the relative volatility between furfural and water changes from 7.5 at 0 percent furfural to about 2.5 at the lower solubility limit (18.4 wt % at 208°F). At the upper solubility limit (84.1 wt % at 208°F), the relative volatility of furfural to water drops to about 0.1. Here furfural becomes less volatile and is even easier to separate from the azeotrope than water.

If the pressure is elevated sufficiently to raise the boiling point of the azeotrope above 250°F, a single phase will condense from the vapor. At this point, the azeo-

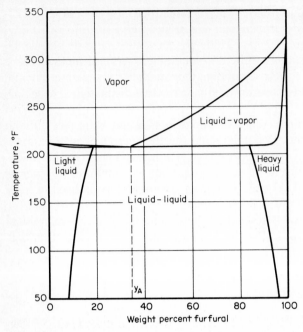

figure 10.1 *The phase diagram for the furfural-water system at atmospheric pressure. (From G.H. Mains, The System Furfural-Water, Chem. Metall. Eng., April 26, 1922. By permission.)*

trope becomes homogeneous. To take advantage of the separation afforded by the limited solubility of the two liquids, the condensate should be cooled to a reasonably low temperature.

Stripping with Decanter Feed   A single column is used when only one pure product is required. This would be the case when excess water is to be removed from a lean mixture while minimizing the loss of furfural. The water product should be virtually pure; the furfural product can be no more concentrated than the upper solubility limit at whatever solution temperature exists. Figure 10.2 illustrates the stripping system with decanter and column feed points indicated.

Consider first the lean furfural feeding the decanter as an aqueous phase saturated with furfural at or above the decanter temperature. In passing through the decanter, it may give up some furfural, leaving it saturated at the decanter temperature. The saturated aqueous phase entering the top of the column will yield a vapor closer to the azeotropic composition. Upon condensing, a furfural-rich layer will form, to be withdrawn as product, along with a water-rich layer recycled with the lean feed. Bottom product from the stripping column will be water that is virtually free of furfural. Open steam is used for stripping.

This system is interesting because of the intertwining of energy and material balances. This property appears in all examples of azeotropic distillation, but furfural stripping is perhaps the simplest illustration. The symbols in Fig. 10.2 will be used for flow rates and weight fractions of furfural in the following analysis.

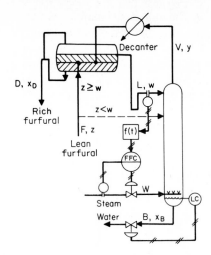

figure 10.2 *Setting steam flow in ratio to reflux can regulate the furfural content of the stripped water, but only when feed is introduced at the decanter.*

Consider first the mass balance for the decanter:

$$F + V = D + L \tag{10.1}$$

Next consider the system balance for furfural only, assuming that x_B approaches zero and is therefore negligible:

$$Fz = Dx_D \tag{10.2}$$

A similar furfural balance may be written for the stripper only, using the same assumption:

$$Lw = Vy \tag{10.3}$$

Solving Eq. (10.2) for D and (10.3) for L, and substituting for D and L in (10.1), yields a solution for the boilup-feed ratio:

$$\frac{V}{F} = \frac{w(x_D - z)}{x_D(y - w)} \tag{10.4}$$

For control purposes, it is desirable to set boilup in ratio to the feed. However, Eq. (10.4) indicates that this ratio varies with feed composition z, which may not be measurable. Instead, boilup could be set proportional to reflux L, eliminating this problem:

$$\frac{V}{L} = \frac{w}{y} \tag{10.5}$$

Reflux composition w can be controlled by decanter temperature, or inferred from it if a minimum rather than a controlled temperature is desirable.

Solving (10.4) for y indicates how it is affected by variations in the V/F ratio:

$$y = w + \frac{w(x_D - z)}{x_D(V/F)} \tag{10.6}$$

Increasing V/F reduces y through the action of the material balance. Yet increasing V/F for a column is known to improve separation, driving y and x_B farther apart. The combination of these two effects is to reduce x_B more than y. Although x_B was neglected as being insignificant compared to the other terms in the material balance, it is nonetheless the primary controlled variable. Although small, x_B is still a finite, measurable number (typically about 0.02 wt %) and is controlled by manipulation of boilup. By contrast, y does not need to be controlled since it is not a final product, and any value between azeotropic composition y_A and w will produce a furfural-rich product. Product concentration x_D is determined solely by decanter temperature.

The relationship between L/V and V/F can be seen by solving Eq. (10.2) for D and substituting it into (10.1):

$$\frac{L}{F} = 1 - \frac{z}{x_D} + \frac{V}{F} \tag{10.7}$$

As can be seen, L/F varies with V/F on a 1:1 basis but is higher for all reasonable operating conditions. Thus increasing V/F will cause L/V to decrease:

$$\frac{L}{V} = 1 + \frac{1 - z/x_D}{V/F} \tag{10.8}$$

If it is desirable to decrease the L/V ratio by increasing V, the operator will find that L will subsequently increase if it is manipulated by a level controller or simply overflows from the decanter. When a new steady state is reached, L/V will be less than during the original steady state but not as low as immediately following the increase in V.

example 10.1

For the furfural stripping column, calculate V/F, D/F, L/F, and L/V on a mass basis for $z = w = 0.9$ weight fraction (at 100°F), $x_D = 0.94$, and $y = 0.25$ weight fraction.

$$\frac{V}{F} = \frac{0.09(0.94 - 0.09)}{0.94(0.25 - 0.09)} = 0.509$$

$$\frac{D}{F} = \frac{0.09}{0.94} = 0.096$$

$$\frac{L}{F} = 1 - \frac{0.09}{0.94} + 0.509 = 1.413$$

$$\frac{L}{V} = \frac{1.413}{0.509} = 2.775$$

In the case of the furfural stripping column, bottom temperature is not a satisfactory measure of composition: there is at most 4°F difference in boiling points between the overhead and bottom product. But if L/V is held constant and decanter temperature is controlled, x_B should not vary appreciably.

Setting boilup in ratio to reflux as in Fig. 10.2 forms a positive-feedback loop: increasing boilup will generate more reflux which will call for more boilup. However, the loop gain is low (0.36), and a lag $f(t)$ is inserted to give the loop dynamic stability; it should be set to model the hydraulic lag between the top and bottom of the column.

Stripping with Column Feed   Figure 10.2 indicates that the lean furfural may be fed either to the column or the decanter. In conventional distillation, feed is not blended with the reflux because it will dilute the reflux and thereby the overhead product as well. In this heterogeneous system, feeding the decanter has no effect on reflux composition (which is determined by temperature). Instead, feed rate and its composition affect only reflux flow. A leaner feed will increase reflux flow, requiring more boilup to hold L/V constant.

A feed which is significantly leaner than the reflux ought to be introduced at the appropriate tray in the column. To illustrate this point, consider a decanter mass balance with feed entering the column:

$$V = D + L \tag{10.9}$$

Next consider a furfural balance on the column only, assuming that x_B approaches zero as before:

$$Vy = Fz + Lw \tag{10.10}$$

Solving (10.9) for L and substituting into (10.10), while substituting for D from (10.2), yields

$$\frac{V}{F} = \frac{z(x_D - w)}{x_D(y - w)} \tag{10.11}$$

Note that this expression is almost (but not quite) identical to (10.4). The difference is the transposed locations of z and w in the numerator. For the case of $z = w$ as in Example 10.1, both will give the same results. But as z approaches zero, Eq. (10.4) approaches a maximum while (10.11) approaches zero.

The explanation for this behavior is related to the reflux-dilution problem. Feed entering the decanter augments the reflux, whose fixed composition increases the furfural entering the column. As the feed grows leaner, reflux flow increases, carrying more furfural to the column. All this furfural must be driven overhead to maintain a uniform bottom product.

When a lean feed enters the column directly, only enough boilup must be provided to drive overhead the furfural it contains, with a proportionate amount for that in the reflux. A leaner feed requires less boilup and hence less reflux. To demonstrate, Eqs. (10.9), (10.10), and (10.2) may be solved for L/F:

$$\frac{L}{F} = \frac{z(x_D - y)}{x_D(y - w)} \tag{10.12}$$

As z approaches zero, L/F approaches zero. But when feed is introduced to the decanter, L/F approaches maximum when z approaches zero.

example 10.2

Compare V/F for decanter feed and column feed when $z = 0.05$, $w = 0.09$, $x_D = 0.94$, and $y = 0.25$.

For decanter feed,

$$\frac{V}{F} = \frac{0.09(0.94 - 0.05)}{0.94(0.25 - 0.09)} = 0.532$$

For column feed,

$$\frac{V}{F} = \frac{0.05(0.94 - 0.09)}{0.94(0.25 - 0.09)} = 0.283$$

Controlling the stripping column when feed is introduced to it rather than to the decanter presents more of a problem because boilup cannot be set in ratio to reflux. Without boilup, there is no reflux. Therefore, boilup must be set in proportion to feed rate and composition as (10.11) indicates. This requires either a feed analyzer to measure z or a bottom-product analyzer to indicate when the V/F ratio is incorrect.

Two-Column Operation   To make a water-free furfural product requires a second column called a *dehydrator*. The stripping column operates essentially as described except that its reflux is augmented by the aqueous portion of the dehydrator overhead stream. As Fig. 10.3 indicates, the two columns may use a common decanter and in fact a common condenser. The dehydrating column requires a reboiler to produce an essentially water-free product.

The material balances are easy to solve if the impurities in the two products are neglected. For feed introduced at the stripping column, a furfural mass balance yields

$$Fz + L_1 w_1 = V_1 y_1 \tag{10.13}$$

In the decanter, there is a net transfer of all the furfural in the feed from column 1 to column 2. Then

$$V_1 - L_1 = Fz \tag{10.14}$$

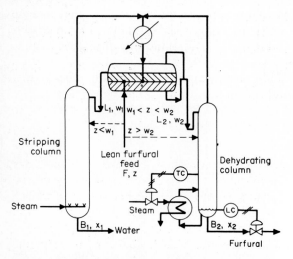

figure 10.3 *Two columns are required to separate a furfural-water mixture into its pure components.*

and

$$L_2 - V_2 = Fz \tag{10.15}$$

Finally, a furfural mass balance on the dehydrating column yields the bottom-product flow Fz as a function of boilup and reflux:

$$Fz = L_2 w_2 - V_2 y_2 \tag{10.16}$$

Equations (10.13) and (10.14) may be combined to give the V/F ratio for the stripping column:

$$\frac{V_1}{F} = z \frac{1 - w_1}{y_1 - w_1} \tag{10.17}$$

Observe that this equation is the same as (10.10) for $x_D = 1$. Combining (10.15) and (10.16) gives the V/F ratio for the dehydrating column:

$$\frac{V_2}{F} = z \frac{1 - w_2}{w_2 - y_2} \tag{10.18}$$

Because w_2 is much larger than w_1, V_2 can be much smaller than V_1.

example 10.3

Estimate the V/F ratios for a two-column system separating a feed of 5 wt % furfural into two essentially pure products. Let $w_1 = 0.09$, $w_2 = 0.94$, $y_1 = 0.25$, and $y_2 = 0.60$.

$$\frac{V_1}{F} = 0.05 \frac{1 - 0.09}{0.25 - 0.09} = 0.284$$

$$\frac{V_2}{F} = 0.05 \frac{1 - 0.94}{0.94 - 0.60} = 0.00882$$

Note that V_2/F appears unreasonably small. Actually, however, the feed to the dehydrator is Fz rather than F, so that V_2/Fz would be a more appropriate index of heat duty. For this cited example, $V_2/Fz = 0.176$, which is comparable to V_1/F.

Whether the low V_2/Fz estimated for the dehydrator is sufficient to produce an acceptably pure bottom stream depends on the relative volatility and the number of trays. As with the furfural stripping column described earlier, increasing boilup will drive y_2 and x_{B2} apart by improving separation and will raise y_2 through the action of the material balance. The latter effect can be seen by rearranging (10.18):

$$y_2 = w_2 - \frac{z(1 - w_2)}{V_2/F} \tag{10.19}$$

As a consequence, increasing V_2/F has more net effect on bottom-product quality than on overhead composition. Furthermore, the actual value of overhead composition achieved at a given boilup rate is inconsequential since it is not a product.

The phase diagram indicated a wide boiling-point range for furfural-rich mixtures. Consequently, column temperature is a sensitive and reliable index of composition and is used successfully to control furfural quality by manipulating heat input.

Energy Balances   The separation in the decanter is not achieved without cost. Sensible heat must be removed from the condensate to enhance separation and must in turn be supplied by the heat input. In the case of the stripping column, condensing the azeotrope without subcooling (at atmospheric pressure) will produce liquid-phase compositions of about 20 and 80 wt % furfural. Purities may be doubled by cooling to 140°F; in fact further cooling is usually carried out simply because of availability of cooling media in the ambient range.

An estimate using compositions given in Example 10.1 indicates that subcooling the condensate to 100°F increases the energy required per pound of feed by nearly 40 percent. Because of the relatively large energy loss associated with this subcooling, its value in reducing the solubility of the components is questionable except for the case where a specified furfural purity (for example, 90 percent) is required with a single column. Subcooling may not be justified at all in two-column operation.

If no subcooling is provided in the two-column system, w_1 will increase, raising y_1. The material balance (10.17) shows that V_1/F will have to increase as w_1 approaches y_1 in the denominator. In addition, V_1/F must increase to maintain control of x_{B1} in the face of a rising y_1. These increases may offset the loss in sensible heat incurred by subcooling. Each particular heterogeneous azeotropic separation should be examined, either in the design stage or in actual operation, to determine whether subcooling raises or lowers the energy required to make acceptable products. An optimum decanter temperature may exist in many systems.

PRESSURE-SENSITIVE HOMOGENEOUS AZEOTROPES

Homogeneous azeotropes lack the fortunate property of liquid-phase separation. If they are to be separated, then, a method must be used to alter somehow the properties of the azeotrope. There are four principal methods presented in this book:
1. Extractive distillation
2. Pressure adjustment
3. Formation of a heterogeneous ternary azeotrope
4. Formation of a heterogeneous binary azeotrope

Of these, extractive distillation was described in the previous chapter. The balance of this chapter is devoted to the other three methods.

A pressure-sensitive azeotrope exhibits a shift in composition of its constant-boiling-point mixture as total pressure is changed. Therefore, the azeotrope formed by distillation at one pressure can be further fractionated at another pressure.

Operation at Two Pressures   Tetrahydrofuran (THF) forms a homogeneous, minimum-boiling-point azeotrope with water. Lean mixtures may be separated by distillation at atmospheric pressure into water and the azeotrope, approximately 95 wt % THF. This mixture may then be fractionated in a second column at 150 lb/in² gage into essentially pure THF bottom product and an overhead vapor approaching the azeotropic mixture of approximately 88 wt % THF. As with other azeotropic separations, this mixture may be recycled to the first column for reconcentration to 95 percent. The flowsheet is shown in Fig. 10.4.

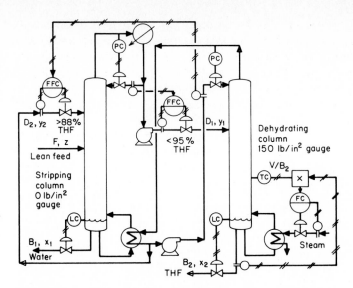

figure 10.4 *Operating the columns at different pressures breaks the azeotrope and allows energy integration.*

The material balances for the two columns are very sensitive to the actual overhead compositions of each. The equations developed below assume that virtually pure products leave the bottom of each column. Symbols for the streams are given in Fig. 10.4, with all flow rates in mass units and compositions in weight fraction THF.

All the THF leaving the first column comes from the feed and the recycle:

$$D_1 y_1 = Fz + D_2 y_2 \tag{10.20}$$

The second column discharges a bottom product at a rate Fz as the difference between feed D_1 and distillate D_2:

$$Fz = D_1 - D_2 \tag{10.21}$$

Substituting (10.21) for Fz in (10.20) yields the ratio of the two distillate flows:

$$\frac{D_2}{D_1} = \frac{1 - y_1}{1 - y_2} \tag{10.22}$$

In the same manner, either distillate flow may be eliminated to yield a solution for the other:

$$\frac{D_1}{F} = z\frac{1 - y_2}{y_1 - y_2} \tag{10.23}$$

$$\frac{D_2}{F} = z\frac{1 - y_1}{y_1 - y_2} \tag{10.24}$$

Observe how the recycle flow increases as y_2 approaches y_1.

example 10.4

Given a feed containing 60 wt % THF, with overhead compositions of 95 and 88 percent, calculate the three flow ratios listed above.

$$\frac{D_2}{D_1} = \frac{1 - 0.95}{1 - 0.88} = 0.417$$

$$\frac{D_1}{F} = 0.6\frac{1 - 0.88}{0.95 - 0.88} = 1.029$$

$$\frac{D_2}{F} = 0.6\frac{1 - 0.95}{0.95 - 0.88} = 0.429$$

Energy Requirements   Each column is essentially a binary distillation where the relationships given in Chap. 3 apply. For example, the feed to column 2, D_1, must command a certain energy input to achieve separation into a controlled bottom product and the recycled distillate. Decreasing column 2 boilup will tend to reduce the purity of the THF product and increase y_2. For an increase of y_2 from 88 to 90 percent, D_1/F will increase from 1.029 to 1.20 according to Eq. (10.23). This increase in column 2 feed will tend further to degrade separation by reducing the boilup-feed ratio. The recycle loop causes column 2 to be especially sensitive to variations in boilup by reinforcing disturbances through positive feedback.

Feed to column 1 is the sum of $D_2 + F$:

$$D_2 + F = F\left(1 + z\frac{1 - y_1}{y_1 - y_2}\right) \tag{10.25}$$

In this column also, separation determines the relative purities of product water and distillate as a function of boilup-feed ratio. An increase in y_2 caused by a reduction in column 2 boilup increases the feed to column 1 and therefore affects the separation there.

example 10.5

Estimate the effect of an increase in y_2 from 88 to 90 percent on total feed to column 1.
  For $y_2 = 0.88$,

$$D_2 + F = 0.429F + F = 1.429F$$

For $y_2 = 0.90$,

$$\frac{D_2}{F} = 0.06\frac{1 - 0.95}{0.95 - 0.90} = 0.6$$

$$D_2 + F = 0.6F + F = 1.6F$$

The increase in D_2 caused by reducing y_1 can further increase D_1 and D_2. So the positive feedback is seen to penetrate both columns.

Energy Integration   Because the columns operate at different pressures, energy integration is possible. Figure 10.4 shows the overhead vapor from the high-pressure column being used to boilup the low-pressure column. Flooded condensers are used to control pressures through manipulation of reflux valves. Separation is established in each by appropriate adjustment of reflux ratios, using the flow ratio controllers (FFC) on each distillate.

Note that the dehydrating column is controlled by manipulation of the V/B ratio. In this way, steam is proportioned to the amount of dehydrated product being made. Presence of water in the product is indicated by column temperature, which is therefore used for feedback control of quality.

No direct control appears on the quality of the water leaving the stripping column because boilup is predetermined by the needs of the dehydrating column. However, if reflux ratios are correctly set, then water quality should be adequately regulated. For example, an increase in L/D for the dehydrating column will reduce the THF content of its distillate, thereby sending less to the stripping column; at the same time, it increases boilup and therefore raises the flow of stripped distillate being recirculated. Both of these responses will reduce the THF loss in the water product. The two reflux ratios need to be coordinated to achieve minimum energy usage consistent with product purity and recovery.

TERNARY HETEROGENEOUS AZEOTROPES

The ternary systems most commonly encountered in industry are those formed to allow separation of binary homogeneous azeotropes. A volatile substance known as an "entrainer" is injected into the column containing the binary azeotrope to form a lower-boiling ternary mixture. Its principal effect is to remove one of the binary components overhead, allowing the other to be concentrated in the bottom of the column. However, the ternary azeotrope must be heterogeneous if the valuable product and the entrainer in the overhead vapor are to be recovered. As with the other applications, an example is used to illustrate the principle.

Dehydration of Ethanol Using Hydrocarbons The most common application of an entrainer to break a homogeneous azeotrope is probably the dehydration of ethanol. Many different processes are actually in use, including adsorption and extractive distillation as well, but the formation of a ternary azeotrope concerns us here.

Benzene was the entrainer most used in the past, but its toxicity has led to the application of other hydrocarbons in recent years. Diethyl either has also been used at elevated pressures, but Black [2] showed that n-pentane at 50 lb/in^2 abs requires fewer trays and less energy per unit ethanol. Hexane has the advantage over pentane of atmospheric operation, along with an aqueous solubility lower than benzene.

Figure 10.5 shows an ethanol dehydrator and stripping column using hexane as an entrainer. The ternary azeotrope has an atmospheric boiling point of 57°C (135°F) and its composition is given in Table 10.1. Upon condensation and cooling to 25°C (77°F), the mixture separates into two liquid layers whose compositions are also given in the table. If the overhead vapor were exactly at the azeotropic composition, the resulting light layer would represent 87.8 wt % of the total condensed vapor.

If sufficient hydrocarbon-rich reflux is added to form the ternary azeotrope with all the water in the feed, then anhydrous ethanol can be produced. The bottom few trays are required to separate the hydrocarbon from the ethanol.

The heavy, aqueous layer must be removed from the dehydrator and stripped of its hydrocarbon and ethanol content elsewhere. Figure 10.5 shows a stripping

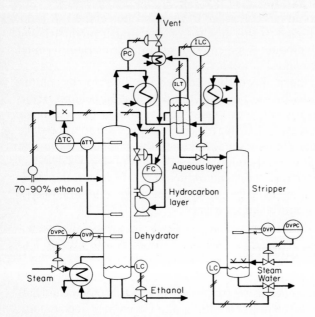

figure 10.5 *The hydrocarbon entrainer is continuously recirculated by the two vapor streams.*

column dedicated to performing this function. Here, overhead composition is unimportant, the principal purpose being to let only water out the bottom.

In a plant where anhydrous ethanol is made from a product of fermentation, weak ethanol (wine or beer) must first be enriched to 70 to 90 percent in a rectifier such as that shown in Fig. 8.4. To eliminate the separate stripping column, processors often recycle the aqueous layer from the decanter in Fig. 10.5 back to the rectifier. Although this practice reduces capital costs, the presence of hydrocarbon in the enriched ethanol affects the indication of the differential-vapor-pressure or temperature measurement in the top of the rectifier, interfering with control of ethanol concentration. The arrangement shown in Fig. 10.5 is much more reliable.

Material Balances   The products leaving the bottoms of the two towers in Fig. 10.5 will be essentially pure. From the standpoint of the material balance, then, it may be assumed that all the water in the feed Fz is transferred from the

TABLE 10.1 Composition of Hexane–Ethanol–Water Mixtures

	Wt %		
	Azeotrope	Upper layer*	Lower layer*
Hexane	85.0	96.6	6.2
Ethanol	12.0	2.9	73.7
Water	3.0	0.5	20.1

*at 25 °C.

dehydrator, through the decanter, and out the stripper. This then is the difference between vapor and reflux flows at the top of the dehydrator:

$$V_1 - L_1 = Fz \tag{10.26}$$

and also at the stripper:

$$L_2 - V_2 = Fz \tag{10.27}$$

(Internal liquid and vapor rates may depart significantly from these values owing to subcooling of the condensate to decanter temperature, yet their differences remain the same.)

A material balance on any component in the decanter gives

$$V_1 y_1 + V_2 y_2 = L_1 w_1 + L_2 w_2 \tag{10.28}$$

where y represents the concentration of that component in the respective vapor stream and w in the respective liquid stream.

There are two independent evaluations of Eq. (10.28) for any two of the components in the ternary mixtures. When these two are combined with (10.26) and (10.27), a solution for the dehydrator reflux can be obtained:

$$L_1 = Fz \frac{(y_{2h} - y_{1h})(w_{2e} - y_{2e}) + (y_{1e} - y_{2e})(w_{2h} - y_{2h})}{(y_{1h} - w_{1h})(w_{2e} - y_{2e}) - (y_{1e} - w_{1e})(w_{2h} - y_{2h})} \tag{10.29}$$

where h denotes hydrocarbon and e ethanol.

The first observation made is that dehydrator reflux must be directly proportional to the flow of water Fz entering the column. The proportionality varies with both vapor and decanter liquid-phase compositions. While the latter can be controlled by temperature, the former are functions of separation factors and bottom-product compositions in the two columns.

The separation factor is determined by the number of trays, the relative volatility, and the L/V ratio. However, material-balance equation (10.26) shows that

$$\frac{V_1}{L_1} = 1 - \frac{Fz}{L_1} \tag{10.30}$$

so that the L/V ratio is related to overhead vapor composition in two different ways. Note, however, that when (10.29) and (10.30) are combined, V_1/L_1 is not affected by feed composition. Similar equations can be derived for the stripping column.

In actual practice, the overhead vapor compositions are unimportant; only bottom compositions require control. Overhead vapor compositions will tend to float, in order to concurrently satisfy material-balance and separation relationships at whatever boilup rates are needed for acceptable product qualities.

Control   Control over ethanol product quality depends on two factors: keeping hydrocarbon out of the product and keeping water out. Excess hydrocarbon will appear at the differential-vapor-pressure transmitter as a high vapor pressure. This response will cause the DVP controller to increase boilup and drive the hydrocarbon back up the column. This is the normal mode of operating the dehydrator.

If water should contaminate the ethanol, it will not elevate its vapor pressure, because the vapor pressure of mixtures between 90 and 100 wt % ethanol are about the same. Therefore the bottom-composition control loop can only function in the absence of water. The bulb of the DVP cell may be filled with any aqueous mixture between 90 and 100 wt % ethanol.

Removal of water is accomplished by maintaining a sufficient ratio of hydro-carbon reflux to water influx, as required by Eq. (10.29). Therefore reflux flow is set proportional to feed rate in Fig. 10.5. But because the proportionality de-pends on many factors, including feed composition (which should be controlled using the system shown in Fig. 8.4), a means of automatically adjusting the ratio is provided. Water removal takes place over the largest portion of the column, where the hydrocarbon content dominates [3]. If insufficient trays are available with the necessary hydrocarbon content, some water will slip through that zone into the bottom product.

The zone of constant hydrocarbon content is identified by a constant tem-perature, i.e., almost complete absence of a temperature gradient. Accordingly, the differential-temperature transmitter in Fig. 10.5 has bulbs positioned near the top of the dehydrator and at a tray where the hydrocarbon content should begin to fall, i.e., where the temperature normally begins to rise. Positioning of the upper bulb allows two or three trays to remove reflux subcooling, common in these col-umns. The upper bulb then acts as a reference for the lower bulb, where tem-perature and composition change with reflux; it therefore provides compensation for changing pressure.

Should reflux flow be insufficient to remove all the water in the feed, the inven-tory of hydrocarbon in the column will become depleted, causing the lower tem-perature to rise. Excess reflux will conversely force hydrocarbon farther down the column, lowering the bottom temperature. The optimum temperature difference is in the range of 5 to 15°F.

Below the lower bulb, the temperature gradient is steeper, where hydrocarbon is stripped from ethanol [3]. The temperature will change from about 145 to 180°F over only a few trays. The DVP measurement should be made at the point where the gradient again begins to diminish, typically four to six trays above the bottom.

The decanter in Fig. 10.5 is equipped with an immersed displacer, whose bouyant force indicates the average density of its surrounding liquid, and therefore the position of the interface between the two phases. The interface-level controller maintains a constant aqueous-phase inventory by feeding the stripping column. The stripping column is controlled simply by admitting enough steam to remove the volatiles overhead. Its DVP transmitter uses water as a reference fluid, as did the lower transmitter in the ethanol-water column in Fig. 8.4. The level of hydro-carbon in the decanter is not controlled at all; it will be lowered by losses which must be made up by the operator.

TERNARY SYSTEMS CONTAINING BINARY AZEOTROPES

Actually, ternary mixtures containing binary azeotropes are much more common than ternary azeotropes. Several combinations are possible—homogeneous systems with one azeotrope, systems with one homogeneous and one heterogeneous azeotrope, etc. But the system described here is most common throughout the chemical, pharmaceutical, and synthetic fiber industries, where it is used in the concentration of organic acids.

These processes are typically carried out in three stages: liquid-liquid extraction, followed by azeotropic distillation, followed by solvent recovery. The process having the most data available is the production of glacial acetic acid using an acetate solvent. Actually, three different solvents are in common use — ethyl, isopropyl, and n-butyl acetate. Of the three, butyl acetate boils over both acetic acid and water and therefore presents some special problems. The other solvents boil lower and are in much more common use.

The Acetic Acid–Water–Ethyl Acetate System Acetic acid does not form an azeotrope with water, but their equilibria depart so much from ideality that the components are very difficult to separate by conventional distillation. The phase diagram for this system is shown in Fig. 10.6. Notice that the bubble-point and dew-point curves bend in the same direction — a most unusual characteristic. Superimposed on the same figure is a phase diagram for the heterogeneous ethyl acetate–water system. The two diagrams have been combined to describe more easily the events taking place in the distillation of this ternary mixture.

The third binary combination in this system, i.e., acetic acid and ethyl acetate, form a somewhat more ideal mixture than either of the other two binaries. Fig-

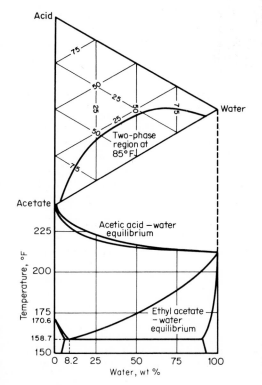

figure 10.6 *Projections of the phase diagram for the acetic acid–water–ethyl acetate system.*

ure 10.6 is actually a projection of the prismatic ternary phase diagram, in which the acid-acetate surface is not visible. The acetate-water phase diagram appears in the foreground while the acid-water diagram is in the background. These two planes meet at the right vertical axis (100 percent water). Looking down on the prism, the viewer will see the three-component liquid-phase diagram projected at the top of Fig. 10.6. As acid concentration is increased, the mutual solubility of acetate and water improves until eventually a single liquid phase is formed. Increasing temperature also improves their mutual solubility, shrinking the two-phase region. Hoffman [4] goes into much detail describing the variable characteristics of these phase diagrams.

The minimum boiling point for the system is that of the acetate-water heterogeneous azeotrope. Acid added to that mixture will distribute itself between the two liquid phases and raise the boiling point of the mixture. However, a minimum boiling point exists for any given concentration of acid as long as there are two liquid phases in equilibrium with the vapor. The combination of increasing acid concentration and boiling point eventually results in a single liquid phase, and the valley, characteristic of lower acid concentrations, disappears. Ellis and Pearce [5] indicate that the azeotrope shifts toward lower water concentrations as acid is added, until above 50 wt % acid it disappears altogether. The valley then apparently curves toward lower water concentration until it flattens into a smooth plane approaching the acid-water equilibrium curve.

The Extraction Process   The first stage of acid concentration is accomplished by liquid-liquid extraction of weak acid with acetate. As in other countercurrent mass-transfer processes, the water leaving the bottom of the extractor approaches equilibrium with the acetate feed. Any acid leaving the bottom of the extractor is lost and also pollutes, so the acetate extractant should be kept free of acid to avoid this loss. Acetate leaving in the aqueous effluent is recovered in a decanter-fed stripping column.

In a similar manner, the concentration of acid leaving the top of the extractor varies with the concentration of the weak-acid feed but also with the solvent-feed ratio. This relationship may be derived from a set of material-balance equations. The overall balance is

$$A + E = F + W \tag{10.31}$$

where the symbols represent mass-flow rates of weak acid, solvent, column feed, and waste, as indicated in Fig. 10.7. An acid balance on the extractor assumes no losses to waste:

$$Ax_a = Fz_a \tag{10.32}$$

where both compositions are in weight fraction of acid in the associated stream. A solvent balance may also be written:

$$Ex_e = Wx_w + Fz_e \tag{10.33}$$

with all compositions in weight fraction of acetate in their respective streams.

When the three balances are combined, a solution may be obtained in terms of the composition of the column feed:

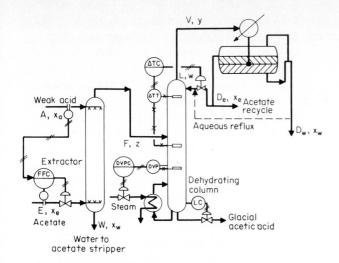

Weak acid
A, x_a

F, z

Extractor

FFC

E, x_e
Acetate

W, x_w

Water to
acetate stripper

V, y

ATC

ATT

L, w

DVPC   DVP

Steam

D_e, x_e Acetate
recycle

Aqueous reflux

D_w, x_w

Dehydrating
column

LC

Glacial
acetic acid

figure 10.7 *Acetate-rich reflux may require adjustment with aqueous reflux to optimize separation in the dehydrator.*

$$\frac{z_e - x_w}{z_a} = \frac{(x_e - x_w)E/A - x_w}{x_a} \tag{10.34}$$

Of the variables given in this equation, x_w is the saturation level of ethyl acetate in water and should only be a function of temperature. If the extractant is taken from a decanter, then its saturation level of acetate x_e is fixed by decanter temperature. Therefore, these two parameters ought to be constant. Then control over the relative proportions of acetate and acid in the column feed can be achieved by adjusting the extractant–weak-acid ratio E/A as a function of weak-acid concentration x_a. If E/A is fixed, however, z_e and z_a will vary with x_a.

Equation (10.34) is a single expression with two unknowns, z_e and z_a. Another relationship is required to determine their absolute concentrations: it is the phase diagram for the ternary system. Solubility limits dictate that the liquid-phase compositions in equilibrium in the extractor lie on the curve in the triangular diagram of Fig. 10.6. Concentrations can then be taken from selected points on the curve until a fit to Eq. (10.34) is found. (Because the stream is acetate-rich, column-feed composition will lie somewhere along the left side of the curve.)

Dehydrating the Acid   The distillation process has three objectives:
1. To make a glacial acetic acid product
2. To minimize acid and solvent losses
3. To minimize energy consumption

Conventional columns have but two degrees of freedom which determine overhead and bottom-product quality — typically the material balance and the separation. This column has a third degree of freedom, however, the ratio of acetate to water.

The lowest boiling point of any mixture in this ternary system belongs to the acetate-water azeotrope; the highest boiling point is that of pure acetic acid. Therefore, one might assume that the average relative volatility in the column would be maximized by adjusting the acetate-water ratio to produce the azeotrope overhead. The merits of this assumption can be evaluated in light of data presented in Ref. 5. Ellis and Pearce have derived values of volatility for ethyl acetate relative to acetic acid, and for water relative to acetic acid, at various concentrations of the ternary mixture. Selected points from this source, converted into molar-average relative volatilities between the acetate-water combination and the acid, are presented in Table 10.2.

Weight percent water on an acid-free basis was chosen as a parameter to illustrate the effect of adjusting the acetate-water ratio: 8.16 percent corresponds to the acid-free azeotrope. Increasing acid content reduces the relative volatility of the 8.16 percent mixture until at 100 percent acid it no longer is maximum. The most effective fractionation would seem to be achieved when the acetate-water ratio is held at the acid-free azeotropic composition, the reduction in volatility at 100 percent acid being slight.

These columns are typically refluxed with only the organic phase from the decanter, however. Since this reflux contains only about 3 wt % water, the relative volatility is not as high as it could be and substantially more energy is required to make the separation. As the liquid proceeds down the column, its acid content increases and so does the temperature, both of which improve the solubility of the solvent and water. The liquid may become richer in water if the feed is rich in water, however. The tray composition and hence the relative volatility is then sensitive to feed composition as well as reflux composition and flow. The relationship is best borne out by the material balance.

Material-Balance Considerations For simplicity, the overhead vapor is assumed to contain no acid and the bottom product no solvent or water. Then all the acetate and water in the feed leave as distillate, giving a decanter balance of

$$V = L + F(z_e + z_w) \tag{10.35}$$

where V and L are mass-flow rates of overhead vapor and reflux and where z_e and z_w are weight fractions of acetate and water in the feed. An acetate balance gives

$$Vy = Lw + Fz_e \tag{10.36}$$

where y and w are weight fractions of acetate in the vapor and reflux.

To meet certain specifications of distillate and bottom-product purity, a certain V/F is required whose value naturally depends on the relative volatility achievable in the column. Since the relative volatility α varies with the acetate-water ratio in the

TABLE 10.2 Molar-Average Volatility of Ethyl Acetate and Water Relative to Acetic Acid

Wt % water in liquid (acid-free basis)	Wt % acid in liquid			
	0	10	50	100
0.0	6.8	5.2	3.0	2.8
3.0	11	8.0	3.3	2.8
8.16	18	10.2	3.7	2.7
100.0	1.2	1.3	1.6	2.6

column, one factor of substantial significance is how overhead composition is affected by the V/F ratio. To determine this relationship, Eqs. (10.35) and (10.36) are combined by eliminating L and solving for y in terms of V/F:

$$y = w - \frac{w(z_e + z_w) - z_e}{V/F} \tag{10.37}$$

For a given V/F ratio, y varies with feed composition. Because of the effect of y on α, this relationship is worth examining in detail. Accordingly, Eq. (10.37) was solved for several values of V/F at 97 wt % acetate in the reflux; the results appear in Fig. 10.8.

Feed-composition values were taken from the solubility curve at the top of Fig. 10.6. As the acid content of the feed increases, its water content also increases, causing a drop in acetate level. Increasing acid concentration thus tends to lower the acetate-water ratio in the column. To demonstrate its effect on column operation, consider an initial condition at $V/F = 1.5$ and $z_a = 12.5$ percent, producing an overhead vapor at the azeotropic composition. An increase in z_a at that V/F will reduce y below the azeotropic composition. The liquid in the column must then become richer in water than the vapor, causing α to fall. If bottom-product composition is controlled by manipulating boilup, the controller will react to deteriorating purity by raising V/F. This action will increase y toward the azeotropic composition, thereby increasing α and returning product purity to its former value. The net result of an increase in z_a is a corresponding increase in V/F, not a normal condition by conventional standards but at least stable.

However, a *decrease* in acid in the feed generates an *unstable* situation. At a given V/F ratio, y increases above the azeotropic composition, developing a liquid phase still richer in acetate. Again α decreases, but deteriorating product quality cannot be corrected by increasing V/F. As Fig. 10.8 indicates, increasing V/F increases y and thereby further lowers α. Some columns have a history of operation in the solvent-rich mode but are prone to be upset easily. In addition, they typically allow significant losses of solvent in the acid product, and use much more energy than should be required to make the separation. Because of their high and variable V/F ratio, their feed-processing capacity also tends to be low.

Solvent-rich operation is characterized by higher acid concentration all across the column, although tray temperatures are lower. A shift from water-rich to solvent-rich operation is detectable as a sharp drop in mid-column temperature from 200 to perhaps 175°F.

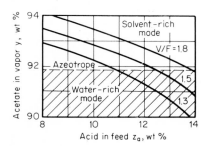

figure 10.8 *Overhead vapor composition as a function of feed composition for reflux containing 97 percent acetate.*

Adjusting the Solvent-Water Ratio   The only way out of the solvent-rich mode is to add water to the column, either in the reflux or the feed. If the solvent-water ratio in the feed is not the same as in the reflux, a composition gradient (acid-free basis) will develop across the column. In fact, the only way a gradient can be avoided is if feed, reflux, and vapor all contain the azeotropic mix of acetate and water. This would seem to provide the maximum overall relative volatility or at least very close to it. By comparison, the initial conditions described earlier had a 97 percent acetate reflux and the feed of 12.5 percent acid contained 88 percent acetate (acid-free basis). A feed this far from azeotropic proportions cannot produce an optimum volatility. The only feed composition on the solubility curve which corresponds to the azeotropic mixture of water and acetate occurs at about 7.5 percent acid, 7.5 percent water, and 85 percent acetate. This particular mix could be obtained from the extractor by increasing its extractant–weak-acid ratio, although adjustment would have to be made for variations in weak-acid concentration.

Regardless of whether feed composition is specifically adjusted to an optimum, reflux composition ought to be adjustable to avoid slipping into the solvent-rich mode. This is best achieved by adding some of the aqueous layer from the decanter to the reflux as shown in Fig. 10.7. Reflux composition w required to develop a specific overhead-vapor composition with varying feed can be calculated by rearranging (10.37):

$$w = \frac{yV/F - z_e}{V/F - (z_e + z_w)} \tag{10.38}$$

Equation (10.38) has been solved for $y = 0.9184$ and three values of V/F, with the results plotted in Fig. 10.9. As observed earlier, when the feed contains the azeotropic ratio, the reflux should also, regardless of the value of V/F. This would seem to be the optimum set of conditions.

Increasing the acid (and water) content of the feed requires that w increase to the point of saturation, i.e., requiring no aqueous reflux. Operation beyond this point is possible at the cost of increasing the V/F ratio. However, adding more solvent to the feed may be more economical.

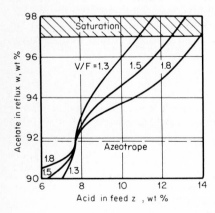

figure 10.9 *Reflux composition should be adjusted as shown here to hold the overhead vapor at azeotropic composition.*

The amount of aqueous reflux L_w required to adjust w to the desired point is actually quite small. It is easily calculated from the compositions of the two phases leaving the decanter:

$$\frac{L_w}{L} = \frac{x_e - w}{x_e - x_w} \tag{10.39}$$

where x_e and x_w represent the weight fractions of acetate in the organic and aqueous phases, respectively. For the case where w is desired at 0.918, given $x_e = 0.97$ and $x_w = 0.07$, $L_w/L = 0.058$. The aqueous stream is seen to contribute relatively little to the reflux flow while at the same time greatly affecting its composition.

Having a two-phase reflux may cause two liquid phases to exist on some of the trays. But this is a natural consequence of operation in either the water-rich mode or precisely at the azeotrope throughout. Only in the solvent-rich mode may a single liquid phase exist, and this is not only suboptimum but demonstrably unstable. Therefore, the acetic acid column must be designed for operation with two liquid phases.

Not being able to analyze a heterogeneous azeotrope creates a control problem. Since reflux composition must be adjusted as a function of feed composition or to hold a certain vapor composition, some sort of analysis is necessary. One suggestion is to measure the flow of each phase leaving the decanter and calculate their ratio. If V_w is that portion of the condensed vapor leaving the decanter in the aqueous phase and V_e is the portion leaving in the organic phase, then

$$\frac{V_w}{V_e} = \frac{x_e - y}{y - x_w} \tag{10.40}$$

To hold y at 0.9184, given that x_e and x_w are 0.97 and 0.07, respectively, V_w/V_e should be 0.061. The calibration of such an "analyzer" appears in Fig. 10.10 as the solution to Eq. (10.40).

The liquid-phase flow rates will lag behind changes in vapor composition because of the large capacity of the decanter. This is to be expected and is in fact typical of product analyses made on reflux or distillate liquid streams. So the flow-ratio calculation would seem to have several of the properties of an analyzer, yet without the sampling and calibration problems. It will be temperature-sensitive, however, so decanter temperature must be controlled. In any case, it may be the

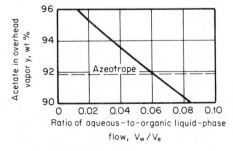

figure 10.10 *Overhead vapor composition may be inferred from the flow ratio of the two liquid phases if their composition is constant.*

only means of determining the overhead vapor composition of a heterogeneous azeotrope.

Control   Figure 10.7 shows ratio control of acetate to weak acid entering the extractor. This will hold column feed composition constant if weak-acid composition is uniform; if it is not, the ratio will require adjustment. In any case, optimum column performance could be realized without aqueous reflux if feed composition is adjusted to 7.5 percent acid, as noted. The ratio adjustment would seem easier and more effective than controlling the addition of aqueous reflux.

A differential-vapor-pressure transmitter is recommended to control acetic-acid quality, as it was for ethanol dehydration. In contrast to the ethanol-water mixture, however, the acetic-acid–water mixture shows significant variation in vapor pressure with composition in that range. Reference 6 indicates that only 0.08 wt % water depresses the atmospheric boiling point of acetic acid by 1°F, which is equivalent to an increase in vapor pressure of 7.3 in H_2O. Furthermore, Ref. 5 reports that water concentration exceeds that of ethyl acetate from two trays below the feed point to the bottom of the column, when operating in the normal, water-rich mode.

A special transmitter is required for acetic-acid service. Although acetic is classified as a weak acid, it may contain impurities of formic, oxalic, and other stronger organic acids. In any case, lower sections of acetic-acid columns must be made of corrosion-resistant alloys because mild steel is severely attacked. The special DVP cell has wetted parts constructed of Hastelloy C; the reference bulb is made of 316 stainless steel but is inserted in a thermowell of Hastelloy C. Products of corrosion — particularly hydrogen — change the vapor pressure of an acetic-acid sample significantly, preventing its use in a reference bulb. Therefore a substitute filling solution is used [7], consisting of 80 percent methyl cellosolve and 20 percent distilled water; it mimics the vapor-pressure curve of a 90 percent acetic-acid–10 percent water mixture.

Figure 10.7 also shows a differential-temperature-control loop manipulating reflux flow, as was done with the ethanol dehydrator. In the latter, it was necessary to maintain an inventory of entrainer across enough trays to ensure removal of all water overhead. Here, however, the normal composition profile [5] shows more water than entrainer below the feed tray, whereas in the ethanol dehydrator, the entrainer was dominant. Consequently, this ΔT control loop will only function properly in the water-rich mode — insufficient water will cause mid-column temperature to fall, thereby decreasing reflux and allowing acid to be carried overhead. If the reflux contains enough water to equal or exceed the azeotropic composition, then this loop should always function properly. The lower temperature bulb is located about two trays below the feed point, where acetate composition shows its sharpest gradient [5]; the upper bulb is located two or three trays below the top, to avoid the influence of subcooled reflux.

REFERENCES

1. Mains, G. H.: "The System Furfural-Water," *Chem. Metall. Eng.,* April 26, 1922.
2. Black, C.: "Distillation Modeling of Ethanol Recovery and Dehydration Processes for Ethanol and Gasohol," *Chem. Eng. Prog.,* September 1980.

3. Black, C., R. A. Golding, and D. E. Ditsler: "Azeotropic Distillation Results from Automatic Computer Calculations," *Adv. Chem.,* ser. 115, 1972, pp. 64–92.
4. Hoffman, E. J.: *Azeotropic and Extractive Distillation,* Wiley-Interscience, New York, 1964, Chaps. 5–8.
5. Ellis, S. R. M., and C. J. Pearce: "Designing Azeotropic Distillation Columns," *Br. Chem. Eng.,* December 1957.
6. Ju Chin Chu: *Distillation Equilibrium Data,* Reinhold, New York, 1950, p. 208.
7. *Custom E13VA-I Vapor Pressure Transmitter for Use with Corrosive Process Solution,* Special Instruction 1-02009, The Foxboro Company, Foxboro, Mass., January 1981.

PART FOUR

Optimization

CHAPTER ELEVEN

Optimizing Control Systems

To optimize a process generally means to operate it in the best possible way consistent with certain objectives and within given constraints. But what constitutes "best" must first be defined by examining those objectives and, from them, deriving an optimum operating policy. It should come as no surprise, however, to find that for most processes optimum conditions lie beyond some constraints. Then the problem reduces to one of constraint control rather than optimization. The distinction here is that the objective function can be driven *through* a maximum or minimum only if that maximum or minimum lies within all constraints.

This chapter on optimization was relegated to the end of the book to place the subject in its proper perspective in order of both time and importance. While optimization is a desirable goal, it cannot be achieved—and, in fact, should not be attempted—until all lower-level control functions are executed. To optimize a process requires that it be moved from some existing state to a more profitable one. Yet the purpose of the product-quality controls was to maintain constant compositions.

Optimization then requires that these compositions be adjusted. For this endeavor to succeed, the compositions must in fact be controlled at their optimum set points and must respond to direction rapidly enough to follow changing plant conditions. If their control is erratic and unresponsive, no optimization program, regardless of its accuracy, can succeed.

PRINCIPLES OF OPTIMIZATION

To optimize a process, the engineer must first select an objective function and a variable to manipulate in a way to maximize or minimize that function. Usually it is not sufficient to solve an optimization problem just once. Any real process is affected strongly by its environment, such that its optimum operating point tends to shift with imposed conditions. Most objective functions are economic in nature; as a result, the optimum point will change with selling prices, costs, and market conditions. Then any optimizing system must be dynamic, capable of readjusting selected variables as plant conditions and economic factors dictate.

Perhaps the first consideration is the scope of the problem—how much of the plant to include. The scope, in fact, determines the nature of the objective function and the complexity of the relationships involved; it also suggests the means necessary for implementing the system.

Levels of Optimization   There are several levels at which optimization may be applied, depending on the scope of plant equipment enclosed within the objective function. To begin, one might optimize the performance of a single column without regard to its impact on the rest of the plant. Since a single column is probably the smallest element in the plant capable of independent operation, this then becomes the lowest level of optimization. It is traditionally designated "local optimization."

Columns in a parallel or series configuration can, in addition, be optimized as a group. This practice avoids the possibility of excessive penalties being absorbed by one column to satisfy the local optimization of another. When the scope of the system is expanded to include all the columns in one distillation unit, we have "unit optimization." Implicit in unit optimization is the effective allocation of limited feedstocks, energy, refrigeration, etc., among the members of that unit.

Complete plant optimization involves coordinating the control of distillation units, utilities, reaction systems, furnaces, compressors, etc., to maximize the profit from the entire operation. Although this is a laudable objective, it may be more of a panacea than a realizable goal. Optimization is the end of the road—all other lower-level controls must function responsively before it can ever be approached. And the larger the segment of a plant enclosed within the scope of the optimization problem, the less the likelihood of reaching a stable optimum state. Expanding the scope of a system to enclose more units of the plant extends the time required to reach an optimum and the exposure to disturbances. Consequently, some level can be reached at which optimization is no longer effective. This limit will naturally differ with the complexity of the plant and its integrity.

The Objective Function   In most cases, the objective function for optimization is monetary in nature. It could be profit or loss, operating cost, or productivity. Occasionally it can be reduced to a singular function specific to one process, such as the yield from a chemical reactor or the efficiency of a boiler or compressor. Although these specific functions are satisfactory for local optimization, their use cannot be extended to combination with other elements.

As an example, consider a stripping column exhausting into a compressor. At a given gas flow rate, there is a certain suction pressure which will maximize the efficiency of the compressor. This pressure may not represent the most efficient operating point for the column, however. Yet if the efficiencies of column and compressor were only reported on a percentage basis, there would be no way to relate the two to optimize their combined performance. If their performance were rated on a Btu basis, they could be compared only if the cost per Btu were equal for both column and compressor. Finally the problem evolves into a cost optimization, since operating cost is the common denominator between these two dissimilar pieces of equipment.

Some will argue that a profit function is most meaningful, since a plant is in operation to make a profit. However, it seems more direct to keep tabs on costs and losses, which, if minimized, will thereby constitute maximum profit within the scope of their influence. Cost functions may then be selected to limit the scope of an optimization problem. For example, one may choose to minimize product losses, energy costs, or a combination of the two, independent of production rate. By contrast, a profit function depends on production rate.

The most useful objective function for most applications is a combination of product losses and utility costs per unit feed. For example, let W represent the flow of a waste or lower-valued stream containing w fraction of the valuable component. Then Ww describes the amount of unrecovered product leaving the process. Given that the value of the component of interest *in the product* is v_P, compared to a value of v_w *in the waste* stream, we have a cost statement per unit feed:

$$\frac{\$_c}{F} = (v_P - v_W)\frac{Ww}{F} + (c_i + c_o)\frac{Q}{F} \tag{11.1}$$

The cost factors for heat input and removal are c_i and c_o, expressed in \$/Btu, where Q is the heat flow required to effect the separation.

By contrast, a profit function includes the sum of the values of all the products, whatever their worth, less the cost of the feed:

$$\$_P = \sum Pv_P - \sum Fv_F - \sum Q(c_i + c_o) \tag{11.2}$$

It is possible that under extreme conditions maximum profit may not be consistent with maximum feed rate. Consider, for example, the separation of a feed into one high-valued product and a very low or negative-valued product. As feed rate is increased, a point can be reached where any additional feed will go unrecovered and profit will begin to decrease. This point constitutes the optimum feed rate.

If specifications must be met on both products from a column, then the maximum amount of feed will be processed when these are just met while under a boilup constraint. This constitutes maximum feed rate under constrained conditions and may maximize the profit function as described by Eq. (11.2). From a control standpoint, however, this is not an optimization problem since profit cannot *pass through* a maximum but can only reach it at a constraint.

Another limitation to the profit function is its assumption of an unlimited market for products and an unlimited supply of feedstock. If, in a given case, these conditions exist, then a maximum or optimum feed rate is desirable. However, the selection of a feed rate that will meet supply and demand limits is a real consideration and, in fact, adds another dimension to the optimization problem.

Finally, in any real plant, feedstock must pass serially through a number of processing elements before a final product emerges. The local optimization of any of these elements must be independent of feed rate, which by necessity interacts with *all* the elements in the plant. Hence the cost function of Eq. (11.1) will be found to apply to the local optimization problem while the profit function may be more appropriate for unit or plant optimization.

Feedforward vs. Feedback   From the presentation on feedforward control in Chap. 6, its vastly superior dynamic responsiveness should be evident. Whereas feedback must search for the correct value of the manipulated variable by trial and error, feedforward arrives at the solution by direct calculation. The cost of this responsiveness is a detailed knowledge of the process, converted into computing controls. By contrast, a feedback controller can arrive at the same solution without such intelligence, but it requires a certain amount of time in an undisturbed condition.

Optimization as well as control may be conducted by either feedforward or feedback strategies. Feedforward optimization relies on a mathematical model of the process generating an off-line solution to the problem through which relationships between measured variables can be established which constitute optimum operation. For example, a cost function may depend on a certain combination of plant variables such as feed rate F, product flow D, product composition y and value v, and heat input Q along with its cost c:

$$\$_c = f_c(F, D, y, v, Q, c) \tag{11.3}$$

A certain relationship among these variables may be found which will minimize $\$_c$. One of the variables is then selected to be manipulated to optimize $\$_c$ in response to variations in the others. Then the established optimum relationship may be solved in terms of that manipulated variable, for example, Q:

$$Q_o = f_o(F, D, y, v, c) \tag{11.4}$$

Here Q_o is the heat input which minimizes $\$_c$ in response to the other variables and f_o defines that relationship which represents the minimum cost. This concept is illustrated by several examples later in the chapter.

If the cost relationship f_c is simple enough, it should be possible to differentiate $\$_c$ with respect to the selected manipulated variable, set the differential to zero, and solve for the manipulated variable:

$$\frac{d\$_c}{dQ} = \frac{df_c}{dQ}(F, D, y, v, Q, c) = 0 \tag{11.5}$$

$$Q_o = \frac{df_c}{dQ}(F, D, y, v, Q, c) + Q \tag{11.6}$$

Then the optimizing relationship f_o is

$$f_o(F, D, y, v, c) = \frac{df_c}{dQ}(F, D, y, v, Q, c) + Q \tag{11.7}$$

For a simple process, this procedure will be found workable assuming that the cost function f_c is linear, quadratic, hyperbolic, or logarithmic, yielding readily to differentiation. Then the optimizing relationship will be soluble by direct calculation.

For more-complex processes, quadratic or higher-order optimizing relationships appear, impeding direct solution. Then an iterative program is needed to increment the manipulated variables in such a way as to arrive at optimum conditions. This can be accomplished automatically with nonlinear programming.

Nonlinear and linear programming operate by feedback mechanisms, as does any iterative or trial-and-error procedure. If conducted directly on the process, time must be allowed for the process to settle out after each trial manipulation. Since many steps are typically required to reach an optimum level for each variable manipulated, regardless of the search method used, the progress toward optimization by feedback is slow indeed. In fact, since upsets may easily develop at any time during the search, the optimum conditions for a real plant are not likely to be reached at all by this means. Although feedback may be used for local optimization, it is not recommended. See Ref. 1 for a discussion of this subject.

Instead, the iterative procedure should be carried out on a steady-state computer *model* of the process, where individual results may be obtained in milliseconds rather than minutes or hours. Then an optimum solution may be found in seconds or minutes and be imposed on the plant with some confidence that it will at least satisfy conditions presently existing.

There are two fundamental limitations to this approach. First, it depends on the accuracy of the plant model. Although the model of the plant may in fact be optimized, there is no assurance that the plant itself is also optimized. Actually this is true of any feedforward system. As subsequent observations indicate a disparity between the plant and the model, adjustments can be made to improve the representation.

The elimination of dynamics from the model reduces its complexity and facilitates a rapid solution. But their absence constitutes a second departure from the true characteristics of the plant itself. Although a manipulation at a given point in time may represent optimum steady-state conditions at that time, its subsequent effect on a changing process could be significantly suboptimum. One constantly changing variable is the environment surrounding the process. Passage of the sun across the sky affects condensers and may therefore cause optimum operating conditions to cycle diurnally. If the plant's serial distribution is extensive, unit optimization under these conditions may be impossible. For example, an increase in feed rate allowable at night could arrive at a downstream column the following noon, at which time it could not be accommodated. While complete plant optimization is desirable, these factors may prevent it from ever being achieved.

Selection of Variables for Manipulation A prerequisite to optimization is the existence of more than the minimum number of manipulated variables needed for regulation of the specified controlled variables. If no constraints are being applied, this requirement can be met if the number of manipulated variables exceeds the number of controlled variables. Naturally a manipulated variable is lost each time a constraint is encountered. Hence constraints place severe limits on the opportunities available for optimization.

The most common application in a fractionating plant is the local optimization of a column making a single final product. In this case, the quality of that product could be controlled by manipulating the material balance. This leaves the heat input free to be manipulated for optimization within the limits of flooding, etc.

When a constraint is encountered, heat input is the variable normally limited. Then encountering a constraint only means suboptimum operation as opposed to loss of control over product quality. But in attempting to follow the command of the optimizing program, heat input will be as close to optimum as the constraint will allow.

Allocation of utilities among several columns is another optimization problem presented later in the chapter. At this juncture, it is only necessary to point out that heating and cooling can be allocated to minimize monetary losses of all the products in the plant. Again, the selected manipulated variable is heating for each individual column.

The exact form of the manipulated variable is open to selection. In a given instance, it could be steam flow, heat flow, refrigerant flow, boilup-feed ratio, separation, or even the composition of the unspecified stream. The choice depends on the significance of the term in the cost function. In the case of a refrigeration-allocation problem, heat flow is probably the most significant variable. For local optimization, either heat flow or the composition of the unspecified product could be used, with different results. However, the use of heat input or reflux can bring about a faster approach to optimum conditions and avoids the need for a second composition-control loop. Again, these various choices will be illustrated by examples.

SINGLE-COLUMN OPTIMIZATION

Confining the optimization effort to a single column facilitates the solution and its implementation. While there is no guarantee that what is best for a single column is best for the plant, single-column optimization is generally preferable to no optimization at all. Furthermore, the opportunities for single-column optimization far exceed those of larger scope. Consequently, these opportunities are examined in detail below since they represent the bulk of the applications which are worth pursuing in distillation units.

Optimization is only possible within constraints, so it is important to identify those conditions where constraints will allow it. The most restrictive specifications are those on product qualities. If *neither* product from a column needs to meet a particular specification, then there is complete freedom to optimize both. The more common situation is where one of the products must meet a specification which allows optimization of the *other* product composition. Although this case is more restrictive, it is also easier to define and so is covered first. A related problem, where *both* products are specified, is treated afterward.

Final Products The most common objective for operating a distillation column is to make a single valuable product meeting some guaranteed specification while at the same time minimizing the loss of that valuable product. In the past, these columns have been operated at the heat-input constraint, which in fact minimizes product losses for any given feed rate. However, rising energy costs have focused attention on total operating costs and have led to an exploration of possibilities for optimization.

In every one of the separations fitting the description above, a heat input exists which will minimize operating costs for any given feed rate. If the cost of energy is low in comparison to the value of the lost product, that optimum heat input may lie beyond the capacity of the column, reboiler, or condenser to accommodate it. But before that can be determined, the exact location of the optimum heat input needs to be found.

Consider the cost function of Eq. (11.1) for a column separating a binary mixture into a more-valuable distillate and a less-valuable bottom product. The cost function is stated as

$$\frac{\$_c}{F} = (v_D - v_B)\frac{Bx}{F} + c\,\Delta H_D\,\frac{V}{F} \tag{11.8}$$

where v_D and v_B are the values of the two products, c is the combined cost of heating and cooling, V is the boilup rate, and ΔH_D is the heat of vaporization of the distillate, all in consistent units.

Because of the interlocking nature of the variables in Eq. (11.8), differentiation with respect to any selected variable is not a simple procedure. If it were desired to find the optimum bottom composition, for example, all terms would have to be differentiated with respect to x at constant y. But B/F and V/F are both functions of x, the latter a rather complex one through the separation factor and reflux ratio. Therefore it is probably better to simply evaluate (11.8) in terms of x at several different compositions and select the least cost. First V/F must be replaced by L/D:

$$\frac{V}{F} = \frac{D}{F}\left(1 + \frac{L}{D}\right)$$

Then L/D must be replaced by the separation factor using Eq. (3.59). Finally, material-balance relationships are substituted for B/F and D/F, giving:

$$\frac{\$_c}{F} = \frac{(v_D - v_B)x(y - z)}{(y - x)} + \frac{c\,\Delta H_D(z - x)}{(y - x)}\left\{1 + \frac{1}{z}\Big/\left[\left(\frac{\alpha}{S^{1/nE}}\right)^2 - 1\right]\right\} \tag{11.9}$$

example 11.1

A splitter separates a 50-50 mixed-butane feed into products each 95 percent pure at an L/D ratio of 7.0. The distillate is \$4.00/bbl more valuable than the bottom product, and the combined cost of heating and cooling is $6.60/10^6$ Btu. The heat of vaporization of isobutane at 100°F is 25,000 Btu/bbl, and relative volatility is 1.35. Calculate the optimum bottom composition, with distillate controlled at 95 percent purity.

First, find nE using Eq. (3.58):

$$nE = \ln S \Big/ \ln\left(\frac{\alpha}{\sqrt{1 + D/Lz}}\right)$$

$$nE = \ln 361 \Big/ \ln\left(\frac{1.35}{\sqrt{1 + 1/3.5}}\right) = 33.76$$

Next, evaluate (11.9) at $x = 0.05$:

$$\frac{\$_c}{F} = \frac{(\$4.00)0.05(0.45)}{0.95} + \frac{(\$6.60/10^6)25,000(0.45)}{0.95}(1 + 7)$$

$$= 0.100 + 0.660 = \$0.760/bbl$$

Then, try $x = 0.04$, to find the slope of the cost function:

$$S = \frac{(0.95)0.96}{(0.05)0.04} = 456$$

$$\frac{\$_c}{F} = \frac{(\$4.00)0.04(0.45)}{0.91} + \frac{(\$6.60/10^6)25,000(0.46)}{0.91}\left[1 + \frac{2}{(1.35/456^{1/33.76})^2 - 1}\right]$$

$$= 0.0791 + 0.7057 = \$0.7849/bbl$$

(A program for evaluating Eq. (11.9) is given in Appendix A.) Evaluation continues above $x = 0.05$, with results appearing in Table 11.1.

An examination of Table 11.1 shows how loss in product value increases as reflux ratio is reduced. The optimum value of x at 0.08 differs considerably from its base value of 0.05, yet the total operating cost is only reduced by about 3 percent. This is typical of optimization problems in distillation. Furthermore, the large value of x which represents minimum cost may be well beyond the tolerance of the downstream process to accommodate. In this event, the butane splitter should be operated at the maximum value of x that is acceptable.

In cases where the cost of heating and cooling is substantially lower or the difference in product values substantially greater, the optimum value of x could be quite low, within the acceptable limit. Then optimization is meaningful, although still perhaps not a source of much revenue.

TABLE 11.1 Operating Costs for the Butane Splitter at $y = 0.95$

x	Loss in product value	Cost of heating and cooling	Total, $/bbl
0.04	0.0791	0.7057	0.7849
0.05	0.1000	0.660	0.7600
0.06	0.1213	0.6245	0.7458
0.07	0.1432	0.5954	0.7386
0.08	0.1655	0.5705	0.7361
0.09	0.1884	0.5487	0.7371

Should a product analyzer not be available or economically justified, it is possible to optimize a flow or flow ratio instead. To do this, the column in Example 11.1 was subjected to feed compositions of 0.30 and 0.40. The optimum values of x were determined as in the example, and from them the corresponding values of V/F, V/B, and L/F were calculated. This information is assembled in Table 11.2.

Observe that the optimum value of V/B changes almost 40 percent over the range of feed compositions, while V/F changes only about 13 percent. The last column shows that L/F changes by only 7.6 percent, indicating that it is the best choice of the three to optimize this column in the presence of feed-composition variations. In other words, holding L/F constant at about 2.90 will produce nearly optimum bottom composition without having to measure either bottom or feed compositions.

Figure 11.1 shows how such a system might be configured. Reflux must be set in ratio to the feed, with the optimum ratio determined as in Table 11.2. Then distillate flow is manipulated in ratio to reflux for top-composition control. This system requires that liquid level in the accumulator (or column pressure, if the accumulator is flooded) be controlled by heat input. Since there is no bottom composition to control, there is no need to consider a relative-gain analysis.

If x must be limited to some specifications, however, then it must be measured and controlled, requiring a bottom-composition loop manipulating V/F or V/B as required for minimum interaction. Figure 7.13 illustrates the type of calculation required to optimize a composition set point within allowable specifications.

Intermediate Products   When neither product from a column is for sale but instead is used elsewhere within the plant, it is possible to optimize the compositions of both. Because neither composition need be controlled at a fixed set point, the material balance is free to be manipulated for optimization. The heat input may then remain fixed or may also be manipulated, in which case the optimization becomes a two-parameter problem.

For the simpler case where V/F is fixed, the cost statement is a sum of the losses of the two products in the opposite streams. For a problem like this to be meaning-

TABLE 11.2 Optimum Conditions vs. Feed Composition for the Butane Splitter at $y = 0.95$

z	x	V/F	V/B	L/F
0.3	0.073	2.983	4.024	2.724
0.4	0.075	3.282	5.221	2.910
0.5	0.082	3.430	6.616	2.949

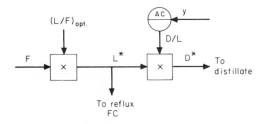

figure 11.1   *Reflux is set in optimum ratio to feed rate, and reflux ratio is used to control top composition; accumulator level must be controlled by boilup.*

ful, *both* products must incur penalties for being lost. In other words, there must be a penalty attached to the heavier component appearing in the distillate as well as to the lighter component appearing in the bottom product. If only a single loss function were used as in the previous examples, there would be no incentive to make the separation—all the feed could leave as the valuable product.

To demonstrate, consider the example of the butane splitter, which has already been introduced. But in addition to the $4.00/bbl penalty for isobutane in the bottom product, consider a $2.00/bbl penalty for n-butane in the distillate. The cost function excluding utilities becomes a loss function only:

$$\frac{\$_L}{F} = \Delta v_B \frac{B}{F} x + \Delta v_D \frac{D}{F}(1 - y) \tag{11.10}$$

where Δv_B and Δv_D represent the bottom and distillate penalties, respectively.

Operation at a fixed V/F ratio with unknown product compositions can be evaluated by selecting various values of D/F and determining reflux ratio:

$$\frac{L}{D} = \frac{V/F}{D/F} - 1 \tag{11.11}$$

Then separation factor S can be found using Eq. (3.58), with compositions determined from the quadratic formula given by Eqs. (3.54) and (3.55). (Appendix A includes a calculator program for finding y and x, given V/F and D/F.)

The column described in Example 11.1 was analyzed according to the cost function and penalties given in Eq. (11.10), at $V/F = 4.0$. The minimum costs found for the three feed compositions are reported in Table 11.3. Note that the bottom product is purer than the distillate as a result of the greater penalty associated with losses there. However, the purity difference decreases as the feed becomes richer in the more-valuable product.

Because neither composition can be constant, another measurable criterion must be found. The only relationship that appears to be useful for on-line optimization is that between the optimum D/F and feed composition. Although the relationship is not exactly linear, a simple multiplication of z by 1.01 to give D/F does not depart significantly from minimum cost.

Although a column model is used to determine the optimum operating conditions off-line, the most effective control system is one that is easy to implement and operate on-line. This is the reason for the design of the system in Fig. 11.1, where the optimum value of L/F was calculated from an off-line model. To optimize the column described in Table 11.3, the feedforward system of Fig. 11.2 would be used. Because neither composition-feedback loop is closed, dynamic response and inter-

TABLE 11.3 Minimum Loss in Product Values for the Butane Splitter at $V/F = 4.0$

z	y	x	D/F	$\$_L/F$
0.3	0.9272	0.0286	0.302	0.1240
0.4	0.9361	0.0366	0.404	0.1390
0.5	0.9419	0.0455	0.507	0.1486

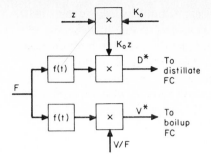

figure 11.2 *Optimization of an intermediate-product column is achieved by setting product-flow rate in proportion to feed rate and composition.*

action are ignored. The desired value of V/F and K_o would be entered manually, with coefficient K_o estimated from the off-line model (here 1.01).

Next the cost of heating and cooling can be entered into the accounting procedure:

$$\frac{\$_c}{F} = \Delta v_B \frac{B}{F} x + \Delta v_D \frac{D}{F}(1 - y) + c \, \Delta H_D \frac{V}{F} \qquad (11.12)$$

Now we have a two-parameter optimization problem. An optimum value of D/F must be found for each selected value of V/F until minimum total cost is achieved. This problem is solved for the butane splitter in Example 11.1, and the results are reported in Table 11.4.

Here again, both compositions are variable, but so is V/F. The ratio L/F shows less variation and can therefore be used for control. The system required for the two-parameter optimization would feature a combination of the schemes in Figs. 11.1 and 11.2. Distillate flow would be set in proportion to Fz as in Fig. 11.2, while reflux would be set in ratio to feed as in Fig. 11.1. The best approximation for optimum conditions across the range of feed compositions calls for K_o to be 0.97 and L/F 2.70. The inaccuracy of the approximation raises costs by 0.01 cents/bbl at $z = 0.4$, and 0.29 cents/bbl at each of the other feed compositions; these are reasonably small fractions of the 6.24 cents/bbl change in minimum cost over the composition range.

Finding the Optimum Feed Rate When the feed to a column is free to be manipulated, its rate may be selected to optimize a profit function. This adds another manipulated variable to those already used. While having the opportunity to select a feed rate for an individual column is rare enough, the ability to reach an optimum profit by adjusting it is still less likely. It requires that no more than one product meet guaranteed specifications. If both products must meet guaranteed

TABLE 11.4 Optimum Conditions vs. Feed Composition for the Butane Splitter with No Specifications

z	y	x	D/F	V/F	L/F	$\$_c/F$
0.3	0.9038	0.0663	0.279	2.83	2.55	0.7120
0.4	0.9175	0.0691	0.390	3.14	2.75	0.7511
0.5	0.9269	0.0782	0.497	3.30	2.80	0.7744

specifications, maximum profit will be achieved at maximum boilup when both products exactly meet their specifications.

With a single guaranteed product, an optimum L/F was found to minimize the combined cost of utilities and losses of product per unit of feed. Actually, that L/F could have been obtained by manipulating L, as described, or F. Assuming that the products from a column are always worth more than the feed, maximum profit can be realized only at maximum boilup.

The optimum feed rate is heavily dependent on the values of the products relative to the feed. If *both* products are worth more than the feed, the optimum feed rate will be higher than if one product is equal to or less than the value of the feed. To determine the optimum feed rate, a profit function must be written:

$$\$_P = v_D D + v_B B - v_F F - \$_o - \$_H \tag{11.13}$$

where v_D, v_B, v_F = values of designated products and feeds
$\qquad \$_o$ = fixed cost of overhead, labor, and capital write-off
$\qquad \$_H$ = cost of heating and cooling at maximum boilup
Feed rate can be factored from the first three terms:

$$\$_P = F\left[v_D \frac{D}{F} + v_B\left(1 - \frac{D}{F}\right) - v_F \right] - \$_o - \$_H \tag{11.14}$$

Then terms may be combined and D/F replaced by compositions to give

$$\$_P = F\left[(v_D - v_B)\frac{z - x}{y - x} + v_B - v_F \right] - \$_o - \$_H \tag{11.15}$$

For a controlled y, x will vary with V/F and hence with F. Although it is possible to differentiate $\$_P$ with respect to F, the resulting derivative contains too many functions of x to be easily solved. It is far simpler to increment F and calculate the profit resulting from its effect on x. The results of such a study are summarized in Table 11.5, normalized to maximum allowable boilup V_M.

The last column in Table 11.5 showing B/V_M is included because its value under optimum conditions does not change greatly. This seems to be characteristic for the waste product. Then if D is manipulated to control y at a specified composition, feed rate may be optimized simply by holding B at an average constant value. Figure 11.3 shows an optimizing control loop to perform this function. Although this configuration may not apply to every situation, it is given as an example of how a simple control system can solve a complex problem. Each individual case should be evaluated as was done here, for all expected load conditions, to arrive at a

TABLE 11.5 Optimum Feed Rate vs. Feed Composition for the Butane Splitter at $y = 0.95$
$(v_D - v_B = \$4.00/bbl$ and $v_B - v_F = -\$1.00/bbl)$

z	F/V_M	x	D/F	$(\$_P + \$_o + \$_H)/V_M$	B/V_M
0.3	0.2156	0.024	0.2981	0.0415	0.1513
0.4	0.2906	0.066	0.3778	0.1486	0.1808
0.5	0.3367	0.123	0.4559	0.2772	0.1832

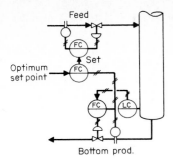

figure 11.3 *The second bottom-flow controller slowly adjusts feed rate to maximize profit.*

workable control system. As in other optimizing applications, the objective function is not terribly sensitive to changes in the manipulated variable, so precise control is not mandatory.

There is an additional economic factor pertaining to Table 11.5 which must not be overlooked. Costs of overhead and heating were not individually specified and were therefore lumped with the profit term. However, the cost of heating and cooling may be inserted using the figures in Example 11.1:

$$\frac{\$_H}{V} = \$6.60/10^6 \text{ Btu} \times 25,000 \text{ Btu/bbl} = \$0.165/\text{bbl}$$

If this figure is deducted from the profit reported in Table 11.5, it will be observed that the leaner two feed conditions result in a net operating loss. The only virtue in optimizing feed rate in these cases is to minimize loss in revenue. Therefore feed conditions are seen to be extremely important in all optimization problems and may determine whether the operation should continue or be shut down.

MULTIPLE-COLUMN OPTIMIZATION

Opportunities for multiple-column optimization are infrequent and diversified. Although it was possible to categorize three classic problems in single-column optimization, the addition of a second element to the process brings with it another dimension. Actually, what is presented here is but a sampling of the possibilities.

There are certainly more two-element optimization problems than those consisting of two columns. Combinations of a column and a compressor, a column and a reactor, etc., are familiar to both the chemical and the petroleum industries. However, reactor characteristics are so specific to the particular process involved that to speculate on such a problem in this text does not seem appropriate. So the discussion that follows is limited to two-column optimization, with a presentation on the optimum policy for allocation of limited resources among multiple users.

Two-Column Optimization The classic single-column program developed the optimum composition for the unspecified key component. By contrast, the two-column problem involves the unspecified off-key component, which by definition cannot be controlled in the column from which the product is withdrawn. An example of this application as shown in Fig. 11.4 was described by the author in an earlier paper [2].

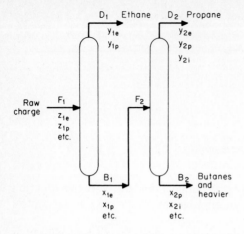

figure 11.4 *The optimum ethane content in the propane product is a function of deethanizer boilup.*

A propane product must meet a purity specification of 95 percent with less than 2 percent isobutane. Without considering the third component — ethane — this could be a single-column optimization if the value of isobutane were sufficiently greater than that of propane. However, the actual value trade-off in this column is not between isobutane and propane but between isobutane and ethane. Typically, the value of ethane is so much lower than that of the other components that its concentration should be maximized. The operating strategy for the depropanizer is then to control propane content as close to its purity specification as possible while optimizing the ethane and isobutane balance.

To reduce the isobutane content, more energy must be applied to the depropanizer. But at the same time, the ethane content is allowed to increase, which requires less energy at the deethanizer. The cost function then contains three terms — isobutane loss, depropanizer boilup, and deethanizer boilup:

$$\$_c = (v_i - v_e)D_2\, y_{2i} + c(\Delta H_2 V_2 + \Delta H_1 V_1) \tag{11.16}$$

where subscripts 1 and 2 refer to the deethanizer and depropanizer and i and e refer to isobutane and ethane. Specifications on the ethane product require that y_{1e}/y_{1p} be constant and similarly the isobutane product requires that x_{2p}/x_{2i} be constant.

To find the optimum, the ratio x_{1e}/x_{1p} may be incremented and the reflux ratio in the deethanizer found. Next, its material balance needs to be developed. Because

$$\frac{y_{1e} - z_{1e}}{y_{1e} - x_{1e}} = \frac{y_{1p} - z_{1p}}{y_{1p} - x_{1p}} \tag{11.17}$$

a solution for x_{1p} is possible:

$$x_{1p} = \frac{(y_{1p} - z_{1p})y_{1e} - (y_{1e} - z_{1e})y_{1p}}{(y_{1p} - z_{1p})(x_{1e}/x_{1p}) - (y_{1e} - z_{1e})} \tag{11.18}$$

Then D_1/F_1 can be found and combined with L_1/D_1 to give V_1/F_1. Note that B_1/F_1 is also F_2/F_1 and is described by Eq. (11.17).

Next, overhead composition in the depropanizer must be determined. First, let

$$y_{2i} = 1 - y_{2p} - y_{2e} \tag{11.19}$$

Next, for simplicity, neglect the propane leaving the bottom, so that y_{2e}/y_{2p} is approximately x_{1e}/x_{1p}. Then

$$\frac{y_{2p}}{y_{2i}} \approx \frac{y_{2p}}{1 - y_{2p} - y_{2p}x_{1e}/x_{1p}} = \frac{1}{1/y_{2p} - 1 - x_{1e}/x_{1p}} \tag{11.20}$$

At this point, ratio y_{2p}/y_{2i} and bottom specification x_{2p}/x_{2i} can be combined to determine L_2/D_2. From the assumption above,

$$\frac{D_2}{F_2} \approx \frac{x_{1p}}{y_{2p}} \tag{11.21}$$

which can then be used to determine V_2/F_2. At this point, enough information is available to evaluate the cost equation in terms of F_2:

$$\frac{\$_c}{F_2} = (v_i - v_e) \frac{D_2}{F_2} y_{2i} + c \left[\Delta H_2 \frac{V_2}{F_2} + \Delta H_1 \frac{V_1}{F_1} \middle/ \frac{F_2}{F_1} \right] \tag{11.22}$$

In the analysis described in Ref. 2, the savings in deethanizer energy were small compared with the cost of depropanizer energy for an incremental increase in x_{1e}/x_{1p}. One factor was the easier separation in the deethanizer. And as ethane content was increased and isobutane content reduced, the difference in energy usage between the two columns became more pronounced. An incremental increase in an already high (4 percent) ethane concentration had much less effect on separation than the same decrease in an already low (0.5 percent) isobutane concentration. As a result, the problem was solved as that of a single column generating an optimum isobutane concentration. The latter was then converted to a set point for x_{1e}/x_{1p} by substitution as in (11.20).

Allocation of Limited Resources In some plants, a single refrigeration unit serves more than one column. Refrigeration may not always be in short supply, but there are certain seasons of the year and hours of the day when it limits plant production. During these times some sort of allocation policy should be applied to optimize the use of this resource.

Consider the separation unit of the ethylene plant, represented schematically in Fig. 8.5. Assume that a common refrigeration unit supplies all the columns. Naturally, ethylene losses from each column will vary with the boilup-feed ratio of each. If refrigeration were apportioned arbitrarily, inequities might cause some columns to sustain high losses while others lost little. Because of the nonlinear relationship between losses and boilup, taking from the "poor" and giving to the "rich" is likely to cause a net increase in total plant losses. The curve of ethylene losses versus V/F for the demethanizer, as shown in Fig. 11.5, bears this out.

Every column has a loss curve of this same shape. Because the slope becomes less steep as V/F increases, any increment applied to a column with a low V/F will have more effect on total plant losses than when applied to a column with a high V/F. Therefore, an optimum policy exists which determines the allocation of addi-

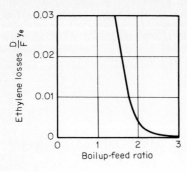

figure 11.5 *Ethylene losses with the distillate vary inversely with the boilup-feed ratio.*

tional utilities resulting in the best net reduction in losses. Naturally, the converse is true—a reduction in resource availability should be applied to minimize the effect on total plant losses.

Reference 3 develops an optimum allocation policy for the demethanizer and ethylene fractionator using simple hyperbolic models of column-loss functions. If the ethylene losses from each column are assumed to be simply inversely proportional to boilup, then it is demonstrated that an optimum percentage of total available energy exists for each column in the unit. Unfortunately, this concept is not so readily proved with the more exact models used here—differentiation of loss functions with respect to boilup yields functions too complex for direct solution.

However, on-line readjustment of allocation does not seem to be necessary. Because the component being lost (ethylene) is the same for all columns in the unit, variations in its value would not affect allocation. Production rate would not need to be considered since all columns would be affected equally by it. Variations in feed composition may call for an adjustment to allocation because the flow of one waste stream (e.g., fuel gas) might increase and thereby increase the losses from that column. A change in composition of the controlled product should also be considered since it will influence ethylene losses at a given V/F ratio.

If these variations are small or infrequent, the optimum allocation can be estimated off-line by an accurate model and implemented as shown in Fig. 11.6. As with the other examples described earlier, the total-loss function can be expected to be rather flat in the optimum region, so that extreme accuracy in predicting the optimum allocation is unwarranted.

The compressor discharge-pressure controller determines the availability of refrigerant vapor to the reboilers. The flow to each column is set in ratio to the total as determined by an off-line optimization program. However, high-flow overrides must be available in the event a flooding situation should develop or to satisfy a local optimization program. Feedback of total flow to an integrator set by the pressure controller prevents an override from upsetting the pressure loop. Should all three flow set points be overridden, the available energy cannot be used by the columns and the integrator will be unable to eliminate the deviation between its two points. However, a subtractor acting on this same deviation will then reduce compressor speed to control pressure. This system combines multiple-output controls with variable structuring as described in Chap. 7.

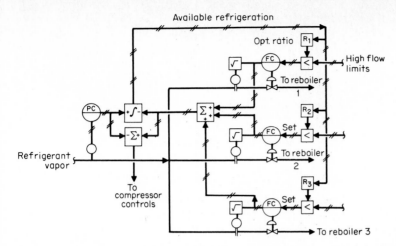

figure 11.6 *A pressure controller can set the total energy available to the columns.*

REFERENCES

1. Shinskey, F. G.: *Process-Control Systems,* 2d ed., McGraw-Hill, New York, 1979, pp. 160–164.
2. Fauth, C. J., and F. G. Shinskey: "Advanced Control for Distillation Columns," *Chem. Eng. Prog.,* June 1975.
3. Shinskey, F. G.: "The Values of Process Control," *Oil Gas J.,* February 18, 1974.

Computer Programs

Calculate y and x from V/F, D/F, α, nE, and z:

Using HP–11C Calculator

Inputs:		
	V/F STO	0
	D/F	1
	α	2
	nE	3
	z	4

Program:	Step	Command
	001–	f LBL A
		RCL 0
		RCL 1
		÷
	005–	1
		– (L/D)
		RCL 4
		×
		1/x

Program:	Step	Command
	010–	1
		+
		$\sqrt{}$
		RCL 2
		x⇆y
	015–	÷
		RCL 3
		y^x
		STO 5 (Sep. S)
		1
	020–	–
		RCL 1
		×
		STO 6 (Coeff. a)
		1
	025–	RCL 5
		–
		RCL 4
		RCL 1
		+
	030–	×
		1
		–
		STO 7 (Coeff. b)
		RCL 4
	035–	RCL 5
		× (Coeff. c)
		RCL 6
		×
		4
	040–	×
		RCL 7
		g x^2
		x⇆y
		–
	045–	$\sqrt{}$
		RCL 7
		+
		CHS
		2
	050–	÷
		RCL 6
		÷
		STO 8 (y)
		RCL 5
	055–	1
		–
		×
		RCL 5
		x⇆y
	060–	–
		RCL 8
		x⇆y
		÷
	064–	g RTN (Read x) RCL 8 (Read y)

Using BASIC

```
1 PRINT "EXPLICIT SOLUTION"
2 INPUT "V/F = ";V
3 INPUT "D/F = ";D
4 INPUT "ALPHA = ";AL
5 INPUT "NE = ";NE
6 INPUT "Z = ";Z
7 RR = V/D − 1
8 S = (AL/SQR(1+1/RR/Z)) ↑ NE
9 A = D*(S − 1)
10 B = −((D + Z) * (S − 1) + 1)
11 C = S * Z
12 Y = (−B−SQR(B↑2 − 4*A*C))/2/A
13 PRINT "Y = " Y
14 PRINT "X = " Y/(S−Y*(S − 1))

RUN
EXPLICIT SOLUTION
V/F = ? 4
D/F = ? .5
ALPHA = ? 1.35
NE = ? 33.76
Z = ? .5
Y = .9500
X = .0500
```

Calculate Slopes and Relative Gains

Using HP–11C Calculator

Inputs:	x	STO	0
	y		1
	z		2
	nE		3
	L/D		4

Slope Program:	Step	Command
(A)	001−	f LBL A
		RCL 2
		RCL 1
		−
	005−	RCL 2
		RCL 0
		−
		÷
		STO 5 (δ)
	010−	1
		RCL 1
		−
		RCL 1
		×
	015−	STO 7
		1
		RCL 0
		−
		RCL 0
	020−	×
		÷
		STO 6 (σ)
		RCL 3
		RCL 7

Slope Program:	Step	Command
	025—	×
		2
		÷
		RCL 1
		RCL 0
	030—	−
		÷
		RCL 2
		RCL 4
		×
	035—	1
		+
		÷
		STO 7 (ϵ)
		RCL 5
	040—	×
		RCL 6
		−
		RCL 7
		1
	045—	−
		÷
		STO 8 (μ_L)
		RCL 4
		1/x
	050—	1
		+
		STO 9 (μ_V)
		RCL 7
		×
	055—	RCL 5
		×
		RCL 6
		−
		RCL 9
	060—	RCL 7
		×
		1
		−
		÷
	065—	STO 9
		RCL 5
		1
		−
		RCL 7
	070—	×
		RCL 6
		−
		1
		RCL 5
	075—	1/x
		−
		RCL 7
		×
		1
	080—	−
		÷
		STO .0 ($\mu_{L/B}$)
		RCL 5
		1

Slope Program:	Step	Command
	085–	–
		RCL 4
		1/x
		1
		+
	090–	×
		STO .1
		RCL 7
		×
		RCL 6
	095–	–
		RCL .1
		RCL 5
		÷
		RCL 7
	100–	×
		1
		–
		÷
		STO .1 ($\mu_{V/B}$)
	105–	g RTN

Slopes Now Stored in Following Locations:

5 (δ)   8 (μ_L)   .0 ($\mu_{L/B}$)
6 (σ)   9 (μ_V)   .1 ($\mu_{V/B}$)

To Calculate Relative Gains from Slopes:

Inputs:	RCL j	(j and k represent the locations
	RCL k	of the input slopes)

RGA Program:	Step	Command
(B)	106–	f LBL B
		÷
		1
		x⇆y
	110–	–
		1/x
		g RTN (Read Λ_{jk})

Calculate Slopes and Relative Gains

Using TI 58C/59

Key	Step/Code		Descr.	Key	Step/Code		Descr.
LBL	00	76	LBL A	STO		42	
A		11		7		07	nE→R7
R/S		91	R/S	R/S		91	
STO		42		STO	15	42	
1		01	x→R1	8		08	L/D→R8
R/S	5	91		+		85	
STO		42		1		01	
2		02	y→R2	=		95	
R/S		91		1/	20	35	
STO		42		STO		42	
3	10	03	z→R3	4		04	D/V→R4
R/S		91		RCL		43	

Key	Step/Code	Descr.	Key	Step/Code	Descr.
2	02		6	06	
–	25 75		1/	35	
RCL	43		*	85 65	
2	02		(	53	
SQ	33		RCL	43	
=	95	$y(1-y)=$	3	03	
PRD	30 49		–	75	
7	07	$R7=R7*y(1-y)$	RCL	90 43	
STO	42		2	02	
5	05	$y(1-y){\rightarrow}R5$	)	54	
RCL	43		=	95	$\delta=$
2	35 02		R/S	91	R/S
–	75		STO	95 42	
RCL	43		6	06	$\delta{\rightarrow}R6$
1	01		*	65	
=	95	$y-x=$	RCL	43	
INV	40 22		7	07	
PRD	49		=	100 95	
7	07	$R7=R7/(y-x)$	STO	42	
1/	35		9	09	$\delta\epsilon{\rightarrow}R9$
*	65		RCL	43	
(	45 53		1	01	
RCL	43		–	105 75	
3	03		RCL	43	
–	75		1	01	
RCL	43		SQ	33	
1	50 01		=	95	$x(1-x)=$
)	54	$z-x=$	INV	110 22	
STO	42		PRD	49	
6	06	$z-x{\rightarrow}R6$	5	05	$R5=R5/x(1-x)$
=	95	$D/F=$	RCL	43	
R/S	55 91	R/S	5	05	RCL σ
+/–	94		R/S	115 91	R/S
+	85		–	75	
1	01		RCL	43	
=	95	$B/F=$	9	09	
STO	60 42		=	95	
0	00	$B/F{\rightarrow}R0$	STO	120 42	
R/S	91	R/S	4	04	$\sigma-\delta\epsilon{\rightarrow}R4$
2	02		–	75	
*	65		RCL	43	
RCL	65 43		9	09	
3	03		÷	125 55	
*	65		RCL	43	
RCL	43		8	08	
8	08		=	95	
+	70 85		STO	42	
2	02		3	130 03	$\sigma-\delta\epsilon(1+$
=	95	$2((zL/D)+1)=$			$D/L){\rightarrow}R3$
INV	22		1	01	
PRD	49		–	75	
7	75 07	$R7=R7/2((zL/D)+1)$	RCL	43	
RCL	43		7	07	
7	07	RCL ϵ	=	135 95	$1-\epsilon=$
R/S	91	R/S	INV	22	
RCL	43		PRD	49	
4	80 04	RCL D/V	4	04	$\mu_L=R4/(1-\epsilon)$
R/S	91	R/S	–	75	
RCL	43		RCL	140 43	

Key	Step/Code	Descr.
7	07	
÷	55	
RCL	43	
8	08	
=	145 95	$1-\epsilon(1+D/L)=$
INV	22	
PRD	49	
3	03	$\mu_v=R3/[1-\epsilon(1+D/L)]$
RCI	43	
5	150 05	
−	75	
RCL	43	
9	09	
÷	55	
RCL	155 43	
0	00	
=	95	
STO	42	
2	02	$\sigma-\delta\epsilon(F/B)\rightarrow R2$
−	160 75	
RCL	43	
9	09	
÷	55	
RCL	43	
0	165 00	
÷	55	
RCL	43	
8	08	
=	95	
STO	170 42	
1	01	$\sigma-\delta\epsilon(F/B)(1+D/L)\rightarrow R1$
1	01	
−	75	
RCL	43	
7	175 07	
÷	55	
RCL	43	
0	00	
=	95	$1-\epsilon(F/B)=$
INV	180 22	
PRD	49	
2	02	$\mu_{L/B}=R2/[1-\epsilon(F/B)]$
−	75	
RCL	43	
7	185 07	
÷	55	
RCL	43	
8	08	
÷	55	
RCI	190 43	
0	00	
=	95	$1-\epsilon(F/B)(1+D/L)=$
INV	22	
PRD	49	
1	195 01	$\mu_{v/B}=R1/[1-\epsilon\cdot(F/B)(1+D/L)]$
RCL	43	RCL μ_L
4	04	
R/S	91	R/S
RCL	43	
3	200 03	RCL μ_v
R/S	91	R/S
RCL	43	
2	02	RCL $\mu_{L/B}$
R/S	91	R/S
RCL	205 43	
1	01	RCL $\mu_{v/B}$
R/S	91	R/S
LBL	76	
Nop	68	LBL Nop
1	210 01	
−	75	
RCL	43	
4	04	
÷	55	
RCL	215 43	
6	06	
=	95	
1/	35	LD=
R/S	91	R/S
LBL	220 76	
Op	69	LBL Op
1	01	
−	75	
RCL	43	
5	225 05	
÷	55	
RCL	43	
6	06	
=	95	
1/	230 35	SD=
R/S	91	R/S
LBL	76	
x≥t	77	LBL x≥t
1	01	
−	235 75	
RCL	43	
6	06	
÷	55	
RCL	43	
3	240 03	
=	95	
1/	35	DV=
R/S	91	R/S
LBL	76	
Σ+	245 78	LBL +
1	01	
−	75	
RCL	43	
4	04	
÷	250 55	
RCL	43	
3	03	
=	95	
1/	35	LV=
R/S	255 91	R/S
LBL	76	
x̄	79	LBL x

Key	Step/Code		Descr.	Key	Step/Code		Descr.
1		01		RCL		43	
−		75		2	300	02	
RCL	260	43		=		95	
5		05		1/		35	S,L/B=
÷		55		R/S		91	R/S
RCL		43		LBL		76	
3		03		Dsz	305	97	LBL Dsz
=	265	95		1		01	
1/		35	SV=	−		75	
R/S		91	R/S	RCL		43	
LBL		76		6		06	
IfFlg		87	LBL If Flg.	÷	310	55	
1	270	01		RCL		43	
−		75		1		01	
RCL		43		=		95	
6		06		1/		35	D, V/B
÷		55		R/S	315	91	R/S
RCL	275	43		LBL		76	
2		02		ADV		98	LBL Adv
=		95		1		01	
1/		35	D, L/B=	−		75	
R/S		91	R/S	RCL	320	43	
LBL	280	76		4		04	
DMS		88	LBL D.MS	÷		55	
1		01		RCL		43	
−		75		1		01	
RCL		43		=	325	95	
4	285	04		1/		35	L, V/B=
÷		55		R/S		91	R/S
RCL		43		LBL		76	
2		02		Prt		99	LBL Prt
=		95		1	330	01	
1/	290	35	L, L/B=	−		75	
R/S		91	R/S	RCL		43	
LBL		76		5		05	
π		89	LBL π	÷		55	
1		01		RCL	335	43	
−	295	75		1		01	
RCL		43		=		95	
5		05		1/		35	S, V/B=
÷		55		R/S	339	91	R/S

Notes

1. Call program with program key A, advance by R/S

2. Enter variables in order x, y, z, nE, L/D,

3. Results sequence
   D/F
   B/F
   ϵ
   D/V
   Slopes: constant D − R6   V − R3
                 S − R5   L/B − R2
                 L − R4   V/B − R1

4. For relative gains, key
   sequence GTO (LBL), R/S
   according to the table below:

	N_{OP} Λ_{LD}	O_P Λ_{SD}
$x \geq t$ Λ_{DV}	$\Sigma +$ Λ_{LV}	$\overline{x}$ Λ_{SV}
IfFlg $\Lambda_{D,L/B}$	D.MS $\Lambda_{L,L/B}$	π $\Lambda_{S,L/B}$
Dsz $\Lambda_{D,V/B}$	Adv $\Lambda_{L,V/B}$	Prt $\Lambda_{S,V/B}$

or continue to advance by
R/S, which fills the relative
gain table by rows.

Using BASIC

```
 1 PRINT "DISTILLATION RGA"
 2 INPUT "X = "; X
 3 INPUT "Y = "; Y
 4 INPUT "Z = "; Z
 5 INPUT "NE = "; NE
 6 INPUT "L/D = "; RR
 7 DL = (Z − Y)/(Z − X)
 8 PRINT "DELTA = " DL
 9 SG = Y*(1 − Y)/X/(1 − X)
10 PRINT "SIGMA = " SG
11 EP = NE*Y*(1 − Y)/2/(Z*RR + 1)/(Y − X)
12 PRINT "EPSILON = " EP
13 LF = (SG − DL*EP)/(1 − EP)
14 PRINT "SLOPE L/F = " LF
15 VF = (SG − DL*EP*(1 + 1/RR))/(1 − EP*(1 + 1/RR))
16 PRINT "SLOPE V/F = " VF
17 FB = (Y − X)/(Y − Z)
18 LB = (SG − DL*EP*FB)/(1 − EP*FB)
19 PRINT "SLOPE L/B = " LB
20 VB = (SG − DL*EP*(1 + 1/RR)*FB)/(1 − EP*(1 + 1/RR)*FB)
21 PRINT "SLOPE V/B = "VB
22 PRINT "Λ DV = " 1/(1 − DL/VF)
23 PRINT "Λ DL/B = "1/(1 − DL/LB)
24 PRINT "Λ DV/B = "1/(1 − DL/VB)
25 PRINT "Λ LB = "1/(1 − LF/DL)
26 PRINT "Λ LV = " 1/(1 − LF/VF)
27 PRINT "Λ LL/B = " 1/(1 − LF/LB)
28  PRINT "Λ LV/B = " 1/(1 − LF/VB)
29  PRINT "Λ SB = " 1/(1 − SG/DL)
30 PRINT "Λ SV = " 1/(1 − SG/VF)
31 PRINT "Λ SL/B = " 1/(1 − SG/LB)
32 PRINT "Λ SV/B = " 1/(1 − SG/VB)

RUN
DISTILLATION RGA
X = ? .05
Y = ? .95
```

Z = ? .5
NE = ? 30
L/D = ? 2.5
DELTA = −1
SIGMA = 1
EPSILON − .35185
SLOPE L/F = 2.0857
SLOPE V/F = 2.9416
SLOPE L/B = 5.7500
SLOPE V/B = 134.000
ΛDV = .7463
ΛDL/B = .85185
ΛDV/B = .99259
ΛLB = .32407
ΛLV = 3.4369
ΛLL/B = 1.5692
ΛLV/B = 1.0158
ΛSB = .5
ΛSV = 1.5150
ΛSL/B = 1.2105
ΛSV/B = 1.0075

Calculate Cost Given *x* and *y*
Using HP–11C Calculator

Inputs:

x	STO	0
y		1
z		2
nE		3
α		4
$v_D - v_B$		5
$c\Delta H_D$		6

Program:

Step	Command
001–	f LBL A
	RCL 1
	RCL 0
	÷
005–	1
	RCL 0
	−
	×
	1
010–	RCL 1
	−
	÷
	RCL 3
	1/x
015–	y^x
	RCL 4
	x⇆y
	÷
	g x^2
020–	1
	−
	RCL 2
	×
	1/x

```
025–    1
        +
        RCL 2
        RCL 0
        –
030–    RCL 1
        RCL 0
        –
        ÷
        f PSE (Read D/F)
035–    ×
        f PSE (Read V/F)
        RCL 6
        ×
        RCL 1
040–    RCL 2
        –
        RCL 1
        RCL 0
        –
045–    ÷
        RCL 5
        ×
        RCL 0
        ×
050–    +
        g RTN (Read $/F)
```

Using BASIC

```
1 PRINT "COST GIVEN X & Y"
2 INPUT "X = "; X
3 INPUT "Y = "; Y
4 INPUT "Z = "; Z
5 INPUT "NE = "; NE
6 INPUT "ALPHA = "; AL
7 INPUT "VALUE = "; VU
8 INPUT "COST = "; C
9 INPUT "HEAT = "; H
10 D = (Z – X)/(Y – X)
11 PRINT "D/F = " D
12 S = Y*(1 – X)/X/(1 – Y)
13 RR = 1/Z/((AL/S↑(1/NE))↑2 – 1)
14 V = D*(RR + 1)
15 PRINT "V/F = " V
16 PRINT "$/F = " VU*(1 – D)*X + C*H*V

RUN
COST GIVEN Y & X
X = ? .05
Y = ? .95
Z = ? .5
NE = ? 33.76
ALPHA = ? 1.35
VALUE = ? 4
COST = ? 6.6E-6
HEAT = ? 25000
D/F = .5
V/F = 4.000
$/F = .7599
```

Table of Symbols

UPPERCASE

A Area; acid flow
B Bottom-product flow
C Heat capacity; correction factor; coefficient
D Distillate flow; derivative time
E Extractant flow; efficiency
F Feed flow; flow
G Gas flow; gain
H Enthalpy; inverse gain
$\mathbf{H}$ Inverse gain matrix
I Integral time
J Decoupler gain
K Equilibrium constant; steady-state gain
$\mathbf{K}$ Steady-state gain matrix
L Reflux flow; liquid flow
M Molecular weight

P Product flow; proportional band
Q Heat flow
R Recovery; gas constant
S Separation
T Temperature
U Heat-transfer coefficient
V Vapor flow
W Mass; mass flow
X Excess air; stream flow; composition
Y Stream flow; composition
Z Van Laar's constant; stream flow

LOWERCASE

a Coefficient; valve opening; accumulator level
b Coefficient; base level
c Controlled variable; cost
d Differential
e Deviation; 2.178
f Fractional flow; function
g Gravitational acceleration
$\mathbf{g}$ Dynamic gain
h Head; differential pressure
i Component
j Component
k Constant
l Length
m Manipulated variable
n Number of trays; number
p Pressure
$p°$ Vapor pressure
q Feed enthalpy
r Set point; rangeability
s Entropy
t Time
u Velocity
v Value; volume, volume fraction
w Composition; work; weight fraction
x Liquid composition; mol fraction
y Vapor composition
z Feed composition

GREEK

α Relative volatility
γ Activity coefficient; specific-heat ratio
Δ Difference
∂ Partial differential
δ Slope
ϵ Column characterization factor
η Thermodynamic efficiency
Λ Relative-gain subset
Λ Relative-gain array
λ Relative gain

μ Joule-Thompson coefficient; slope
π 3.1416
ρ Density; specific gravity
Σ Sum
σ Surface tension; slope
τ Time constant; period

APPENDIX C

Conversion of Compositions

Converting mole fraction to liquid-volume fraction,

$$v_i = \frac{x_i M_i/\rho_i}{\Sigma x_i M_i/\rho_i} \tag{C.1}$$

where v_i = liquid volume fraction of component i
x_i = mole fraction of component i
M_i = molecular weight of component i
ρ_i = liquid density of component i
  Converting liquid-volume fraction to weight fraction,

$$w_i = \frac{v_i \rho_i}{\Sigma v_i \rho_i} \tag{C.2}$$

where w_i = weight fraction of component i.
  Converting liquid-volume fraction to mole fraction,

$$x_i = \frac{v_i \rho_i / M_i}{\Sigma v_i \rho_i / M_i} \tag{C.3}$$

Converting mole fraction to weight fraction,

$$w_i = \frac{x_i M_i}{\Sigma x_i M_i} \tag{C.4}$$

TABLE C.1 Molar Volumes of Light Hydrocarbons in bbl/mol at 60°F

	M, lb/mol	ρ, lb/bbl	M/ρ, bbl/mol
Ethane	30	131	0.229
Propane	44	161	0.248
Isobutane	58	197	0.295
n-Butane	58	205	0.284
Isopentane	72	218	0.330
n-Pentane	72	221	0.326
n-Hexane	86	232	0.370

APPENDIX D

Summary of Important Equations

MATERIAL BALANCE

$$F = D + B$$
$$Fz_i = Dy_i + Bx_i$$
$$V = L + D$$

$$\frac{D}{F} = \frac{z_i - x_i}{y_i - x_i} \qquad y_i = \frac{z_i - (B/F)x_i}{D/F}$$
$$\frac{B}{F} = \frac{y_i - z_i}{y_i - x_i} \qquad x_i = \frac{z_i - (D/F)y_i}{B/F}$$

SEPARATION

Definition

Binary
$$S \equiv \frac{y(1-x)}{x(1-y)}$$
$$y = \frac{xS}{1+x(S-1)}$$
$$x = \frac{y}{y + S(1-y)}$$

Multicomponent
$$S \equiv \frac{y_L}{x_L}\frac{x_H}{y_H}$$
$$y_{LL} = \frac{z_{LL}}{D/F} \qquad y_L = 1 - y_H - y_{LL}$$
$$x_{HH} = \frac{z_{HH}}{1 - D/F} \qquad x_H = 1 - x_L - x_{HH}$$

Model
$$S = \left[\sqrt{\frac{\alpha}{\sqrt{1 + \frac{D}{LZ}}}}\right]^{nE} \qquad \frac{D}{L} = z\left[\left(\frac{\alpha}{S^{1/nE}}\right)^2 - 1\right]$$

PRODUCT QUALITY

Binary
$$y = \frac{-b - \sqrt{b^2 - 4ac}}{2a}$$
$$a = \left(\frac{D}{F}\right)(S-1)$$
$$b = (1-S)\left(z + \frac{D}{F}\right) - 1$$
$$c = zS$$

$$x = \frac{-b - \sqrt{b^2 - 4ac}}{2a}$$
$$a = \left(\frac{B}{F}\right)(S-1)$$
$$b = \left(z + \frac{B}{F}\right)(1-S) + S$$
$$c = -z$$

Multicomponent
$$y_H = \frac{-b + \sqrt{b^2 - 4ac}}{2a}$$
$$a = \frac{D}{F}(S-1)$$
$$b = \frac{D}{F} - z_{LL} + S\left(1 - \frac{D}{F} - z_{HH}\right) - z_H(S-1)$$
$$c = -z_H\left(1 - \frac{z_{LL}}{D/F}\right)$$
$$x_L = \frac{y_L(1 - x_{HH})}{y_H S + y_L}$$

Assumptions
1. $\partial y_H = -\partial y_L$
2. All LL keys overhead, all HH keys in bottom product

SENSITIVITY TO FEED COMPOSITION DISTURBANCES

$$\lambda \equiv \frac{1}{1 + \frac{(y-z)x(1-x)}{(z-x)y(1-y)}}$$
$$\frac{dy_i}{dz_i} = \frac{\lambda}{D/F} \qquad \frac{dx_i}{dz_i} = \frac{1-\lambda}{1-D/F}$$

DUAL PRODUCT QUALITY CONTROL

Definitions
$$\delta = -\frac{y_i - z_i}{z_i - x_i}$$

Binary
$$\epsilon = \frac{nEy_i(1-y)}{2\left(\frac{zL}{D}+1\right)(y-x)}$$

Multicomponent
$$\epsilon = \frac{nEy_Ly_H}{2\left[\frac{z_L + z_{LL}}{D}L + 1\right](y_L - x_L)}$$

$$\sigma = \frac{y(1-y)}{x(1-x)} \qquad \sigma = \frac{1/x_L + 1/x_H}{1/y_L + 1/y_H}$$

Slopes
$$\frac{\partial y}{\partial x}\Big|_D = \delta \qquad \frac{\partial y}{\partial x}\Big|_S = \sigma$$

$$\frac{\partial y}{\partial x}\Big|_L = \frac{\sigma - \delta\epsilon}{1 - \epsilon} \qquad \frac{\partial y}{\partial x}\Big|_V = \frac{\sigma - \delta\epsilon(1 + D/L)}{1 - \epsilon(1 + D/L)}$$

$$\frac{\partial y}{\partial x}\Big|_{L_B} = \frac{\sigma - \delta\epsilon(F/B)}{1 - \epsilon(F/B)} \qquad \frac{\partial y}{\partial x}\Big|_{V_B} = \frac{\sigma - \delta\epsilon(1 + D/L)(F/B)}{1 - \epsilon(1 + D/L)(F/B)}$$

Relative gains

$$\Lambda_{m_y m_x} = \frac{y}{x}\sqrt{\frac{\lambda_{ym_y}(\Lambda_{m_y m_x})}{1 - \lambda_{ym_y}(\Lambda_{m_y m_x})}}\quad \frac{1 - \lambda_{ym_y}(\Lambda_{m_y m_x})}{\lambda_{ym_y}(\Lambda_{m_y m_x})}$$

$$\lambda_{ym_y}(\Lambda_{m_y m_x}) = \frac{1}{1 - \dfrac{\partial y/\partial x|_{m_y}}{\partial y/\partial x|_{m_x}}}$$

Subsets

	m_y	m_x
$\Lambda_{DS} = \frac{y}{x}$	D	S
$\Lambda_{DL} = \frac{y}{x}$	D	L
$\Lambda_{DV} = \frac{y}{x}$	D	V
$\Lambda_{D,L_B} = \frac{y}{x}$	D	L_B
$\Lambda_{D,V_B} = \frac{y}{x}$	D	V_B
$\Lambda_{SV} = \frac{y}{x}$	S	V
$\Lambda_{SL} = \frac{y}{x}$	S	L
$\Lambda_{S,L_B} = \frac{y}{x}$	S	L_B
$\Lambda_{S,V_B} = \frac{y}{x}$	S	V_B
$\Lambda_{LV} = \frac{y}{x}$	L	V
$\Lambda_{L,L_B} = \frac{y}{x}$	L	L_B
$\Lambda_{V,V_B} = \frac{y}{x}$	V	V_B
$\Lambda_{V,L_B} = \frac{y}{x}$	V	L_B
$\Lambda_{L_B,V_B} = \frac{y}{x}$	L_B	V_B

NOTES
1. For feed and reflux as liquids at bubble point
2. B may be substituted for D
3. S may be manipulated through L/D, L/V, or D/V

figure D.1 *Equations for binary and multicomponent separations.*

COLUMN SPECIFICATIONS

x = _____ y = _____ z = _____ nE = _____ L/D = _____

$$\frac{D}{F} = \frac{z_i - x_i}{y_i - x_i} = \underline{\hspace{2cm}}$$

$$\epsilon = \frac{nEy(1-y)}{2(zL/D+1)(y-x)} = \underline{\hspace{2cm}}$$

$$\frac{B}{F} = 1 - \frac{D}{F} = \underline{\hspace{2cm}}$$

$$\frac{D}{V} = \frac{1}{L/D+1} = \underline{\hspace{2cm}}$$

ENTER CALCULATED RELATIVE GAINS INTO OPEN AREAS

$$\left.\frac{\partial y}{\partial x}\right|_D = -\frac{y-z}{z-x} = \underline{\hspace{1cm}} \equiv \delta$$

$$\Lambda_{m_y m_x} = \cfrac{1}{1 - \cfrac{\left.\partial y/\partial x\right|_{m_y}}{\left.\partial y/\partial x\right|_{m_x}}}$$

$$\left.\frac{\partial y}{\partial x}\right|_S = \frac{y(1-y)}{x(1-x)} = \underline{\hspace{1cm}} \equiv \sigma$$

$$\left.\frac{\partial y}{\partial x}\right|_L = \frac{\sigma - \delta\epsilon}{1-\epsilon} = \underline{\hspace{1cm}}$$

$$\left.\frac{\partial y}{\partial x}\right|_V = \frac{\sigma - \delta\epsilon(1+D/L)}{1-\epsilon(1+D/L)} = \underline{\hspace{1cm}}$$

$$\left.\frac{\partial y}{\partial x}\right|_{L/B} = \frac{\sigma - \delta\epsilon(F/B)}{1-\epsilon(F/B)} = \underline{\hspace{1cm}}$$

$$\left.\frac{\partial y}{\partial x}\right|_{V/B} = \frac{\sigma - \delta\epsilon(1+D/L)(F/B)}{1-\epsilon(1+D/L)(F/B)} = \underline{\hspace{1cm}}$$

m_y \ m_x	D	L	$S(L/D)$	$S(D/V)$	$S(L/V)$
D					
(B)					
V					
L/B					
V/B					

figure D.2 *Relative-gain worksheet for binary distillation.*

Index

ABOUT THE AUTHOR

F. G. Shinskey, chief application engineer with the Foxboro
Company, has also worked as a process engineer for E. I. Du
Pont de Nemours and Company and Olin Mathieson
Chemical Company. He pioneered in the application of
feedforward controls to distillation, evaporation, heat transfer,
and effluent-treating processes. He also developed inferential
control systems based on the models of drying principles,
which have been applied to many types of industrial dryers.
His experience with the unusual problems encountered in
treating industrial wastes led to the development of nonlinear
and self-tuning controllers, many of which have been
patented. Mr. Shinskey has lectured at several universities,
has given seminars on process control at many locations
around the world, and has written numerous papers and five
books, including *Process Control Systems*, a McGraw-Hill
publication.